MW01257900

OPTICS

This is a rerun of the Fifth Printing. The only change is that all the color plates now appear in the frontispiece.

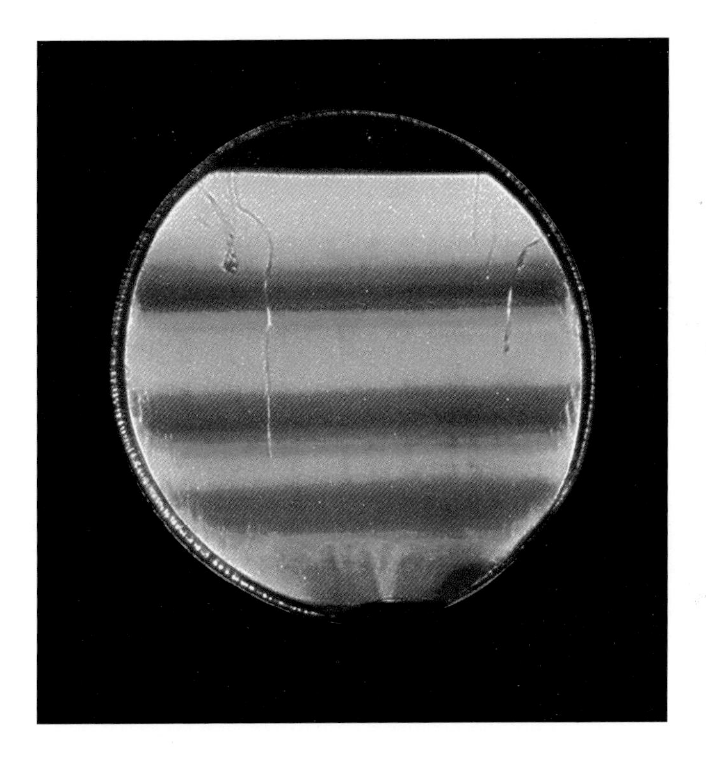

Interference colors in the light reflected
from a thin soap film.

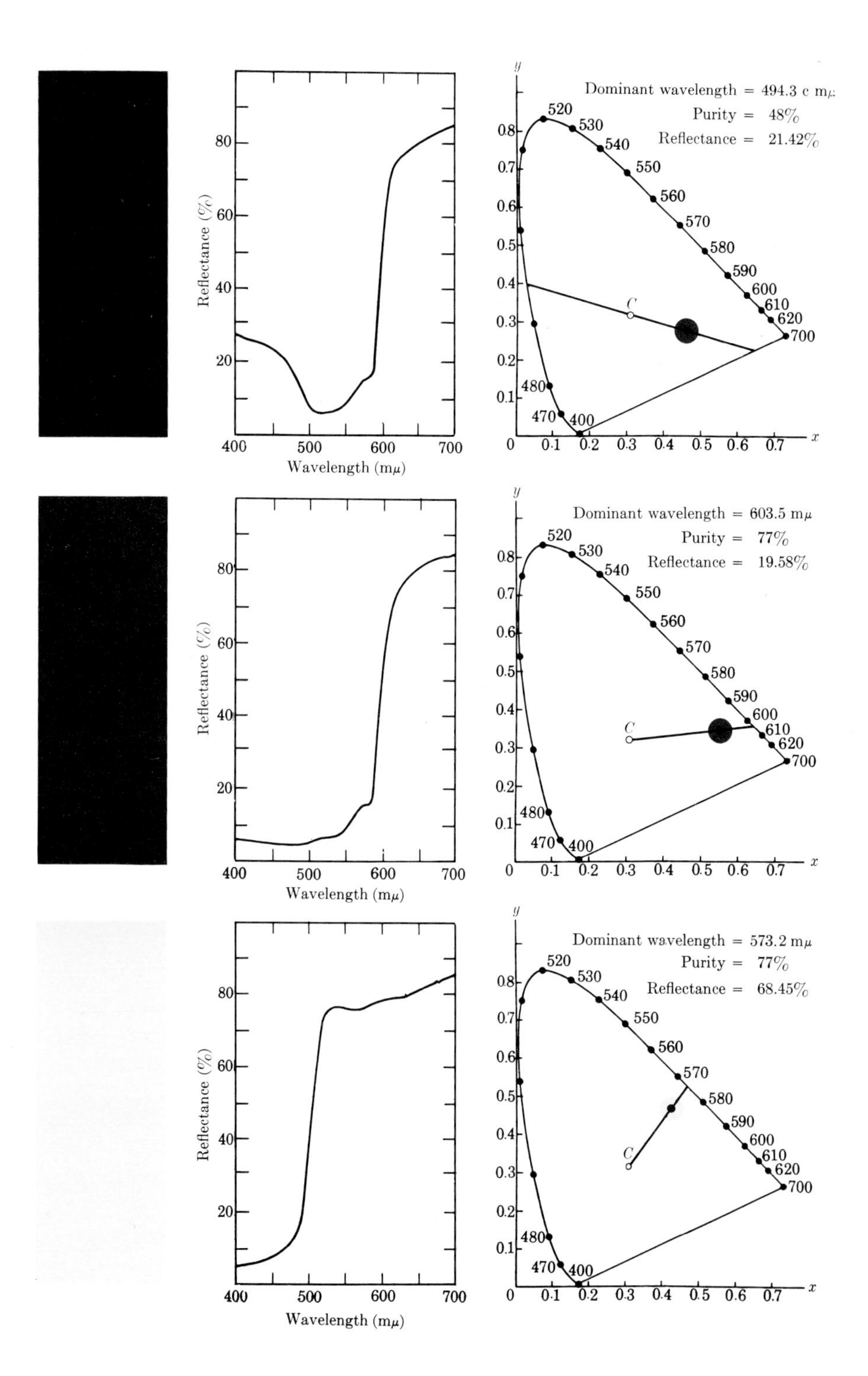

Reflectance (%)
Wavelength (mμ)
Dominant wavelength = 494.3 c mμ
Purity = 48%
Reflectance = 21.42%
Dominant wavelength = 603.5 mμ
Purity = 77%
Reflectance = 19.58%
Dominant wavelength = 573.2 mμ
Purity = 77%
Reflectance = 68.45%

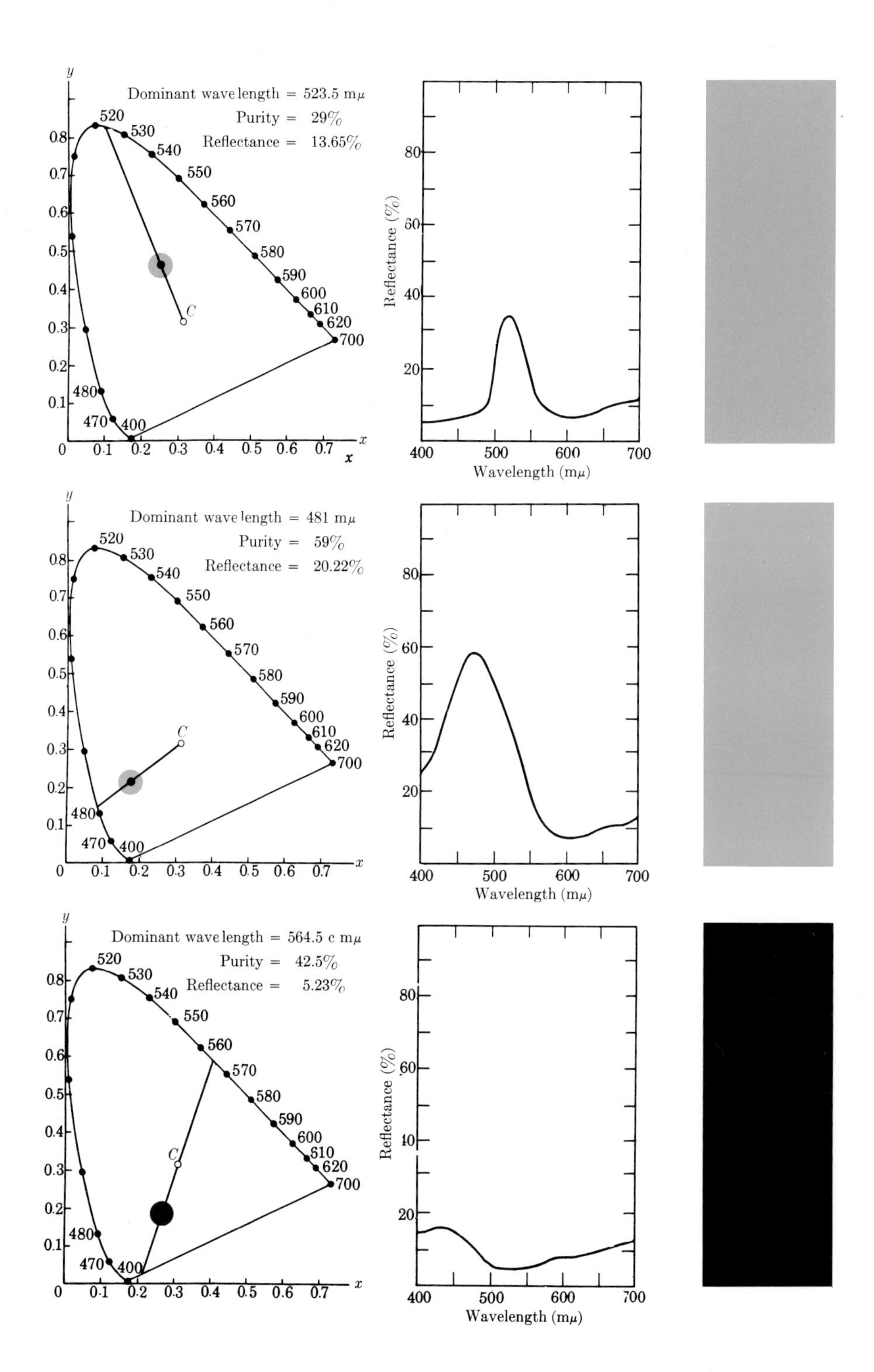
Dominant wave length = 523.5 mμ
Purity = 29%
Reflectance = 13.65%
Reflectance (%)
Wavelength (mμ)
Dominant wave length = 481 mμ
Purity = 59%
Reflectance = 20.22%
Reflectance (%)
Wavelength (mμ)
Dominant wave length = 564.5 c mμ
Purity = 42.5%
Reflectance = 5.23%
Reflectance (%)
Wavelength (mμ)

PRINCIPLES OF PHYSICS SERIES

OPTICS

by

FRANCIS WESTON SEARS

Department of Physics
Massachusetts Institute of Technology

ADDISON-WESLEY PUBLISHING COMPANY

READING, MASSACHUSETTS · MENLO PARK, CALIFORNIA
LONDON · AMSTERDAM · DON MILLS, ONTARIO · SYDNEY

Printed in the United States of America

Third Edition

Fifth printing — April 1958

ISBN 0-201-06915-6
RSTUVWXYZ-AL-8987654321

PREFACE

This book is the third volume of a series of texts written for a two-year course in general physics. Several changes have been made in the present edition as a result of suggestions by teachers and students. The author is particularly grateful to his colleagues Professors Arthur C. Hardy and Siebert Q. Duntley.

The order of presentation in the chapter on lenses has been rearranged so that, without sacrificing generality of approach, a detailed discussion of thick lenses may be omitted if desired. The chapters on physical optics have been largely rewritten. Polarization precedes interference and diffraction, and Fraunhofer diffraction is discussed before Fresnel diffraction. The limit of resolution of an optical instrument has been given a more unified treatment.

Many of the diagrams are new. A number of figures illustrating image formation by lenses and optical instruments have been printed in reverse, and will awaken memories of *Light for Students* in the minds of those who first studied Optics from Edser's admirable series of texts. The assistance of Mr. Joseph S. Banks in making these diagrams is gratefully acknowledged.

The letter symbols have been revised to accord with those recommended by the Committee on Letter Symbols and Abbreviations of the American Association of Physics Teachers, as listed in the American Standard, ASA–Z10, published in 1947.

Cambridge, Mass.
September 1948

F.W.S.

CONTENTS

CHAPTER 1

THE NATURE AND PROPAGATION OF LIGHT

1-1 The nature of light. Until about the middle of the 17th century, it was generally believed that light consisted of a stream of corpuscles. These corpuscles were emitted by light sources, such as the sun or a candle flame, and traveled outward from the source in straight lines. They could penetrate transparent materials and were reflected from the surfaces of opaque materials. When the corpuscles entered the eye, the sense of sight was stimulated.

If the test of the adequacy of any theory is its ability to account for known experimental facts with a minimum of hypotheses, we must admit that the corpuscular theory was an excellent one. The theory was called on to explain why light appeared to travel in straight lines, why it was reflected from a smooth surface such as a mirror with the angle of reflection equal to the angle of incidence, and why and how it was refracted at a boundary surface such as that between air and water or air and glass. For all of these phenomena, a corpuscular theory provides a simple explanation.

By the middle of the 17th century, while most workers in the field of optics accepted the corpuscular theory, the idea had begun to develop that light might be a wave motion of some sort. Christian Huygens, in 1670, showed that the laws of reflection and refraction could be explained on the basis of a wave theory and that such a theory furnished a simple explanation of the recently discovered phenomenon of double refraction. The wave theory failed of immediate acceptance, however. For one thing, it was objected that if light were a wave motion one should be able to see around corners, since waves can bend around obstacles in their path. We know now that the wave lengths of light waves are so short that the bending, while it does actually take place, is so small that it is not ordinarily observed. As a matter of fact, the bending of a light wave around the edges of an object, a phenomenon known as diffraction, was noted by Grimaldi as early as 1665, but the significance of his observations was not realized at the time.

It was not until 1827 that the experiments of Thomas Young and Augustin Fresnel on interference, and the measurements of the velocity of light in liquids by Leon Foucault at a somewhat later date, demonstrated the existence of optical phenomena for whose explanation a corpuscular theory was inadequate. The phenomena of interference and diffraction will be discussed further in Chaps. 8 and 9, where it will be shown that they are only what would be expected if light is a wave motion. Young's experiments enabled him to measure the wave length of the waves and Fresnel showed that the rectilinear propagation of light, as well as the diffraction effects observed by Grimaldi and others, could be accounted for by the behaviour of waves of short wave length.

The exact nature of light waves, and of the medium through which they were transmitted, remained an unsolved problem. The "ether," which had been invented by Huygens as a medium of transmission and which was supposed to fill all of otherwise empty space and to permeate the pores of transparent materials, turned out to have some strikingly inconsistent properties. If waves of light were elastic waves similar to sound waves, then in order to account for the observed large velocity of propagation of light, the ether was required to be extremely rigid. Nevertheless, it offered no opposition to the motion of a body, since the planets moved through it with no appreciable diminution of speed.

The next great forward step in the theory of light was the work of the Scotch scientist, James Clerk Maxwell. In 1873, Maxwell showed that an oscillating electrical circuit should radiate electromagnetic waves. The velocity of propagation of the waves could be computed from purely electrical and magnetic measurements and it turned out to be very nearly 3×10^8 m/sec. Within the limits of experimental error, this was equal to the measured velocity of propagation of light. The evidence seemed inescapable that light consisted of electromagnetic waves of extremely short wave length. Fifteen years after this discovery of Maxwell, Heinrich Hertz, using an oscillating circuit of small dimensions, succeeded in producing short wave length waves (we would speak of them today as microwaves) of undoubted electromagnetic origin and showed that they possessed all the properties of light waves. They could be reflected, refracted, focussed by a lens, polarized, and so on, just as could waves of light. Maxwell's electromagnetic theory of light and its experimental justification by Hertz constituted one of the triumphs of physical science. By the end of the 19th century it was the general belief that little, if anything, would be added in the future to our knowledge of the nature of light. Such was not to be the case.

The classical electromagnetic theory failed to account for the phenomenon of photoelectric emission, that is, the ejection of electrons from a conductor by light incident on its surface. In 1905, Einstein extended an idea proposed five years earlier by Planck and postulated that the energy in a light beam, instead of being distributed through space in the electric and magnetic fields of an electromagnetic wave, was concentrated in small packets or *photons*. A vestige of the wave picture was retained, in that a photon was still considered to have a frequency, and the energy of a photon was proportional to its frequency. The mechanism of the photoelectric effect consisted in the transfer of energy from a photon to an electron. Experiments by Millikan showed that the kinetic energies of photoelectrons were in exact agreement with the formula proposed by Einstein.

Still another striking confirmation of the photon nature of light is the Compton effect. A. H. Compton, in 1921, succeeded in determining the motion of a photon and a single electron, both before and after a "collision" between them, and found that they behaved like material bodies having kinetic energy and momentum, both of which were conserved in the collision. The photoelectric effect and the Compton effect, then, both seem to demand a return to a corpuscular theory of light.

The present standpoint of physicists, in the face of apparently contradictory experiments, is to accept the fact that light appears to be dualistic in nature. The phenomena of light propagation may best be explained by the electromagnetic wave theory, while the interaction of light with matter, in the processes of emission and absorption, is a corpuscular phenomenon.

In the preceding discussion, the term "light" has been used in a purely objective or physical sense, with reference to electromagnetic waves or photons. The same word is used in a psychological or subjective sense to refer to the sensation in the consciousness of a human observer when electromagnetic waves, or photons, strike the retina of his eye. A committee of the Optical Society of America has proposed a third definition of the term which combines both its objective and subjective aspects and is described as *psychophysical*. According to this definition, "Light is that aspect of radiant energy of which a human observer is aware through the visual sensations which arise from stimulation of the retina of the eye." For the present, we shall be concerned only with the first, or objective, meaning of the word light.

1-2 Wave fronts and rays. It is convenient to represent a train of waves of any sort by means of *wave fronts*. A wave front is defined as the locus of points, all of which are in the same phase. Thus in the case of sound waves spreading out in all directions from a point source, any spherical surface concentric with the source is a possible wave front. Some spherical surfaces are the loci of points at which the pressure is a maximum, others where it is a minimum, and so on, but the phase of the pressure waves is the same over any spherical surface. It is customary to draw only a few wave fronts, usually those which pass through the maxima and minima of the disturbance. Such wave fronts are separated from one another by one-half a wave length.

If the wave is a light wave, the quantity which corresponds to the pressure in a sound wave is the electric or magnetic intensity. It is usually unnecessary to indicate in a diagram either the magnitude or direction of the intensity, but simply to show the shape of the wave by drawing the wave fronts or their intersections with some reference plane. For example, the electromagnetic waves radiated by a small light source may be represented by spherical surfaces concentric with the source or, as in Fig. 1-1 (a), by the intersections of these surfaces with the plane of the diagram. At a sufficiently great distance from the source, where the radii of the spheres have become very large, the spherical surfaces can be considered planes and we have a train of plane waves as in Fig. 1-1 (b).

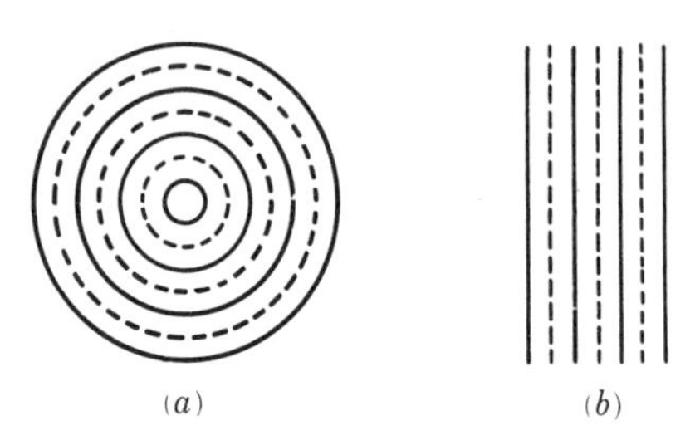

FIG. 1-1. Wave fronts.

Fig. 1-2 (a) illustrates the electric and magnetic field distributions in a train of plane electromagnetic waves at any one instant. The waves are advancing along the x-axis toward the right. The electric lines of force are shown in gray, and the magnetic lines in black. Where the lines are close together the intensities are large; where they are far apart the intensities are small. The directions of the electric and magnetic fields are shown by arrows and by the usual convention of dots and crosses. A more conventional diagram of the wave form is shown in Fig. 1-2 (b).

The wave illustrated in Fig. 1-2 is of a relatively simple type known as *linearly polarized* (also as plane polarized). As the wave advances, the vector representing the electric or magnetic intensity at any fixed point oscillates along a straight line. Other more complicated types of polarization are discussed in Chap. 7.

A train of light waves may often be represented more simply by means of *rays* than by wave fronts. In a corpuscular theory, a ray is simply the path followed by a light corpuscle. From the wave viewpoint, a ray is an imaginary line drawn in the direction in which the wave is traveling. Thus in Fig. 1-1 (a) the rays are the radii of the spherical wave fronts and in Fig. 1-1 (b) they are straight lines perpendicular to the wave fronts. In fact, in every case in which the waves are traveling in a homogeneous

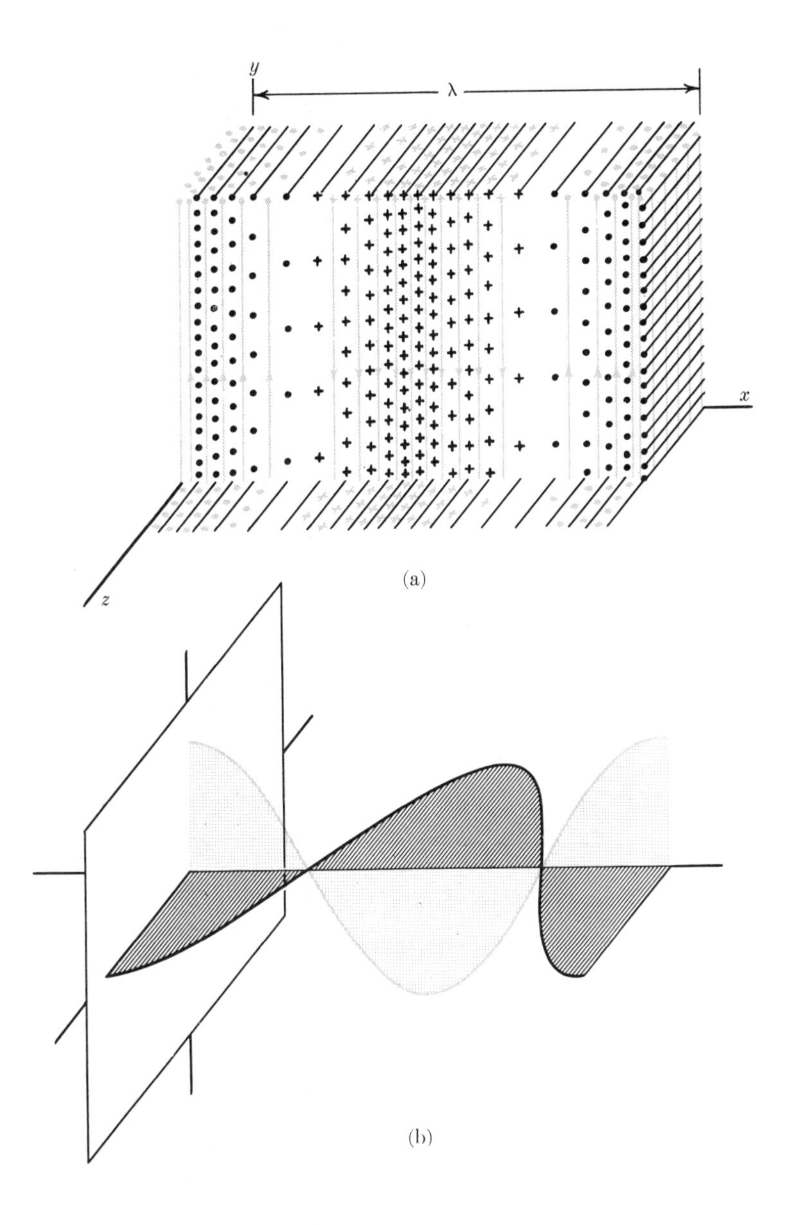

FIG. 1-2. (a) Electric and magnetic field distribution of a plane electromagnetic wave at any one instant. The electrostatic lines of force are shown in gray and the magnetic lines of force are shown in black. (b) A conventionalized diagram of the wave form at the same instant.

isotropic medium, the rays are straight lines, normal to the wave fronts. At a boundary surface between two media, such as the surface between a glass plate and the air outside it, the direction of a ray may change suddenly but it is a straight line both in the air and in the glass. If the medium is not homogeneous, for instance, if one is considering the passage of light through the earth's atmosphere where the density and hence the velocity vary with elevation, the rays are curved but are still normal to the wave fronts. If the medium is anisotropic, as is the case in certain crystals, the direction of the rays is not always normal to the wave fronts. This problem will be considered in more detail in Chap. 7.

A narrow cone of rays diverging from a common point is called a *pencil.* (Strictly, a *homocentric* pencil. Other types will be described later.) The entire group of pencils originating at all points of a surface of finite extent is called a *beam.*

1-3 Huygens' principle. Huygens' principle is a geometrical method for finding, from the known shape of a wave front at some instant, what the shape will be at some later instant. The principle states that every point of a wave front may be considered as the source of small "secondary" wavelets, which spread out in all directions from their centers with a velocity equal to the velocity of propagation of the wave. The new wave front is then found by constructing a surface tangent to the secondary wavelets or, as it is called, the *envelope* of the wavelets. If the velocity of propagation is not the same at all portions of the wave front, the appropriate velocity must be used for the various wavelets.

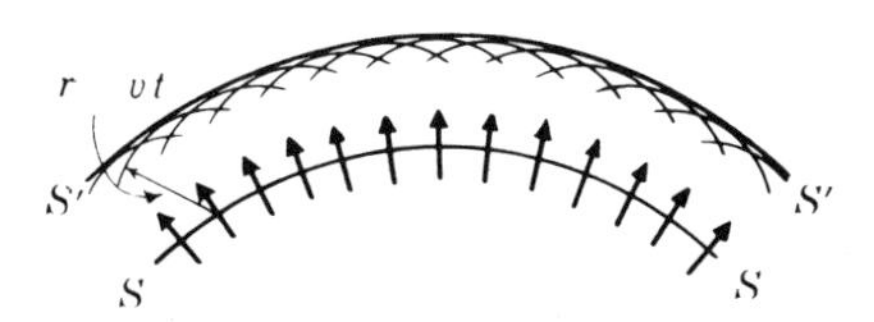

Fig. 1-3. Huygens' principle.

Huygens' principle is illustrated in Fig. 1-3. The original wave front, S–S, is traveling as indicated by the small arrows. We wish to find the shape of the wave front after a time interval t. Let v represent the velocity of propagation. Construct a number of circles (traces of spherical wavelets) of radius $r = vt$, with centers along S–S. The trace of the envelope of these wavelets, which is the new wave front, is the curve S'–S'. The velocity v has been assumed the same at all points and in all directions.

In the simple form given above, Huygens' principle is not fully satisfactory. For instance, the secondary wavelets, if they spread out in *all* directions, should also combine to give a "back" wave which is not observed. The complete significance of Huygens' method, which in fact was not fully appreciated by Huygens himself, is only to be understood when the principles of interference are applied to the secondary wavelets as will be done later in Chap. 9. It will be shown then that the absence of a back wave can be satisfactorily explained.

1-4 Atmospheric refraction. The velocity of light in all material substances is less than its velocity in free space and in a gas the velocity decreases as the density increases. The density of the earth's atmosphere is greatest at the surface of the earth and decreases with increasing elevation. As a result, light waves entering the earth's atmosphere are continuously deviated as shown in Fig. 1-4 (a). The line A–A' represents a

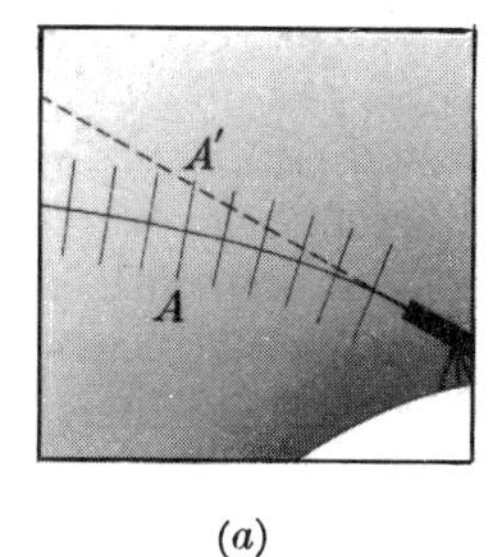

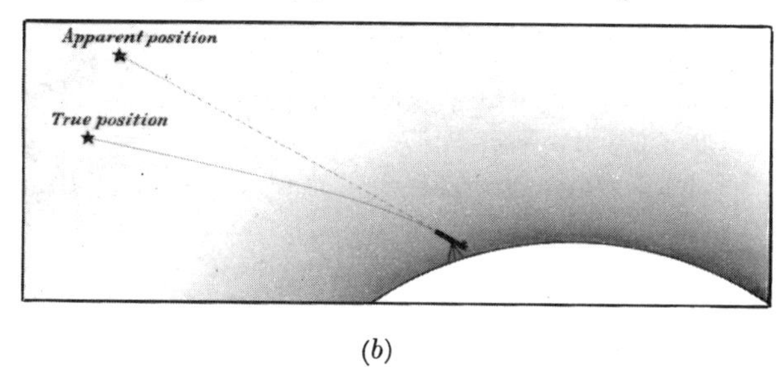

(*a*) (*b*)

FIG. 1-4. Deviation of a ray by the earth's atmosphere.

wave front in the light from the sun or a star. The density of the air at the lower portion of the wave front is greater than that at the upper portion. Hence the lower portion of the wave always travels more slowly than the upper portion and Huygens' construction leads to the shift in the direction of the wave front as shown. An observer at the earth's surface sees the light source in the direction of the tangent to the rays when they reach the earth and concludes that the object is nearer the zenith than its true position.

Rays entering the earth's atmosphere horizontally are "lifted" by atmospheric refraction through about 0.5°. This is very nearly equal to the angle subtended by the sun's disk, so that when the sun appears to be just above the horizon at sunrise or sunset, it is, geometrically, just below it. Furthermore, since the sun requires about two minutes to move (apparently) a distance equal to its own diameter, the day (at the equator) is lengthened by about two minutes at both sunrise and sunset. At higher latitudes the increase is even greater. The necessary correction for atmospheric refraction must be made by every navigator in the process of "shooting" the sun or any other heavenly body.

The deviation of light by atmospheric refraction decreases with increasing angle of elevation of the light above horizontal, falling to zero for light incident normally on the earth's surface. Since rays from the upper portion of the sun's disk are incident at a slightly greater angle than those from the lower part, they are refracted to a smaller extent. This accounts for the slightly flattened appearance of the sun at sunset or sunrise, the lower portion being lifted more than the upper.

Another phenomenon produced by atmospheric refraction is the mirage, illustrated in Fig. 1-5. The conditions necessary for its production require that the air nearer the surface of the ground shall be less dense than that above, a situation which is sometimes found over an area intensely heated by the sun's rays. Light from the upper portion of an object may reach the eye of an observer by the two paths shown in the figure, with the

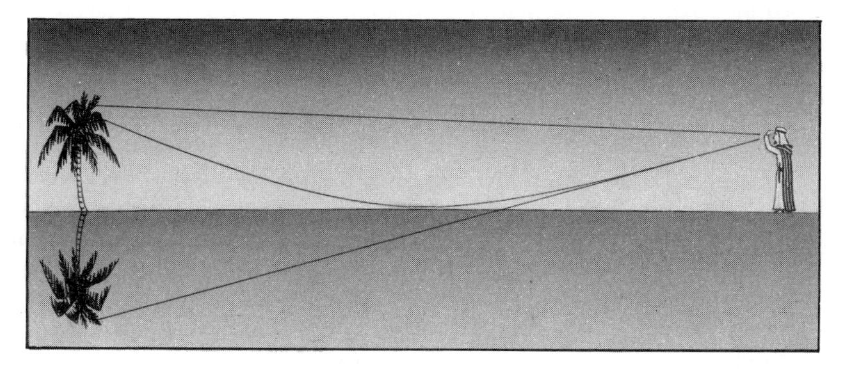

FIG. 1-5. The mirage.

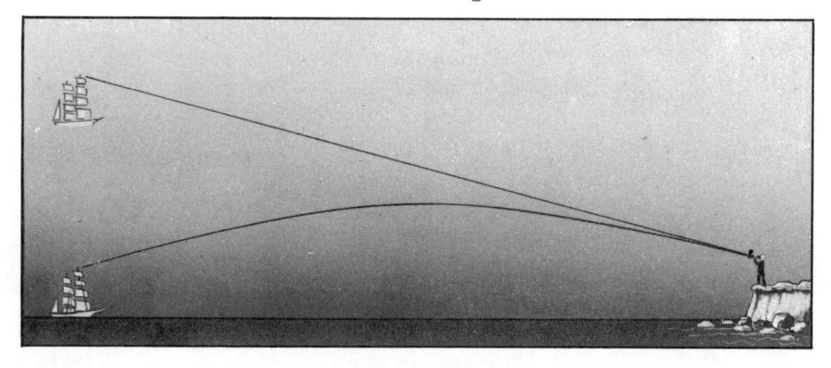

FIG. 1-6. Looming.

result that the object is seen in its actual position, together with its inverted image below it, as though a reflecting surface lay between the object and observer. The weary traveler in the desert interprets the reflecting surface as a body of water. This same phenomenon accounts for the "wet" appearance of the surface of a smooth highway under a hot sun, when a rise in the road ahead permits it to be seen at a glancing angle. Mirages are also produced when the reverse conditions obtain, which is sometimes the case over large bodies of water. Objects at a distance appear to be lifted above their true positions. This phenomenon is known as "looming." (Fig. 1-6.)

1-5 Shadows. Probably one of the first optical phenomena to be noted was that the shadow of an object illuminated by a source of small dimensions has the same shape as the object and that the edges of the shadow are the extensions of straight lines from the source tangent to the edges of the object. Apart from diffraction effects, which will be discussed in Chap. 9, the formation of shadows can be treated satisfactorily in terms of a ray picture.

Point O in Fig. 1-7 (a) represents a *point source* of light. That is, the dimensions of the source are small in comparison with other distances involved. S is a screen and P is a circular obstacle between source and screen. The area of the screen, bounded by rays from the source tangent to the edges of the obstacle, is called the *geometrical shadow* of the obstacle.

If the source is not sufficiently small to be considered a point, as in Fig. 1-7 (b), the shadow consists of two portions. The region behind the obstacle which receives no light from the source is called the *umbra*. This is surrounded by the *penumbra*, within which a part of the source is screened by the obstacle. The fuzzy appearance of the edges of a shadow cast by a frosted bulb incandescent lamp is due to the penumbra. An observer

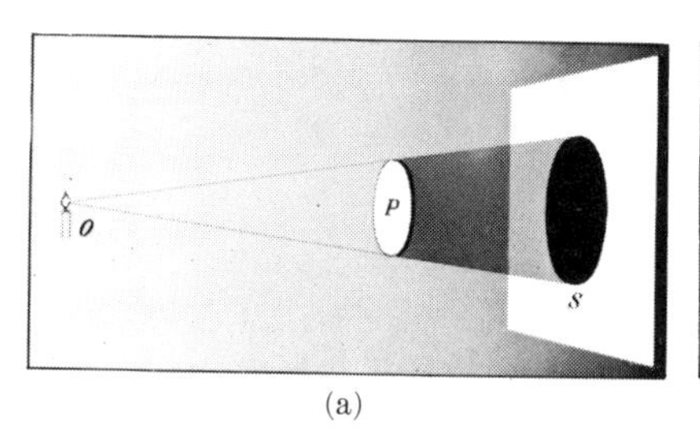

(a)

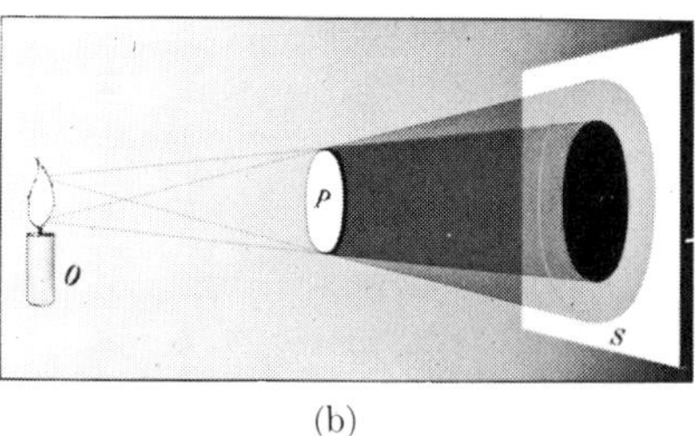

(b)

FIG. 1-7. (a) A point source of light casts a sharply-defined shadow. (b) If the source is not a point, the shadow consists of a central umbra surrounded by a penumbra.

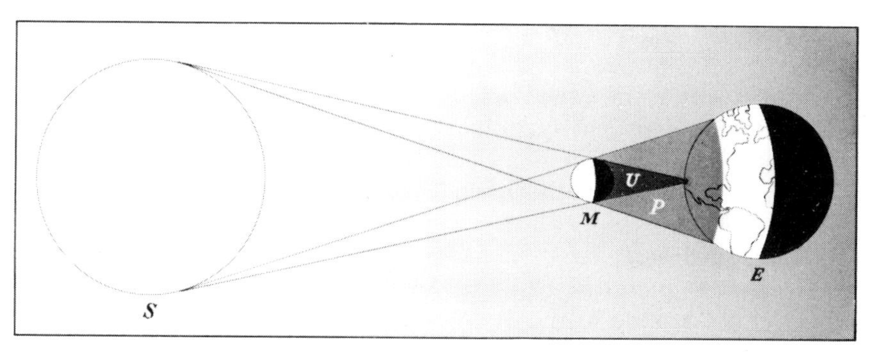

FIG. 1-8. A solar eclipse. The eclipse is total for an observer within the umbra U, partial for one within the penumbra P.

within the umbra cannot see any part of the source, one within the penumbra can see a portion of the source, while from points outside the penumbra the entire source can be seen.

The phenomenon of a partial or total solar eclipse is caused by the passage of a portion of the earth's surface within the penumbra or umbra of the shadow of the moon, cast by the sun. In Fig. 1-8 (obviously not to scale) S, M, and E represent the sun, moon, and earth. The moon's shadow in space consists of a conical umbra U surrounded by a penumbra P. When a portion of the umbra near the tip sweeps over the earth's surface, the solar eclipse will be total for all observers within it. Within a band on either side lying in the penumbra, the eclipse will be only a partial one. Eclipses of the moon arise in a similar manner when the relative positions of sun, earth, and moon are such that the moon lies within the shadow of the earth.

1-6 The velocity of light. The magnitude of the velocity of propagation of light in free space is one of the fundamental constants of nature. The velocity is so great (about 186,000 mi/sec or 3×10^8 m/sec) that it evaded experimental measurement until 1675. Up to that time it was generally believed that light traveled with an infinite velocity.

The first attempts to measure the velocity of light were made in 1667, using a method proposed by Galileo. Two experimenters were stationed on the tops of two hills about a mile apart. Each was provided with a lantern, the experiment being performed at night. One man was first to uncover his lantern and, observing the light from this lantern, the second was to uncover his. The velocity of light could then be computed from the known distance between the lanterns and the time elapsing between the instant when the first observer uncovered his lantern and when he observed the light from the second. While the experiment was entirely correct in principle, we know now that the velocity is too great for the time interval to be measured in this way with any degree of precision.

Eight years later, in 1675, the Danish astronomer Olaf Roemer, from astronomical observations made on one of the satellites of the planet Jupiter, obtained the first definite evidence that light is propagated with a finite velocity. Jupiter has eleven small satellites or moons, four of which are sufficiently bright to be seen with a moderately good telescope or a pair of field glasses. The satellites appear as tiny bright points at one side or the other of the disk of the planet. These satellites revolve about Jupiter just as does our moon about the earth and, since the plane of their orbits is nearly the same as that in which the earth and Jupiter revolve, each is eclipsed by the planet during a part of every revolution.

Roemer was engaged in measuring the time of revolution of one of the satellites by taking the time interval between consecutive eclipses. He found, by a comparison of results over a long period of time, that while the earth was receding from Jupiter the periodic times were all somewhat longer than the average, and that while it was approaching Jupiter the times were all somewhat shorter. He concluded rightly that the cause of these variations was the varying distance between the earth and Jupiter.

Fig. 1-9, not to scale, illustrates the case. Let observations be started when the earth and Jupiter are in the positions E_1 and J_1. Since Jupiter requires about 12 years to make one revolution in its orbit, then by the time the earth has moved to E_2 (about five months later) Jupiter has moved only to J_2. During this interval the distance between the planets has been continually increasing. Hence at each eclipse, the light from the

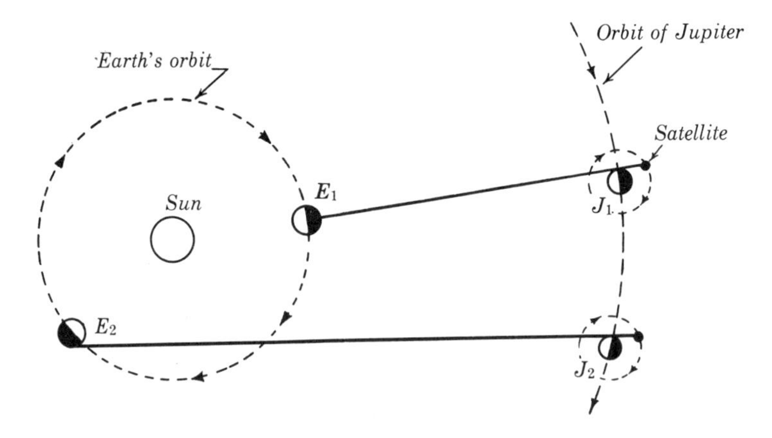

FIG. 1-9. Roemer's method of deducing the velocity of light.

satellite must travel a slightly greater distance than at the preceding eclipse and the observed time of revolution is slightly larger than the true time.

Roemer concluded from his observations that a time of about 22 minutes was required for light to travel a distance equal to the diameter of the earth's orbit. The best figure for this distance, in Roemer's time, was about 172,000,000 miles. Although there is no record that Roemer actually made the computation, had he used the data above he would have found a velocity of about 130,000 mi/sec or 2.1×10^8 m/sec.

Another means of determining the velocity of light from astronomical observations makes use of the *aberration* of light, first discovered by the English astronomer Bradley in 1728. The principle involved may be made clearer by means of an analogy. Suppose rain is falling vertically. Then in order for the drops to pass through a long tube, the axis of the tube must be held in a vertical direction as in Fig. 1-10 (a). If, however, one is walking along and carrying the tube, the drops entering at the upper end would strike the sides if the tube were held vertically. In order that they shall not, it is necessary to incline the tube forward in the direction of its motion as in Fig. 1-10 (b). The angle of inclination must be such that the lower end of the tube, at the time a falling drop reaches it, is directly under the point which was occupied by the upper end of the tube at the time the drop entered it. This angle may readily be found from the construction of Fig. 1-10 (c). Let the downward velocity of the drops be c, and the forward velocity of the tube be v. Then during the time t required for a falling drop to pass through the tube, or to cover a vertical distance ct, the lower end of the tube has moved a horizontal distance vt. If θ is the angle of inclination of the tube away from the vertical, evidently

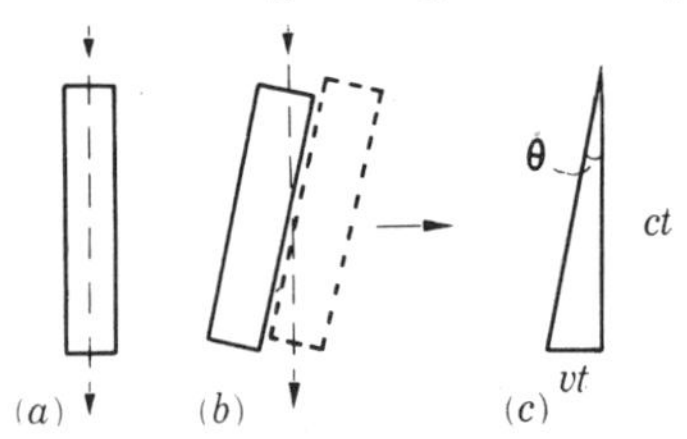

FIG. 1-10. Aberration of light.

$$\tan \theta = v/c.$$

Now let the tube in Fig. 1-10 represent a telescope and the drops of rain the light coming from a star. If the earth were at rest, the axis of the telescope would have to be aimed directly at the star, but because of the earth's motion in its orbit the telescope must be inclined forward by the angle θ.

In Fig. 1-11, the circle $ABCD$ represents the earth's orbit, with the sun S at its center. Point s is a star, on a line through the sun S at right

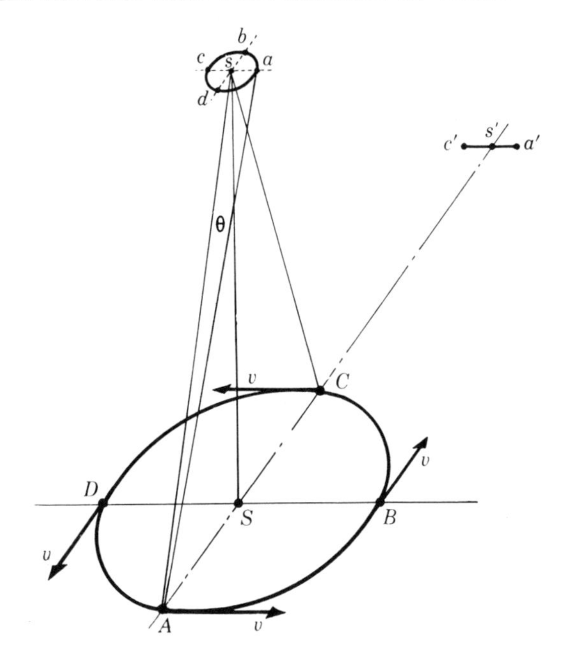

FIG. 1-11. Aberration of light from a star. $ABCD$ is the earth's orbit. The star at s appears to move in a circle; that at s' appears to oscillate along a line.

angles to the plane of the earth's orbit. When the earth is at A, the star appears to be at a, where the angle sAa corresponds to the angle θ in Fig. 1-10. Three months later, when the earth is at B, the star appears to be at b, and so on. Hence during the course of a year the star appears to describe a small circle subtending an angle 2θ (the measured value of 2θ is about 41 seconds) and from this angle and the known velocity v of the earth in its orbit, the velocity of light, which corresponds to c in Fig. 1-10, can be found.

A star such as s', lying in the plane of the earth's orbit, appears to oscillate along a line, being apparently at a' when the earth is at A, at c' when the earth is at C, and at its true position s' when the earth is at B or C and is moving directly toward the star. Stars in a direction making some angle other than 90° or 0° with the plane of the earth's orbit appear to move in ellipses.

Fig. 1-11 is not drawn to scale. The distances from the sun to the nearest stars are tremendous compared to the diameter of the earth's orbit.

In addition to the aberration due to the motion of the earth in its orbit, there is a small additional aberration due to its rotation about its axis, which amounts at the maximum to about 0.31 second.

The first determination of the velocity of light from purely terrestrial measurements was made by the French scientist Fizeau in 1849. A schematic diagram of his apparatus is given in Fig. 1-12. Lens L_1 forms an image of the light source S at a point near the rim of a toothed wheel T, which can be set into rapid rotation. G is an inclined plate of clear glass. Suppose first that the wheel is stationary and the light passes through one of the openings between the teeth. Lenses L_2 and L_3, which were separated by about 8.6 km, form a second image on the mirror M. The light is reflected from M, retraces its path, and is in part reflected from the glass plate G through the lens L_4 into the eye of an observer at E.

If the wheel T is set in rotation, the light from S is "chopped up" into a succession of wave trains of limited length. If the speed of rotation is such that by the time the front of one wave train has traveled to the mirror and returned, an opaque segment of the wheel has moved into the position formerly occupied by an open portion, no reflected light will reach the observer E. At twice this angular velocity, the light transmitted through any one opening will return through the next and an image of S will again be observed. From a knowledge of the angular velocity and radius of the wheel, the distance between openings, and the distance from wheel to mirror, the velocity of light may be computed. Fizeau's measurements were not of high precision. He obtained a value of 3.15×10^8 m/sec. Fizeau's apparatus was modified by Foucault, who replaced the toothed wheel with a rotating mirror. By introducing between the wheel and the mirror a tube filled with water, he proved that the velocity of light in water was less than in air. A corpuscular theory demands that it shall be greater and at the time these measurements were made they were taken as conclusive proof that a corpuscular theory was untenable.

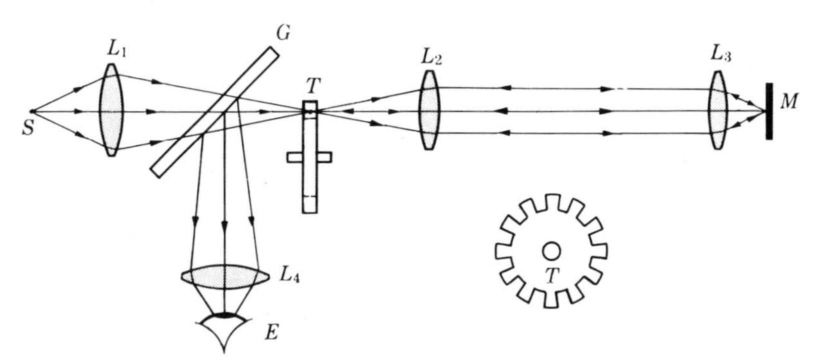

FIG. 1-12. Fizeau's toothed wheel method for measuring the velocity of light. S is a light source, L_1, L_2, and L_3 are lenses, T is the toothed wheel, M is a mirror, and G a glass plate.

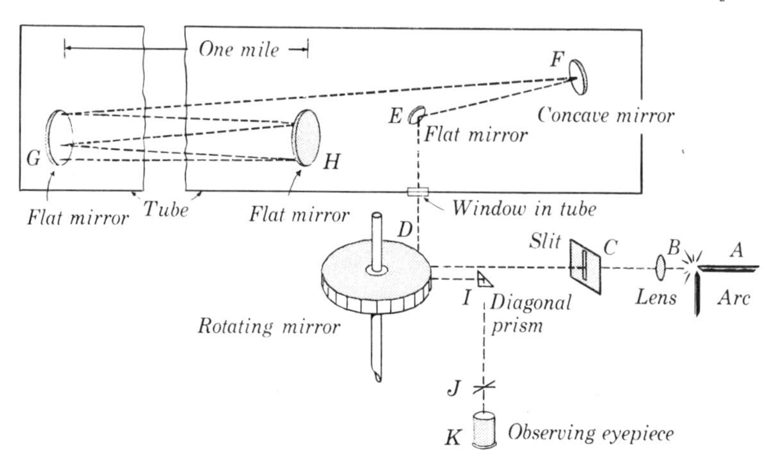

FIG. 1-13. Rotating mirror apparatus used by Michelson, Pease, and Pearson for measuring the velocity of light. A is an arc lamp, C a slit, B a lens, and D the rotating mirror. E, F, G, and H are other mirrors within a tube about 1 mile long. Diagonal prism I serves as a mirror to reflect the returned beam into the ocular K provided with measuring cross hairs J.

The most precise measurements by the Foucault method were made by the American physicist Albert A. Michelson (1852–1931). His first experiments were performed in 1878 while he was on the staff of the Naval Academy at Annapolis. The latest, which were under way at the time of his death, were completed in 1935 by Pease and Pearson. Michelson's latest setup is shown diagrammatically in Fig. 1-13. Light from an arc lamp A is imaged on slit C by lens B. It then strikes the upper half of one face of the rotating mirror D, which has 32 plane polished faces, and is reflected through a glass window set in the wall of a corrugated sheet steel tube about 1 mile in length. After reflection from the flat mirror E, the light strikes a concave mirror F. It next passes above the flat mirror H and, after repeated reflections between G and H, a magnified image of the slit is formed on the surface of G. The light then retraces its original path and emerges from the tube, striking this time the lower half of mirror D. The diagonal prism I (equivalent to a plane mirror) reflects the beam into the observing eyepiece K, provided with reference cross hairs J.

The repeated reflections between mirrors G and H resulted in an effective path length of 8 or 10 miles. The tube could be evacuated to a pressure of about 1 mm of mercury, so that only a small correction was required to obtain the velocity in free space.

If the mirror is not rotating, an observer sees an image of the slit S in a certain direction. When the mirror is rotating, light enters the tube after being reflected from some one of the mirror faces, but by the time the light has traveled the 8 or 10 miles and returned, this face will have turned through a small angle and the beam emerges in a direction slightly different from that when the mirror is at rest. The angle through which it is deflected can be measured and the velocity can be computed from a knowledge of this angle, the angular velocity of the mirror, and the effective path length of the system. The average value of a large number of measurements was

$$c = 2.99774 \times 10^8 \text{ m/sec.}$$

(The velocity of light in free space is always represented by the letter c.)

The most recent measurements of the velocity of light were made by W. C. Anderson at Harvard University between 1937 and 1941. Anderson's method is indicated in Fig. 1-14, which is reproduced from his report. The principle is best described in his own words:

"A light beam is passed through a modulator, where it is made to vary sinusoidally in intensity about some steady value. From the modulator, the beam passes through a half-silvered mirror, a portion being reflected from the surface over to a movable mirror. From this mirror the beam is returned, passing through the half-silvered mirror to a photoelectric cell.

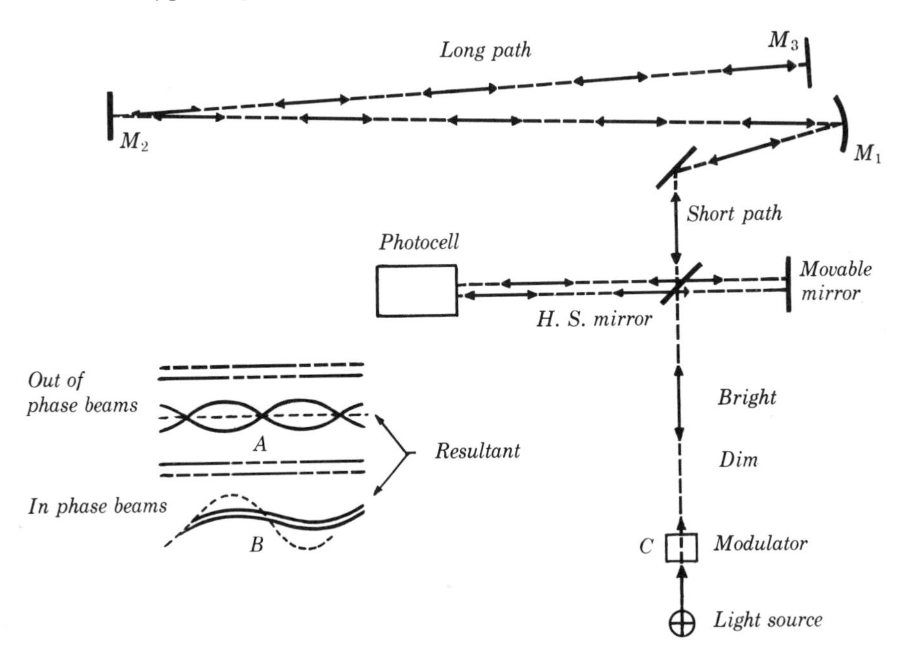

FIG. 1-14. Apparatus used by Anderson to measure the velocity of light.

The other portion of the original beam transmitted by the half-silvered mirror passes over the much longer path ($M_1M_2M_3$) and is returned along the same path, being reflected this time from the half-silvered mirror over to the same photoelectric cell. A tuned circuit converts these photoelectric currents into voltages which are amplified and recorded. It can be shown that the resultant voltage is dependent upon the phase relation of the original light modulations and this voltage will be either a maximum when the beams are in phase or a minimum when one beam is an odd number of half-cycles behind the other. By noting the path difference for a given minimum position, the velocity of light is readily computed by the relation:

$$c = 2fs/n,$$

where n = the number of half-cycles phase difference, s = the optical path difference between the two light beams, f = the frequency of modulation of the beams, and c = the velocity of light."

Anderson reports as his best value a velocity of 2.99776×10^8 m/sec.

In an exhaustive analysis of all work since 1928, E. N. Dorsey of the National Bureau of Standards concludes that the best value to date is

$$c = 2.99773 \times 10^8 \text{ m/sec},$$

which he believes correct within $\pm 0.00010 \times 10^8$ m/sec.

Electromagnetic theory predicts that the velocity of electromagnetic waves in free space is given by

$$c = \sqrt{\frac{1}{\mu_0 \epsilon_0}} \quad \text{(rationalized mks units)},$$

where, by definition

$$\mu_0 = 4\pi \times 10^{-7} \text{ newton-sec}^2/\text{coul}^2,$$

and the factor ϵ_0, found by an experiment equivalent to that of measuring the force between two point charges in vacuum, is

$$\epsilon_0 = (1/4\pi \times 8.9875 \times 10^9) \text{ coul}^2/\text{newton-m}^2.$$

(The most precise value is from measurements made by Rosa and Dorsey at the National Bureau of Standards.) Hence from the equation above

$$c = \sqrt{\frac{4\pi \times 8.9875 \times 10^9}{4\pi \times 10^{-7}} \times \frac{\text{newton-m}^2}{\text{coul}^2} \times \frac{\text{coul}^2}{\text{newton-sec}^2}}$$
$$= 2.9979 \times 10^8 \text{ m/sec}.$$

This is in excellent agreement with the measured velocity of light.

1-7 Index of refraction. The velocity of light stated in the preceding section is that in empty space. The velocity in material substances, with a few exceptions, is smaller. Furthermore, while light of all wave lengths travels with the same velocity in empty space, the velocity in material substances is different for different wave lengths. This effect is known as *dispersion* and will be discussed in more detail in Chap. 2. The ratio of the velocity of light in a vacuum to the velocity of light of a particular wave length in any substance is called the *index of refraction* of the substance *for light of that particular wave length.* We shall designate index of refraction by the letter n, stating, if necessary, the particular wave length to which it refers. If no wave length is stated, the index is usually assumed to be that corresponding to the yellow light from a sodium flame, of wave length 0.0000589 cm. Index of refraction is evidently a pure number (the ratio of two velocities) and in most instances is numerically greater than unity.

$$n = \frac{c}{v}.$$

The velocity of light in a gas is nearly equal to its velocity in free space and the dispersion is small. For example, the index of refraction of air at standard conditions, for violet light of wave length 0.00004359 cm, is 1.0002957, while for red light of wave length 0.00006563 cm the index is 1.0002914. It follows that for most purposes the velocity of light in air can be assumed equal to its velocity in free space or the index of refraction of air can be assumed unity. The index of refraction of a gas is directly proportional to its density.

The index of refraction of most of the common glasses used in optical instruments lies between 1.46 and 1.96. There are only a very few substances having indices larger than these values, diamond being one with an index of 2.42.

INDICES OF REFRACTION

(For light of wave length 0.0000589 cm)

Substance	Index
Glass	1.46–1.96
Iceland spar ($CaCO_3$)	1.658
Quartz (SiO_2)	1.544
Rock salt (NaCl)	1.544
Fluorite (CaF_2)	1.434
Carbon disulphide	1.629
Ethyl alcohol	1.361
Water	1.333

1-8 The wave length of light waves. Newton, in 1690, made a study of the colored fringes which surround the point of contact between a convex spherical surface, such as that of a lens, and a plane glass plate. These fringes are still called "Newton's rings." They are caused by interference effects between the light waves reflected from the convex and plane surfaces and are discussed further in Sec. 8-3. From his measurements of the diameters of the rings, Newton actually had sufficient data from which he could have computed the wave lengths of the light producing the rings. However, although Newton appears to have suspected the wave nature of light, he did not carry the calculation through. His own theory was that the surfaces were subject to "fits of easy reflection or transmission."

The first experimental measurements of the wave lengths of light waves were made about 1827 by Young, Fresnel, and Fraunhofer. The methods used, which involve the phenomena of interference and diffraction, will be described in Chaps. 8 and 9. The wave lengths turned out to be extremely small, much smaller than anyone had hitherto supposed. The shortest waves capable of affecting the sense of sight are violet and their wave length is about 0.00004 cm. The longest visible waves are red, of wave length about 0.00007 cm. (These limits are somewhat different for different observers.) The range of electromagnetic waves between these limits constitutes the *visible spectrum.*

Since wave lengths in the visible spectrum are so small, it is convenient to make use of a small unit in which to express them. The units in common use are (1) the micron, (2) the millimicron, and (3) the Angstrom unit. One micron, abbreviated 1 μ, is equal to 10^{-6} meter. One millimicron, abbreviated 1 mμ, is equal to 10^{-3} μ or 10^{-9} meter. One Angstrom unit (1 A) equals 10^{-10} meter.

$$\begin{aligned} &1 \text{ micron } (1\ \mu) &&= 10^{-6}\text{ m} = 10^{-4}\text{ cm.} \\ &1 \text{ millimicron } (1\ \text{m}\mu) &&= 10^{-9}\text{ m} = 10^{-7}\text{ cm.} \\ &1 \text{ Angstrom } (1\text{ A}) &&= 10^{-10}\text{ m} = 10^{-8}\text{ cm.} \end{aligned}$$

Most workers in the field of optical instrument design, color, and physiological optics, express wave lengths in millimicrons. In the field of spectroscopy, the Angstrom unit is more widely used.

For example, the wave length of the yellow light from a sodium flame, which is 0.0000589 cm, would be written

$$\lambda = 0.589\ \mu = 589\ \text{m}\mu = 5890\text{ A.}$$

One of the fundamental relations applying to any sort of wave motion is that the velocity of propagation, v, equals the product of wave length λ and frequency f.

$$v = f\lambda.$$

It was stated earlier that the velocity of propagation of light in a transparent medium is not equal to its velocity in free space. When light passes from a vacuum into a transparent medium, or from one medium into another having a different index, the frequency is the same in both media, since the number of waves leaving the boundary surface in any time interval must equal the number arriving. Hence the wave length changes whenever the velocity changes. Thus the wave length of a light wave in water, where the velocity is but $\frac{3}{4}$ of that in free space, is $\frac{3}{4}$ of the wave length of the same wave in free space. The wave length of sodium light, which is 589 mμ in free space, is $\frac{3}{4} \times 589$, or 478 mμ, in water.

1-9 The electromagnetic spectrum. As far as their fundamental nature is concerned, there is no difference between light waves and other electromagnetic waves such as those from an oscillating electrical circuit. It will be worth while to survey briefly the electromagnetic spectrum, using the term "spectrum" to designate the entire range of electromagnetic waves, just as the visible spectrum includes those waves capable of stimulating the sense of sight.

There is no limit to the longest electromagnetic waves that can be produced. By operating an A.C. generator sufficiently slowly, the frequency f may be made as small as desired. The wave length of the waves radiated by a 60-cycle transmission line is

$$\lambda = \frac{c}{f}$$

$$= \frac{3 \times 10^8}{60}$$

$$= 5 \times 10^6 \text{ m} = 5000 \text{ km}.$$

By increasing the generator speed and the number of poles, the frequency may be increased up to about 100,000 cycles/sec, corresponding to a wave length of 3 km, but mechanical difficulties set a limit to this method. Higher frequencies may be developed by oscillating electrical circuits, the frequency of such an oscillation being given by

$$f = \frac{1}{2\pi}\sqrt{\frac{1}{LC}},$$

where L and C are respectively the inductance and capacitance of the circuit. Electromagnetic waves of the order of a centimeter in length (so-called microwaves) can be produced by vacuum tube oscillators. Even shorter waves have been produced by spark-excited circuits. Using electrodes of thin metal foil, 0.2 mm by 0.2 mm, Nichols and Tear produced waves as short as 1.8 mm.

The shortest microwaves are shorter than the longest waves that have been detected in the radiant energy from infrared sources. Waves as long as 4 mm have been observed in the energy radiated by a mercury vapor lamp. Radiant energy originating in infrared sources is identical in all its properties with that of the same frequency produced by electrical oscillations.

The waves from light sources are emitted by molecules and atoms and extend from the relatively long infrared waves through the visible spectrum and into the ultraviolet. Some sort of excitation of a molecule or atom is required before it can emit radiant energy. This excitation may be brought about by the thermal agitation of the molecules or it may be acquired by collision processes in an electrical discharge. The molecules store up energy temporarily and release it in the form of electromagnetic waves. The emission is a quantum phenomenon and the greater the energy associated with the process, the higher the frequency, or the shorter the wave length, of the waves emitted. For moderate amounts of energy, only the outer electrons of an atom take part in the process, but when atoms are bombarded with electrons of large kinetic energy, their inner electrons may be displaced and radiant energy of extremely short wave length is emitted when the atom returns to its normal state. The deceleration of the bombarding electrons also gives rise to electromagnetic radiation. Waves produced in this way are called *x-rays*, and their wave lengths extend from about 10^{-5} cm to about 10^{-10} cm. The only limit at the short wave length end seems to be set by the difficulties of obtaining high-speed electrons for the bombarding process.

Waves of even shorter wave length accompany the spontaneous disruptions of atomic nuclei in the processes of radioactive disintegration. These waves are called *gamma rays*. Electrons and positively charged helium nuclei are also ejected in radioactive disintegrations. Before their corpuscular nature was recognized, they were given the names of beta rays and alpha rays, respectively, and these names are still used, but beta particles and alpha particles are to be preferred.

There are today no gaps in the electromagnetic spectrum. All frequencies, from those of gamma rays at one end of the spectrum to radio waves at the other, may be produced and studied. Each portion of the spectrum overlaps the adjacent portions at both the long and short wave length ends. That is, the shortest waves produced by x-ray methods are shorter than the longest gamma rays, and so on. No sharp dividing lines can be drawn between various portions of the spectrum, which are all alike in their fundamental nature and differ only in wave length or frequency.

A chart of the electromagnetic spectrum is given in Fig. 1-15. Note the relatively small portion occupied by the visible spectrum.

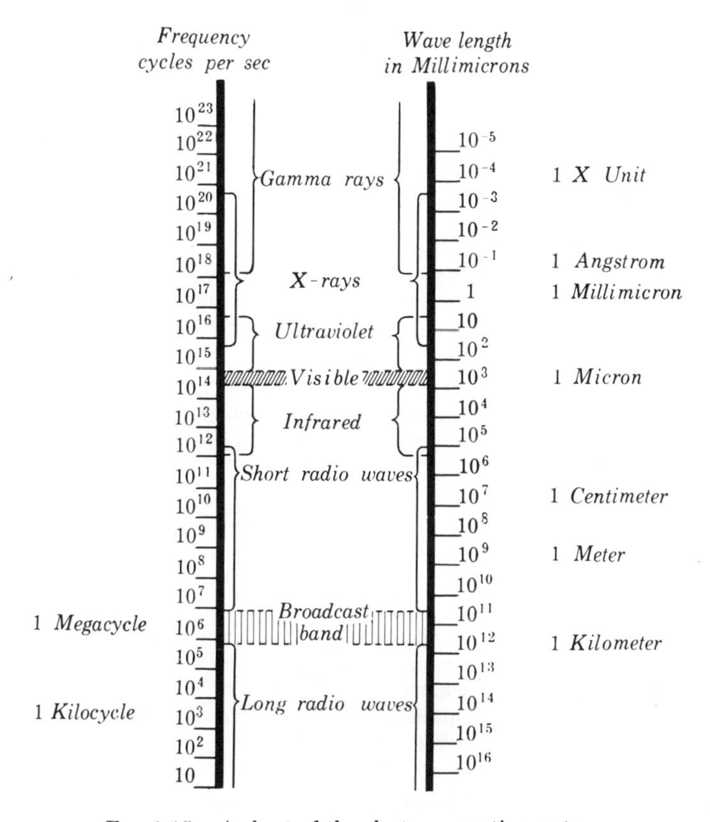

Fig. 1-15. A chart of the electromagnetic spectrum.

Problems—Chapter 1

(1) The diameters of the sun, earth, and moon are respectively 864,000 mi, 7920 mi, and 2160 mi. The distances from the earth to the sun and to the moon vary somewhat, but suppose an eclipse of the sun takes place when the distance from earth to sun is 92,900,000 mi and the distance from earth to moon is 226,000 mi. Compute the length of the conical umbra of the moon's shadow and compare with the distance from the moon to the earth's surface.

(2) Use the figure for the radius of the earth's orbit in Prob. 1, and the best value of the velocity of light, to compute the time required for light to travel a distance equal to the diameter of the earth's orbit. Compare with Roemer's value of 22 minutes.

(3) Compute the velocity of light from the diameter of the earth's orbit, the time of one revolution, and the measured value of the aberration of light from the stars, which is 20.47 seconds.

(4) Fizeau's measurements of the velocity of light were continued by Cornu, using Fizeau's apparatus but with the distance between mirrors increased to 22.9 km. One of the toothed wheels used was 40 mm in diameter and had 180 teeth. Find the angular velocity at which it should rotate in order that light transmitted through one opening will return through the next.

(5) (a) What is the velocity of light of wave length 500 mμ (in vacuum), in glass whose index at this wave length is 1.50? (b) What is the wave length of these waves in the glass?

(6) A glass plate 3 mm thick, of index 1.50, is placed between a point source of light of wave length 600 mμ (in vacuum) and a screen. The distance from source to screen is 3 cm. How many waves are there between source and screen?

(7) A beaker 10 cm deep is filled with alcohol ($n = 1.361$) while an identical beaker contains a layer of water ($n = 1.333$) upon which floats a layer of mineral oil ($n = 1.473$) sufficient in thickness to fill the beaker. The thickness of the layer of mineral oil is such that each beaker contains the same number of waves when light is passed vertically through them. How thick is the layer of mineral oil?

(8) Find the ratio of the thickness of a layer of water ($n = 1.33$) to the thickness of a layer of mineral oil ($n = 1.47$) if the minimum time required for light to traverse the layer is the same for each.

(9) A glass rod 3 m long, of index 1.50, conducts a pulse of light started simultaneously with another pulse of light which covers the same distance through air. What is the difference in the time required for the pulses to reach an observer, and which one arrives first?

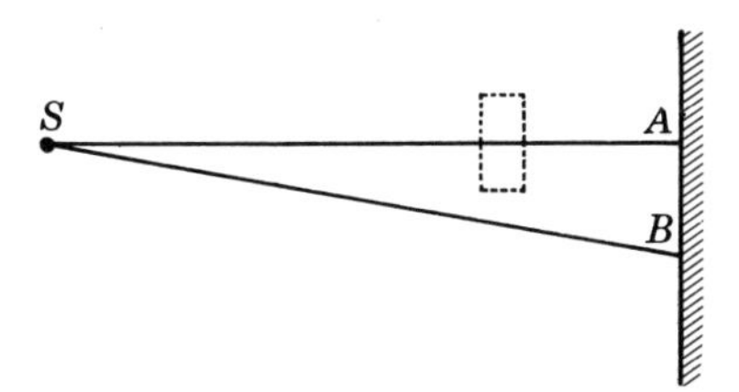

FIG. 1-16.

(10) A point light source S in Fig. 1-16 emits light of wave length 500 mμ in air. A and B are two points 1 cm apart on a screen 100 cm from S. (a) How many more waves are there in the path SB than in the path SA? (b) A plate of glass of index 1.50 is inserted in the path SA, with its faces normal to SA. What thickness is required, if the number of waves in the path SA is equal to the number in the path SB?

(11) The visible spectrum includes a wave length range from about 400 mμ to about 700 mμ. Express these wave lengths in inches.

(12) (a) What is the frequency of light waves of wave length 500 mμ? (b) What is the frequency of x-rays of wave length 1A?

CHAPTER 2

REFLECTION AND REFRACTION AT PLANE SURFACES

2-1 Reflection and refraction at plane surfaces. As the first step in the study of the reflection and refraction of light by mirrors, prisms, and lenses, we consider the general problem of a train of plane electromagnetic waves, traveling in one medium and incident on a plane surface bounding a second medium in which the velocity of propagation differs from that in the first. It might be expected that if both media are transparent the incident wave train will merely continue on into the second medium. Common experience, however, tells us that this is not what happens. Everyone has seen the image of the sun reflected from the surface of a body of water, or an image formed by reflection from the surface of a pane of clear glass. Furthermore, the broken appearance of an oar dipped in water and the bending of light rays by a prism show that, in general, a train of light waves changes direction when it crosses a boundary surface. Fig. 2-1 illustrates what actually happens to the train of incident waves. A *reflected* wave train and a transmitted or *refracted* wave train, originate at the boundary surface. That is, except in certain special cases, only a part of the incident light passes into the second medium, the remainder being reflected. Furthermore, the directions of travel of the reflected and transmitted waves (again except in special cases) are different from that of the incident wave.

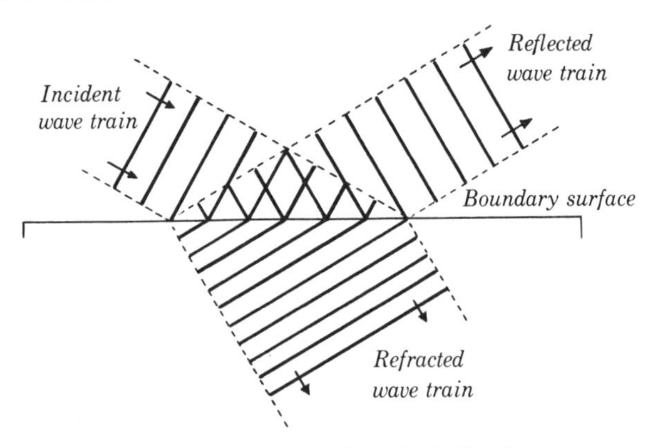

FIG. 2-1. Reflection and refraction of a train of plane waves.

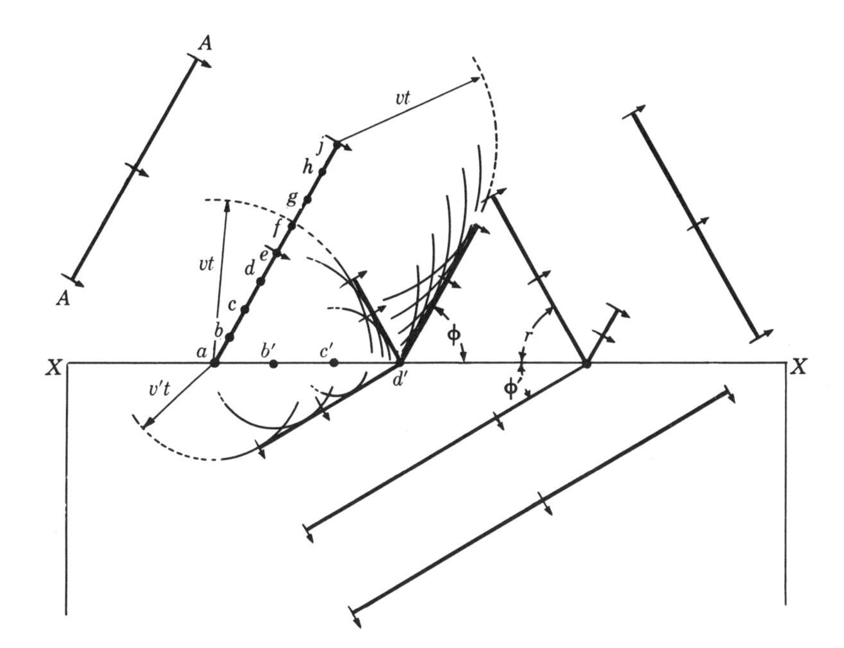

FIG. 2-2. Use of Huygens' construction to find the angles of reflection and refraction of a plane wave front.

2-2 The laws of reflection and refraction. Fig. 2-2 traces the course of a plane wave front A–A, traveling as indicated by the short arrows, as it is incident on a plane surface represented by the line X–X. The wave front and the surface are at right angles to the plane of the diagram. The velocity of propagation in the medium above the boundary plane is v and the index of refraction is $n = c/v$. The velocity in the lower medium is v' and the corresponding index is $n' = c/v'$. The diagram has been drawn for the case in which $v > v'$ and $n' > n$.

Consider an instant at which the lower edge of the incident wave front is just making contact with the boundary surface along a line through point a. Let us make use of Huygens' construction to find the shape of the wave fronts after a time interval t, equal to that required for the incident wave to advance a distance dd'. At the end of this interval the incident wave is making contact with the boundary surface along a line through d'.

Since the incident wave gives rise to both a reflected and a refracted wave, we must consider that two Huygens wavelets originate at point a, one spreading out in the upper medium with a velocity v, the other in the lower medium with a velocity v'. After a time interval t, the radii of these wavelets are respectively vt and $v't$, as shown. As the incident wave advances, point b makes contact with the surface at point b', and again two wavelets originate at this point. These wavelets start out at a somewhat later time than do those from point a, and hence at the end of the time interval t their radii are smaller than those of the wavelets centered at a. In the same way, two more wavelets start out from point c' at a still later time. Tangent lines to the reflected and refracted wavelets with centers at points a, b', and c', give the shapes of the reflected and refracted wave fronts. As indicated, these are planes at right angles to the diagram, intersecting one another, the boundary surface, and the incident wave front, along a line through d'. The Huygens wavelets from points d to j all travel in the upper medium with velocity v, and their common tangent is a plane parallel to that of the incident wave front.

The angle ϕ between the incident wave front and the surface is called the *angle of incidence*; the angle ϕ' between the refracted wave front and the surface is the *angle of refraction*; and the angle r between the reflected wave front and the surface is the *angle of reflection*. We next derive the relations between these angles.

Fig. 2-3 is the same as a portion of Fig. 2-2, redrawn for clarity, with the Huygens wavelets omitted. Point A in Fig. 2-3 corresponds to point a in Fig. 2-2, point B to point d, and point C to point d'. Evidently, from the construction of Fig. 2-2,

$$BC = vt, \quad AD = vt, \quad AE = v't. \tag{2-1}$$

Also, since the triangles ABC, ADC, and AEC are right triangles,

$$AC = \frac{BC}{\sin \phi} = \frac{AD}{\sin r} = \frac{AE}{\sin \phi'}. \tag{2-2}$$

Inserting in Eq. (2-2) the expressions for BC, AD, and AE from Eqs. (2-1), we get

$$\frac{vt}{\sin \phi} = \frac{vt}{\sin r} = \frac{v't}{\sin \phi'},$$

or

$$\frac{\sin \phi}{v} = \frac{\sin r}{v} = \frac{\sin \phi'}{v'}.$$

From the first two terms,

$$\phi = r, \tag{2-3}$$

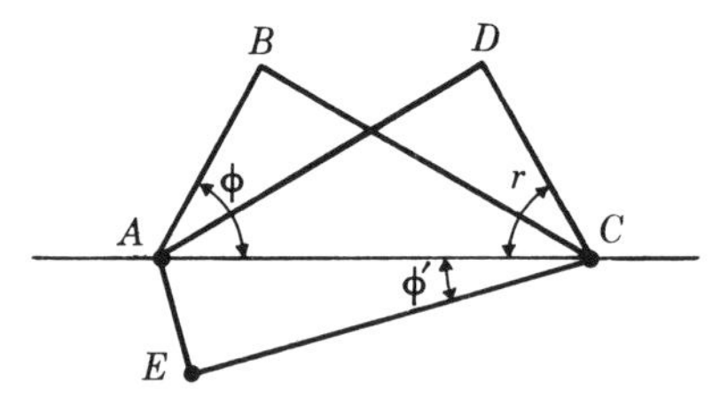

FIG. 2-3.

and from the first and third,

$$\frac{\sin \phi}{\sin \phi'} = \frac{v}{v'} \cdot \tag{2-4}$$

Eq. (2-3) is the law of reflection: *a plane wave is reflected from a plane surface with the angle of reflection equal to the angle of incidence*, and Eq. (2-4) is the law of refraction: *the ratio of the sine of the angle of incidence to the sine of the angle of refraction is equal to the ratio of the velocities in the two media.*

Since by definition

$$n = c/v \quad \text{and} \quad n' = c/v',$$

it follows that

$$v/v' = n'/n,$$

and Eq. (2-4) may be written

$$\frac{\sin \phi}{\sin \phi'} = \frac{n'}{n},$$

or

$$\boxed{n \sin \phi = n' \sin \phi'.} \tag{2-5}$$

This is a more useful equation, since the index of refraction of a substance is the property which is usually tabulated, rather than the velocity of light in that substance. Since for any pair of substances the ratio n'/n is a constant, Eq. (2-5) is equivalent to

$$\frac{\sin \phi}{\sin \phi'} = \text{constant.}$$

The discovery that the sines of the angles of incidence and refraction stand in a constant ratio to one another is usually credited to Willebrord Snell, in 1621, although there seems to be some doubt whether it was original with him. In accord with common usage, however, we shall refer to Eq. (2-5) as *Snell's law*.

The preceding discussion gives only the relation between the *directions* of the incident, reflected, and refracted wave trains. To find what fraction of the incident light is reflected and what fraction is refracted, one must set up the boundary conditions that have to be satisfied by the electric and magnetic fields in the respective wave trains. The problem is worked out in some detail in Chap. 7, where it is shown that the proportions reflected and refracted depend on the state of polarization of the incident light, on the indices of refraction of the two media, and on the angle of incidence. At this point it will suffice to give the results in

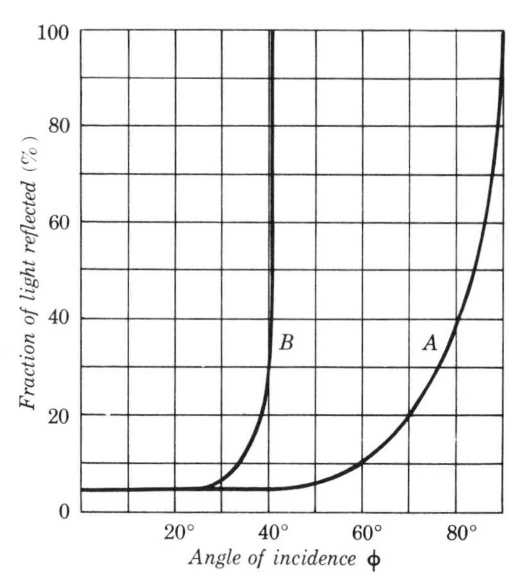

FIG. 2-4. Fraction of incident light reflected as a function of angle of incidence. Curve A, index of second medium greater than that of first; Curve B, index of first medium greater than that of second.

graphical form only, and for so-called *natural* or *unpolarized* light (see Sec. 7-1). The graphs are shown in Fig. 2-4, for the case in which one medium is air, of index 1.000, while the index of the other medium is 1.523, a typical value for optical glass. Curve A gives the fraction of the incident light reflected, as a function of the angle of incidence ϕ, when the first medium is air and the second glass. Curve B applies when the first medium is glass and the second is air. Both curves practically coincide for angles of incidence less than 30°, and up to this angle the fraction reflected is of the order of 5%. Curve A (light traveling in air before reflection) then gradually rises to 100% at an angle of incidence of 90°, while curve B (light traveling in glass before reflection) rises sharply to 100% at an angle of incidence of about 41°, and remains at 100% from that point on. This effect, known as *total reflection*, is discussed further in Sec. 2-8.

2-3 Ray treatment of reflection and refraction. It is possible to analyze the passage of light through any optical system by successive applications of Huygens' principle to the wave fronts, as was done in the preceding section at a single surface. However, it is often much simpler to trace a few rays through the system. The wave fronts, if desired, may then be constructed perpendicular to the rays.

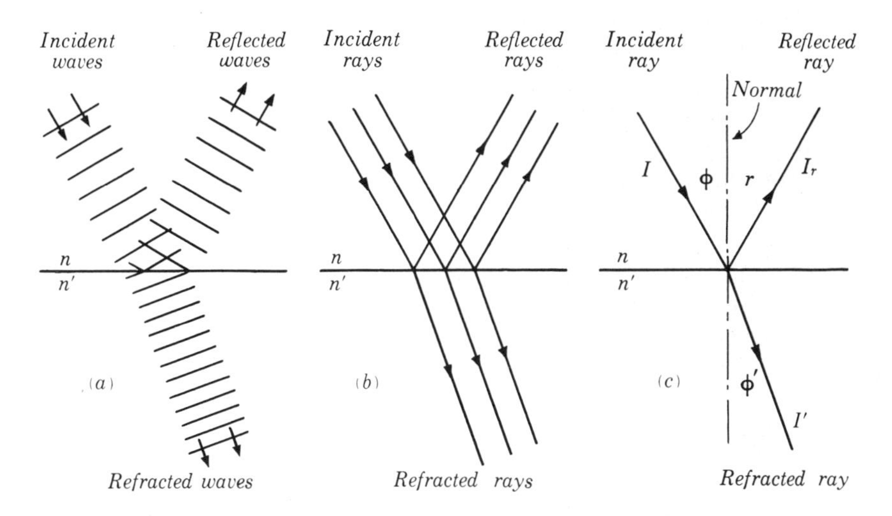

FIG. 2-5. (a) A train of plane waves is in part reflected and in part refracted at the boundary between two media. (b) The waves in (a) are represented by rays. (c) For simplicity, only a single incident, reflected, and refracted ray are drawn.

The simple case of reflection or refraction of plane waves at a plane surface is illustrated in Fig. 2-5. In Fig. 2-5 (a) there is shown a train of plane wave fronts incident on a plane boundary surface, together with the reflected and refracted wave fronts. In Fig. 2-5 (b) the same wave trains are represented by rays. Fig. 2-5 (c) shows a single incident ray I, together with the refracted ray I' and the reflected ray I_r. The angle between any *wave front* and the *surface* is equal to the angle between the corresponding *ray* and the *normal* to the surface, since the sides of the angles are mutually perpendicular. Thus in Fig. 2-5 (c), ϕ is the angle of incidence, ϕ' is the angle of refraction, and r is the angle of reflection. The planes of the incident and refracted waves and the plane of the boundary surface are all perpendicular to the plane of the diagram. It follows that the corresponding rays and the surface normal, which are perpendicular to these planes, *all lie in the same plane,* that is, the plane of the diagram. The plane determined by an incident ray and the normal to the surface at the point of incidence is called the *plane of incidence.*

It is evident from Snell's law that the angle of refraction is always less than the angle of incidence for a ray passing from a medium of smaller into one of larger index, as from air into glass. In such a case the ray is bent *toward* the normal. If the light is traveling in the opposite direction, the reverse is true and the ray is bent *away from* the normal.

To summarize the laws of reflection and refraction in terms of rays:

When a ray of light is reflected, the angle of reflection is equal to the angle of incidence. The incident ray, the reflected ray, and the normal to the surface at the point of incidence, all lie in the same plane.

When a ray of light is refracted, $n \sin \phi = n' \sin \phi'$. *The incident ray, the refracted ray, and the normal to the surface at the point of incidence, all lie in the same plane.*

Examples. (1) In Fig. 2-5, let the upper medium be water of index 1.33 and the lower glass of index 1.50. Let the ray I be incident at an angle of 45° with the normal. The reflected ray then makes an angle of 45° with the normal also. To find the angle of refraction, we have from the given data,

$$n = 1.33,\ n' = 1.50,\ \sin \phi = \sin 45° = 0.707.$$

Application of Snell's law gives

$$1.33 \times .707 = 1.50 \sin \phi',$$
$$\sin \phi' = 0.627,$$
$$\phi' = 38.5°.$$

(2) Suppose that light is incident from *below* on the same boundary surface as in the preceding example, at an angle of incidence of 38.5°. Find the angle of refraction. We now have

$$n = 1.50, \quad n' = 1.33, \quad \sin \phi = \sin 38.5° = 0.627.$$

That is, unprimed indices and angles refer to the medium in which the light is traveling *before* refraction, primed indices and angles to the medium in which it is traveling *after* refraction. Application of Snell's law gives

$$\sin \phi' = 0.707,$$
$$\phi' = 45°.$$

Hence if the direction of a ray is reversed, the ray retraces its original path.

Just as in the preceding example, a part of the incident light is reflected at the surface with the angle of reflection equal to the angle of incidence.

2-4 Fermat's principle of least time. The laws of reflection and refraction may be shown to follow from a general principle first stated by Fermat in 1658. Fermat's principle states that the path of a light ray from one point to another is that which requires the least time.[1]

If the medium in which the light is traveling is homogeneous, the path of least time is the path of least distance. Hence in such a medium the

[1] Strictly speaking, the time along the actual path is an extremum. That is, if the path is varied slightly the difference between the times along the actual path and the varied path is an infinitesimal of higher order than the displacement between the paths. In most instances, the actual path is the path of minimum time.

rays are straight lines. If the light passes through several such media in succession, the path in each medium must be straight or the time in that medium would not be the shortest possible.

To derive the law of reflection, consider a plane reflecting surface MM (Fig. 2-6). It will first be shown that the incident ray, the reflected ray, and the normal to the surface at the point of incidence, all lie in the same plane. Suppose a light ray from A is reflected at point C, and passes through B after reflection. Construct the plane through A and B perpendicular to the plane MM, and draw CO from C perpendicular to this plane. Now, unless C and O coincide, $AC > AO$ and $CB > OB$. Hence the time along ACB is greater than the time along AOB, which is contrary to Fermat's principle. Therefore C and O must coincide and the rays AO, OB, and the normal to MM at O all lie in the same plane (AOB), which was to be proved.

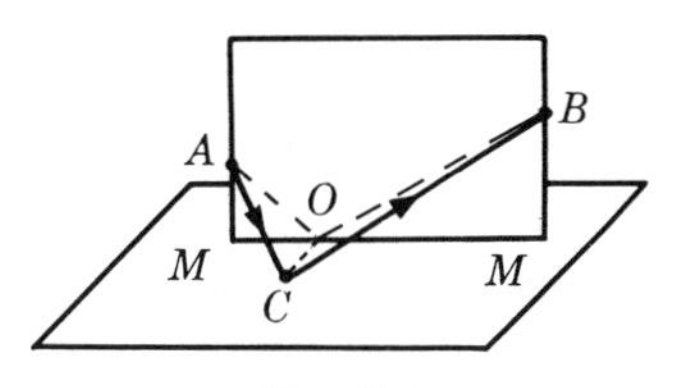

FIG. 2-6.

We now locate the point O so that the time from A to B is a minimum. In Fig. 2-7, the plane of the diagram represents the vertical plane in Fig. 2-6 and the points lettered A, O, and B have the same meaning as in Fig. 2-6. We assume that O may lie anywhere along the line MM. Drop perpendiculars from A and B to MM, and erect the normal at point O. Angles ϕ and r are, respectively, the angles of incidence and reflection. Let v represent the velocity of propagation. The length of path is $s + s_1$, and the time t along the path is

$$t = \frac{s + s_1}{v}.$$

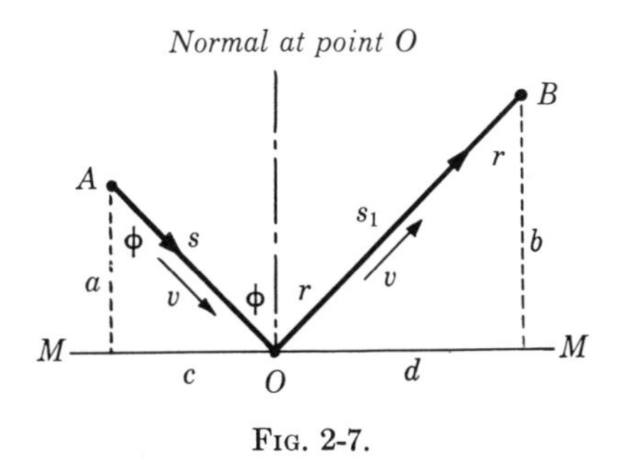

FIG. 2-7.

It is easily seen from the diagram that

$$s = a \sec \phi, \quad s_1 = b \sec r.$$

It follows from the two preceding equations that

$$t = \frac{1}{v}(a \sec \phi + b \sec r).$$

If the point O is displaced slightly, the angles ϕ and r will change by $d\phi$ and dr, and the corresponding change dt in the time is

$$dt = \frac{1}{v}(a \sec \phi \tan \phi \, d\phi + b \sec r \tan r \, dr).$$

If the time is a minimum, $dt = 0$ and

$$a \sec \phi \tan \phi \, d\phi = -b \sec r \tan r \, dr.$$

The differentials $d\phi$ and dr are not independent however. It follows from Fig. 2-7 that

$$c + d = \text{const} = a \tan \phi + b \tan r.$$

Taking differentials of both sides, we obtain

$$0 = a \sec^2\phi \, d\phi + b \sec^2 r \, dr.$$

When the condition for minimum time is combined with this equation, we obtain

$$\sin \phi = \sin r,$$

and hence

$$\phi = r.$$

That is, the light path *AOB* that is traversed in the shortest time is that for which the angle of reflection equals the angle of incidence.

Snell's law may be derived in a similar manner. The proof that the incident ray, the refracted ray, and the normal to the refracting surface all lie in the same plane will not be given, as it is essentially the same as the corresponding proof for the case of reflection. In Fig. 2-8, *MM* represents the boundary plane between two substances having indices of refraction n and n', in which the corresponding velocities are v and v'.

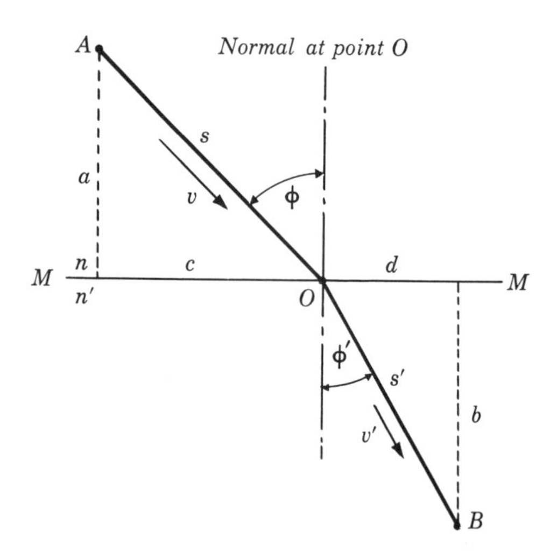

FIG. 2-8.

AOB is the path of a ray from A to B, and ϕ and ϕ' are the angles of incidence and of refraction. The time from A to B is

$$t = \frac{s}{v} + \frac{s'}{v'} = \frac{a \sec \phi}{v} + \frac{b \sec \phi'}{v}.$$

If the point O is displaced slightly,

$$dt = \frac{a \sec \phi \tan \phi \, d\phi}{v} + \frac{b \sec \phi' \tan \phi' \, d\phi'}{v'}.$$

If the time is a minimum, $dt = 0$ and

$$\frac{a \sec \phi \tan \phi \, d\phi}{v} = -\frac{b \sec \phi' \tan \phi' \, d\phi'}{v'}.$$

Also, since $c + d = \text{const}$,

$$a \sec^2\phi \, d\phi = -b \sec^2\phi' \, d\phi'.$$

Dividing one of the preceding equations by the other, we obtain

$$\frac{\sin \phi}{v} = \frac{\sin \phi'}{v'}.$$

Since $v = c/n$ and $v' = c/n'$, this reduces to

$$n \sin \phi = n' \sin \phi',$$

the familiar form of Snell's law.

Fermat's principle is one of a number of "minimal" principles in physics. For example, all of the laws of mechanics may be derived from Hamilton's principle of least action, which states that the motion of any mechanical system is such that the integral of the "action" is a minimum along the path taken by the system, the "action" being the energy of the system multiplied by the differential of time.

2-5 Reflection of a spherical wave at a plane surface. The wave fronts from a point source in a homogeneous medium are spherical surfaces concentric with the source. Consider such a source P, in Fig. 2-9 (a), above a plane reflecting surface MM'. Let AVA' be the trace of a wave front just making contact with the surface at V. It will be seen from an application of Huygens' construction that the wave front AA' takes on successively the shapes $BDED'B'$ and CC', and is reflected as another spherical wave whose center is at P', where PP' is perpendicular to the plane MM' and the distances from P and P' to the plane are equal.

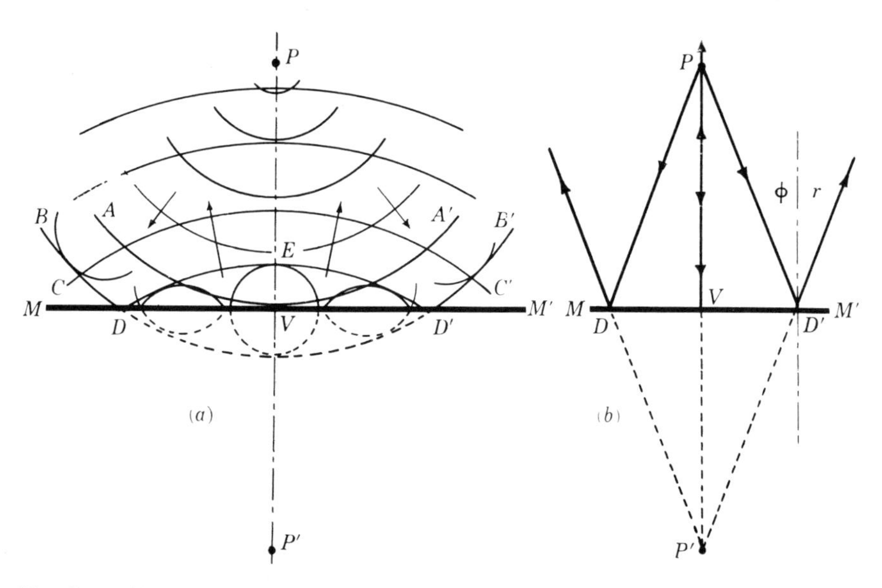

FIG. 2-9. (a) A spherical wave is reflected at a plane surface. (b) The wave in (a) is represented by a few rays.

The same result is obtained by considering the rays from P, as in Fig. 2-9 (b). The ray striking the surface at D' is reflected with the angle r equal to the angle ϕ. Construct the perpendicular from P to the surface and extend it below the surface. Extend the reflected ray backward from D' until it intersects the perpendicular at P'. Then the triangles PVD' and $P'VD'$ are equal, and $PV = P'V$. Since this is true whatever the angle ϕ, then all rays from P which strike the surface *appear* to diverge from P' after reflection. The reflected wave fronts, which are the surfaces normal to the rays, are then spheres with P' as a center.

2-6 Images in plane mirrors. Virtual and real images. The wave fronts or rays which actually originate at point P in Fig. 2-9, *appear* to diverge from the point P' after reflection from the mirror MM'. Point P' is called the *image* of the point P. It follows from the preceding section that *the image of a point object formed by a plane mirror lies on the normal to the mirror and is as far behind the mirror as the object is in front of it.*

The image of a finite object, such as the arrow AB in Fig. 2-10, may be found from the rule above. To every point of the arrow there corresponds an image point behind the mirror. Thus the image of A is at A', the image of B is at B', etc. The totality of all these point images is the image of the arrow.

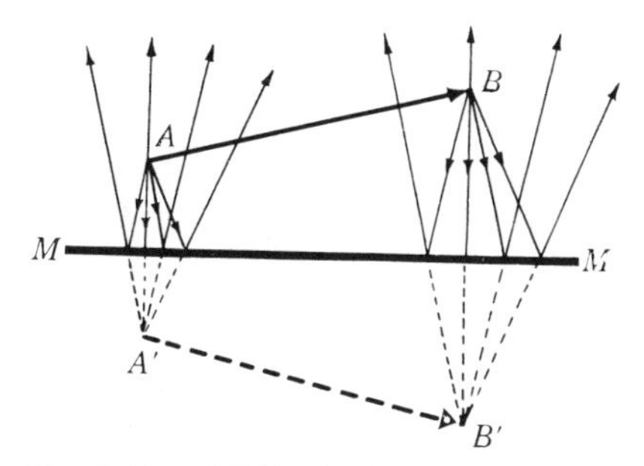

FIG. 2-10. $A'B'$ is the image of the arrow AB, formed by reflection at a plane mirror.

Images such as these, in which the wave fronts do not actually originate at the image but only appear to do so, are called *virtual*. Later on, in connection with spherical mirrors and lenses, we shall find cases in which rays diverging from a point source are made to converge and pass through a common point, beyond which they again diverge just as from an actual object. Images of this sort are called *real*.

In Fig. 2-11, E represents the eye of an observer and P is a point source. A portion of the wave fronts and some of the rays from P are shown. The observer sees the source only by means of the small portion of the wave fronts included within the narrow pencil of rays entering the pupil of his eye.

Suppose the eye of an observer is placed at E in Fig. 2-12, which corresponds to Fig. 2-9. The observer sees the virtual image P' only by means of the small portion of the reflected wave fronts which enter his eye. However, since these wave fronts have precisely the same curvature as would those which originated at an actual point source at P', the eye treats these waves just as it would those from an actual source at P'.

It is not necessary that the reflecting surface extend over the entire plane MM' for an image of P to be formed, or even that the surface inter-

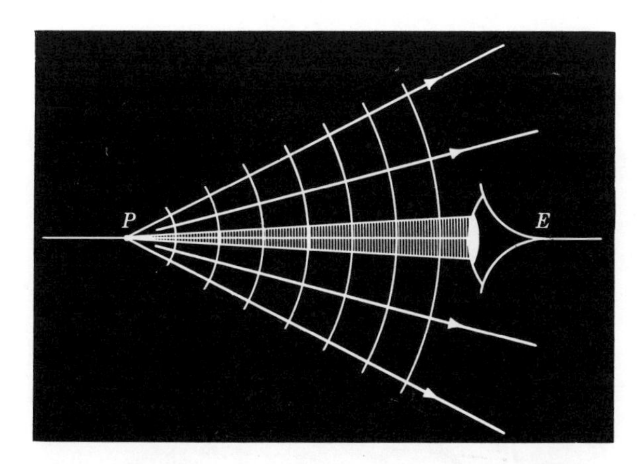

FIG. 2-11. Observer E sees source P by means of the small shaded pencil of rays.

sect the line PP'. Thus, in Fig. 2-12, suppose that the mirror extends only over the region indicated by the heavy line AB. The rays from the source to the eye will still proceed as shown and an image of P can still be seen at P', but only by an observer within the cone of rays limited by AQ and BR. Of course, the observer must look toward the image P' in order to see it. When we say, then, that the image of an object formed by a plane mirror lies on the normal to the mirror, it is implied that one means the normal to the *plane* of the mirror, even though the normal may not intersect the mirror itself.

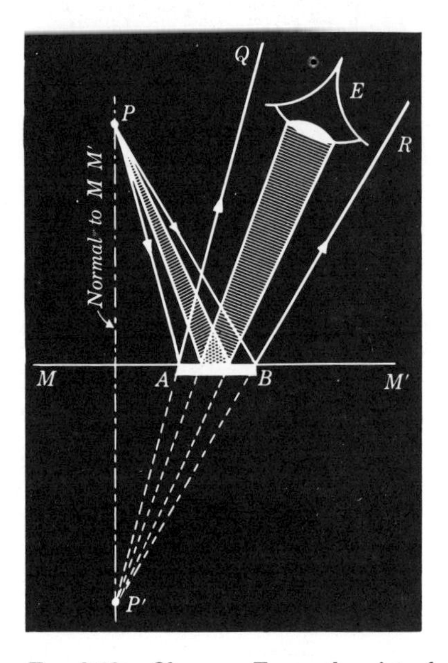

FIG. 2-12. Observer E sees the virtual image P' by means of the shaded pencil of rays.

Example. Find the position of the image of the arrow OP in Fig. 2-13, formed by the plane mirrors MV and $M'V$.

Extend the plane of the mirror MV to m. Construct lines from P and O perpendicular to this plane, and make $P'a = Pa$, $O'b = Ob$. Then $O'P'$ is the image of OP, formed by the first mirror. The second mirror, $M'V$, forms an image of this image. Extend the plane of $M'V$ to m'. Construct lines from O' and P', perpendicular to this plane, and make $P''c = P'c$, $O''d = O'd$. $O''P''$ is the final image.

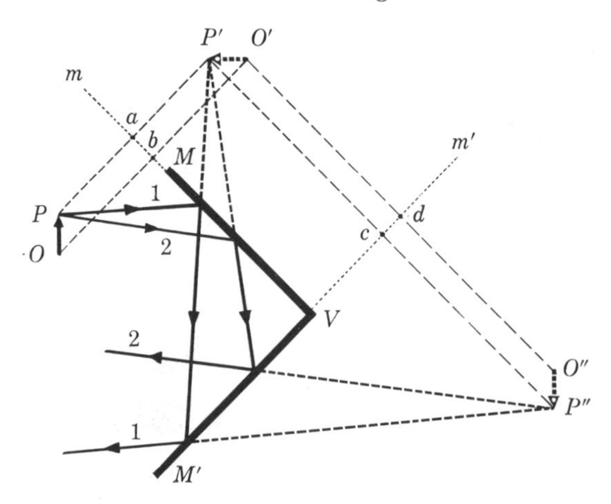

FIG. 2-13. Images formed by successive reflections at two plane mirrors. $O'P'$ is the image of OP, $O''P''$ is the image of $O'P'$.

The construction may be verified by drawing a few rays from P. Ray 1, for example, may be drawn from P in any arbitrary direction to the mirror MV, and its direction after reflection found from the law of reflection. When the reflected ray is projected backward it will be found to pass through P'. The same ray after reflection from $M'V$, when projected backward, will be found to pass through P''.

One type of so-called *retrodirective reflector*, used as a warning signal at danger spots on highways, is a metal cup several inches in diameter. Beginning about an inch from the mouth, the sides are flattened to form three mutually perpendicular reflecting surfaces which meet in a blunt point at the bottom of the cup. The mouth is hermetically sealed with a red cover glass.

The three reflecting surfaces of the inside of the cup have the property that any ray of light which has been reflected successively from all three surfaces is exactly reversed in direction, as illustrated in Fig. 2-14. Thus a headlight beam striking the reflector is projected directly back toward its source, whether or not the source is directly in front of the reflector. The slight spread in the reflected beam that is necessary for it to reach the

FIG. 2-14. A retrodirective reflector, consisting of three mutually perpendicular reflecting surfaces.

eyes of the driver is provided by making the surfaces of the cover glass curved, so as to introduce a small amount of lens action. Another type of retrodirective reflector is described in Sec. 4-4.

In most instances, the object of which an optical system forms an image is three-dimensional. Since to every point in an object there corresponds a point in its image, the image is three-dimensional also. While it often suffices to represent an object and its image by a single arrow as in Fig. 2-13, a complete understanding of image formation (from the geometrical viewpoint) can be attained only through a study of three-dimensional objects and their images.

Consider the three-dimensional image, formed by a plane mirror, of the object represented by the three mutually perpendicular arrows *oa*, *ob*, and *oc*, in Fig. 2-15. The (virtual) image of every point in the object lies on the normal to the mirror and as far behind the mirror as the object is in front of it. Thus, while arrows $o'b'$ and $o'c'$ are parallel to their objects, $o'a'$ is reversed relative to *oa*. When an observer stands at the mirror and faces the object, the arrow *oa* points toward him, *ob* points to his right, and *oc* points upward. In order to see the image, the observer must turn around and look toward the mirror. The arrow $o'a'$ then points toward him as did *oa*, arrow $o'c'$ points upward as did *oc*, but $o'b'$ points toward the left instead of the right. Thus, although the arrow *ob* and its image, $o'b'$, are parallel *in space*, the reversal in direction of light at the mirror surface makes it necessary for an observer who has been looking at the object to turn around in order to see the image and, because he must turn around, the image, to him, appears reversed from left to right with respect to the object. It is evident also that since he turns about an axis parallel to *oc* and $o'c'$, these arrows point in the same direction in both object and image, and since *oa* and $o'a'$ are opposite in direction, they point toward him whichever way he faces.

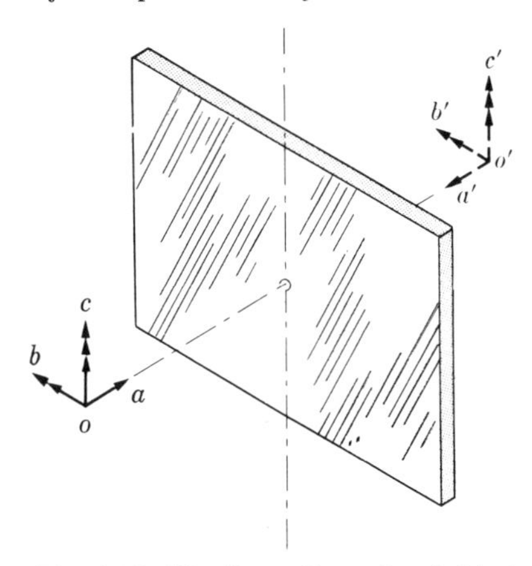

FIG. 2-15. The three-dimensional object *oabc* has a three-dimensional image $o'a'b'c'$. The image is said to be perverted.

The object and its image are related to one another in the same way as are a right hand and a left hand. The reader may easily verify this by applying the "left-hand rule" to the object and the "right-hand rule" to its image. That is, point the thumbs of the two hands along *oa* and $o'a'$, the forefingers along oc and $o'c'$, and the center fingers along ob and $o'b'$. This is in agreement with the familiar fact that the image of a left hand in a plane mirror is a right hand.

When the transverse dimensions of an image are parallel to and in the same sense as the corresponding dimensions in the object, as in Fig. 2-15, the image is said to be *erect*. When image and object are related as are a right and left hand, the image is called *perverted*. Thus a plane mirror forms an erect but perverted image.

2-7 Refraction of a spherical wave at a plane surface. In Fig. 2-16, the X–Z plane represents the boundary surface between two media having different indices of refraction. Suppose light waves originate at a point P in the lower medium at a distance y below the X–Y plane, and that the index n of the medium below the X–Z plane is greater than the index n'

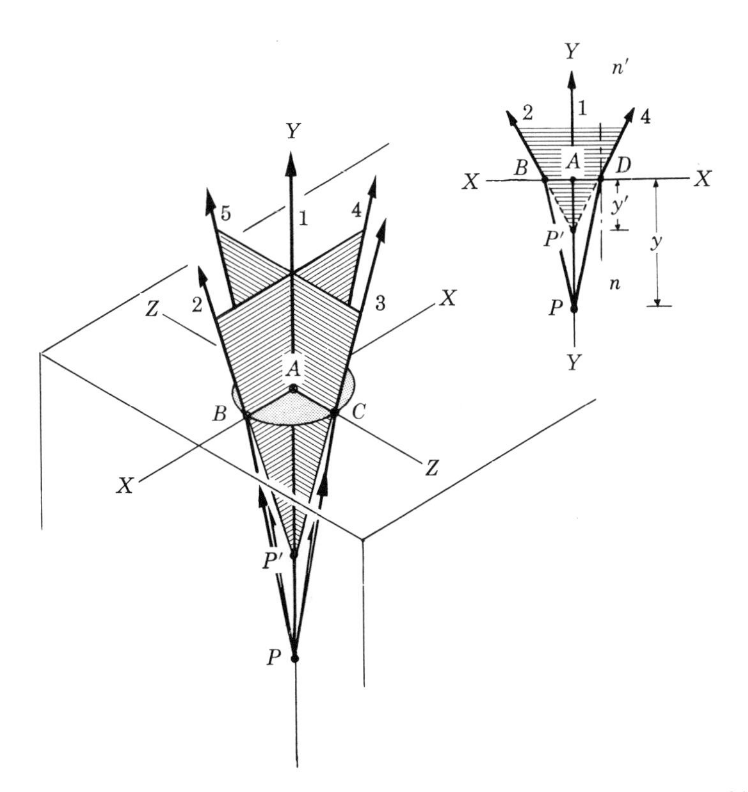

FIG. 2-16. Refraction at a plane surface. Rays near the normal form a virtual image of P at P'

of the medium above this plane. For example, P might be a point beneath the surface of a calm body of water. The wave fronts originating at P, and traveling in the lower medium, are spherical surfaces with centers at P. If Huygens' principle is applied to find the shape of the wave fronts after refraction at the boundary plane, it will be found that these are *not* spherical surfaces. That is, the refracted waves do not appear to diverge from a common point and the surface does not form a point image of a point object. The mathematical expression for the shape of the refracted wave front is so complex as to make it impracticable for purposes of computation. It is here that the simplifications afforded by the ray method are first apparent.

Let us consider a hollow circular cone of rays diverging from point P and intersecting the boundary surface in a circle with center at A, directly above P. Two cross sections of the cone, in the X–Y and Z–Y planes, are indicated by shading. Four rays in the cone are shown in the diagram, numbered 2, 3, 4, and 5, and intersecting the surface at points B, C, D, and E respectively. After refraction, these rays are deviated away from the normal and if projected backward appear to diverge from point P', directly above P. Ray 1, incident on the surface at point A, is normal to the surface before and after refraction.

We wish to compute the distance of point P' below the boundary surface. Let us call this distance y'. In the small inset figure, the X–Y plane lies in the plane of the diagram. Evidently the angle of incidence ϕ of ray 4 equals the angle APD, and the angle of refraction ϕ' equals the angle $AP'D$. From Snell's law,

$$n \sin \phi = n' \sin \phi'. \tag{2-6}$$

It will be seen from the diagram that

$$y \tan \phi = y' \tan \phi', \tag{2-7}$$

since both of these products are equal to AD.

Dividing Eq. (2-7) by Eq. (2-6) gives

$$\frac{y \tan \phi}{n \sin \phi} = \frac{y' \tan \phi'}{\sin \phi'},$$

or

$$y' = y \times \frac{n'}{n} \times \frac{\cos \phi'}{\cos \phi}. \tag{2-8}$$

The ratio $\cos \phi'/\cos \phi$ varies with the angle of incidence ϕ, so that the distance y' is not the same for rays that diverge from P at other angles than those shown in the diagram. This is equivalent to the statement made earlier that the refracted waves are not spherical and do not appear to spread out from a common point.

Suppose that an observer looks vertically down from a point on the Y-axis above the boundary surface. The small diameter of the pupil of the eye limits the cone of rays received by it to an extremely small angle, so that the angles ϕ and ϕ' are both very small. The cosine of any angle less than 2 degrees is equal to 1, to within an accuracy of one-half of one percent. Hence, for such small angles, we can set $\cos \phi' = \cos \phi$, and Eq. (2-8) becomes

$$y' = y \frac{n'}{n} \cdot \tag{2-9}$$

That is, the distance y' is nearly the same for all rays making small angles with the normal, so that all rays from P which lie in a narrow pencil near the normal appear to diverge from the same point P' after refraction by the surface. This is equivalent to stating that the *central portions* of the emergent wave fronts above the boundary surface may be considered as spherical surfaces with centers at P'. In other words, a narrow pencil of rays near the normal forms a point image of P at P' and an observer looking vertically down into a body of water sees well-defined images of objects beneath its surfacc at an apparent depth given by Eq. (2-9). Since the index of refraction of water is 1.33, the apparent depth of objects beneath a water surface, when viewed normally to this surface, is only about ¾ of the actual depth.

When the line of sight is not at right angles to the refracting surface, the situation is more complicated. Fig. 2-17 shows a hollow cone of rays diverging from point P and intersecting a boundary surface in a circle with center at A. When projected backwards, the emergent rays pass through a short vertical line at P', and also through a second short horizontal line at P''. There is thus no one point from which the refracted rays appear to diverge, and no point image of P is formed. The pencil of refracted rays is said to be *astigmatic*, and the mutually perpendicular *lines* at P' and P'' are the "images" of P. The astigmatic image at P' is directly above P and, if the cone of rays is small, it can be shown that the distance y' below the surface is

$$y' = y \frac{(1 - n^2 \sin^2 \phi)^{1/2}}{n \cos \phi},$$

where ϕ is the angle of incidence of the central ray of the cone.

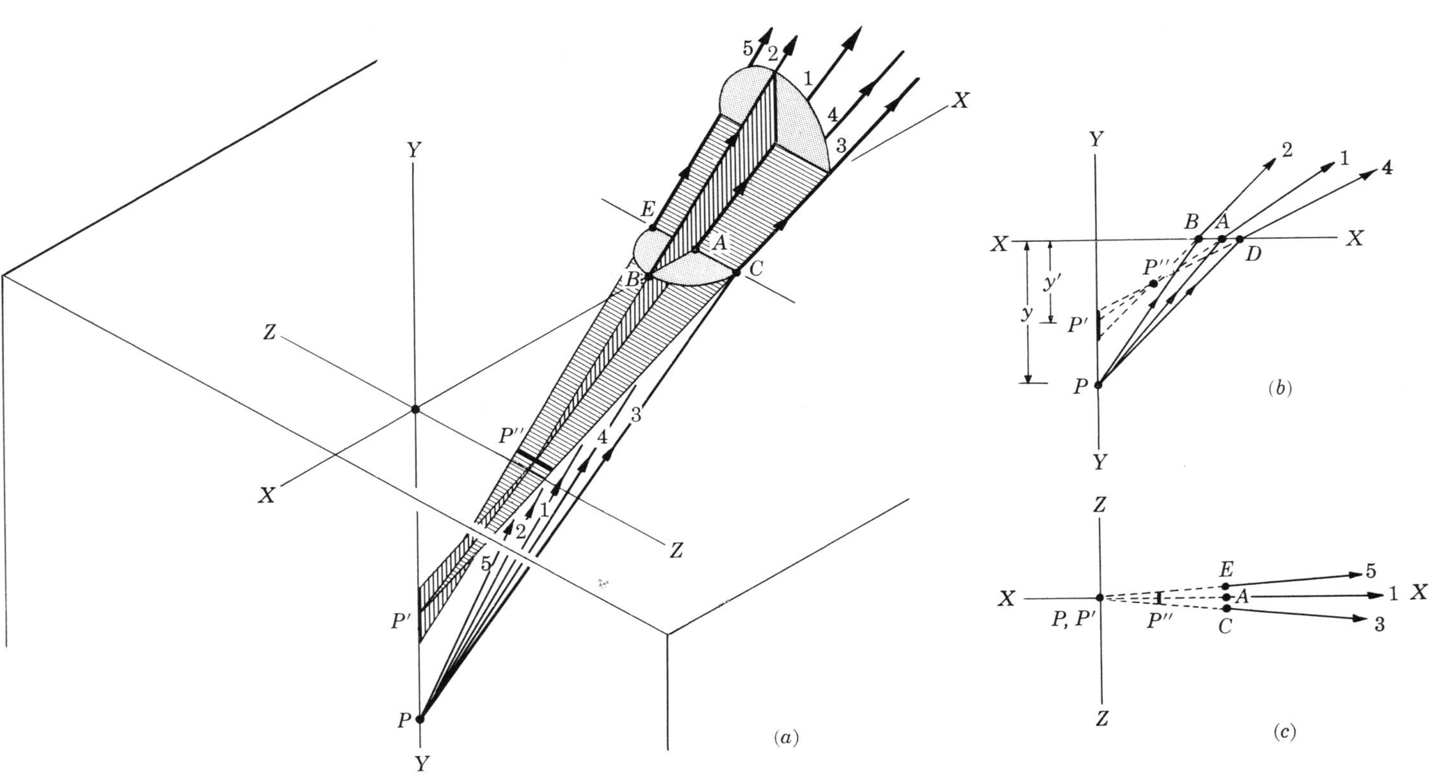

FIG. 2-17. P' and P'' are the astigmatic images of P, formed by refraction at a plane surface.

The image P'' is above and to the right of P. Its coordinates x'' and y'' are

$$x'' = y\,(n^2 - 1)\tan^3\phi,$$

$$y'' = y\,\frac{[1 - (n^2 - 1)\tan^2\phi]^{3/2}}{n}$$

(The positive directions along the X- and Y-axes are taken to the right and downward respectively.)

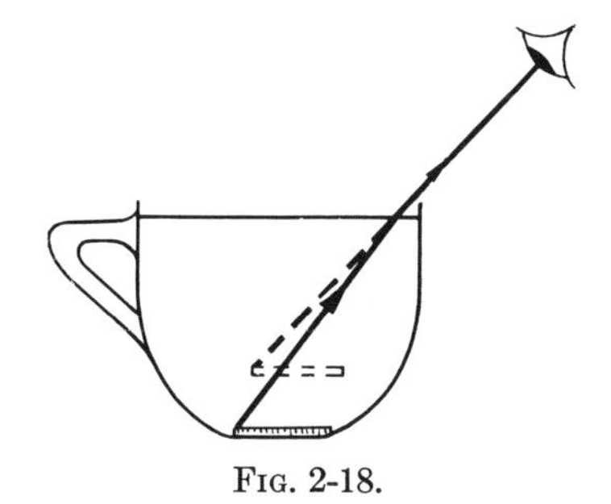

FIG. 2-18.

The astigmatic images of an underwater object are of importance in certain aspects of naval warfare, such as the visual observation of unknown reefs or the observation from the air of submerged submarines.

The "lifting" effect produced by refraction is the basis of one of the earliest recorded experiments in optics, one which was known by the ancient Greeks. A coin is placed in the bottom of an empty vessel as in Fig. 2-18 and the eye of an observer placed in such a position that the coin is just hidden below the edge of the vessel. If water is poured into the vessel the coin appears to rise and come into view. The same effect is responsible for the apparent bend in a straight stick partially immersed in water at an angle with the surface, the immersed portion appearing higher than it actually is.

2-8 Total internal reflection. Fig. 2-19 shows a number of rays diverging from a point source P in a medium of index n and striking the surface of a second medium of index n', where $n > n'$. From Snell's law,

$$\sin\phi' = \frac{n}{n'}\sin\phi.$$

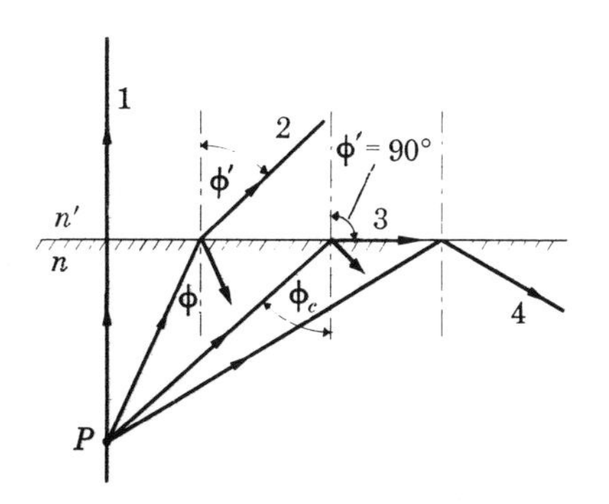

FIG. 2-19. Total internal reflection. The angle of incidence ϕ_c, for which the angle of refraction is 90°, is called the critical angle.

Since n/n' is greater than unity, $\sin\phi'$ is always larger than $\sin\phi$ and evidently equals unity (i.e., $\phi' = 90°$) for some angle ϕ less than 90°. This is illustrated by ray 3 in the diagram, which emerges just grazing the surface at an angle of refraction of 90°. The angle of incidence for which the refracted ray emerges tangent to the surface is called the *critical angle* and is designated by ϕ_c in the diagram. If the angle of incidence is greater than the critical angle, the sine of the angle of refraction, as computed by Snell's law, is greater than unity. This may be interpreted to mean that beyond the critical angle the ray does not pass into the upper medium but is *totally internally reflected* at the boundary surface. Total internal reflection can take place only when a ray is incident on the surface of a medium whose index is *smaller* than that of the medium in which the ray is traveling.

Notice carefully that the reflection of light at a boundary surface does not set in suddenly at the critical angle, but that the approach to total reflection is a gradual one. When a ray strikes the boundary surface between two transparent substances at any angle less than the critical angle, *both* reflection and refraction occur.

When the index of the first medium is greater than that of the second, the fraction of the incident light that is reflected increases with increasing angle of incidence, as in curve B in Fig. 2-4. The angle at which the fraction of light reflected has increased to 100% (about 41° in Fig. 2-4) is the critical angle.

The critical angle for two given substances may be found by setting $\phi' = 90°$ or $\sin \phi' = 1$ in Snell's law. We then have

$$\boxed{\sin \phi_c = \frac{n'}{n}\cdot} \tag{2-10}$$

For a water-air surface,

$$\sin \phi_c = \frac{1}{1.33} = 0.75,$$

$$\phi_c = 48.5°.$$

Only that part of the light from an underwater source which is included within a cone of half-angle 48.5° is refracted into the space above, the remainder being totally internally reflected. Conversely, while light incident on the surface from above at any angle is in part refracted into the water, all of the light after refraction is confined within a cone of this angle. Thus a diver looking upward from beneath the surface can see an object in any position in the space above, but all objects appear to lie within a cone of half-angle 48.5°.

2-9 Reflecting prisms. The critical angle of an air-glass surface, taking 1.50 as a typical index of refraction of glass, is

$$\sin \phi_c = \frac{1}{1.50} = 0.67, \quad \phi_c = 42°.$$

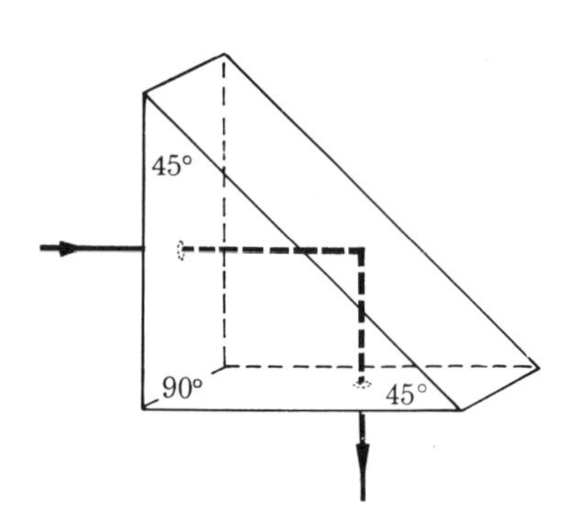

Fig. 2-20. A totally reflecting prism.

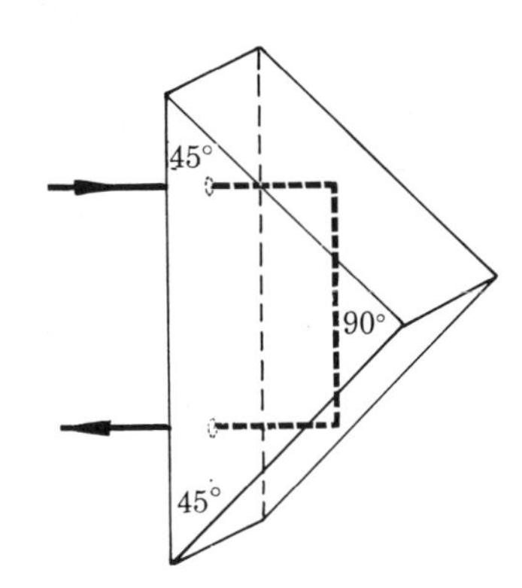

Fig. 2-21. The Porro prism.

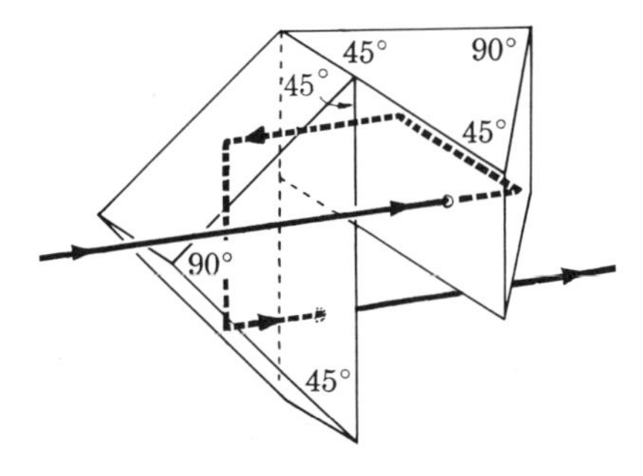

Fig. 2-22. A combination of two Porro prisms.

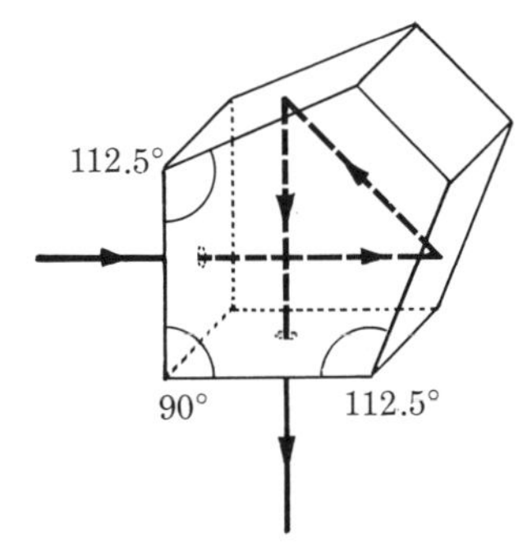

Fig. 2-23. The penta prism.

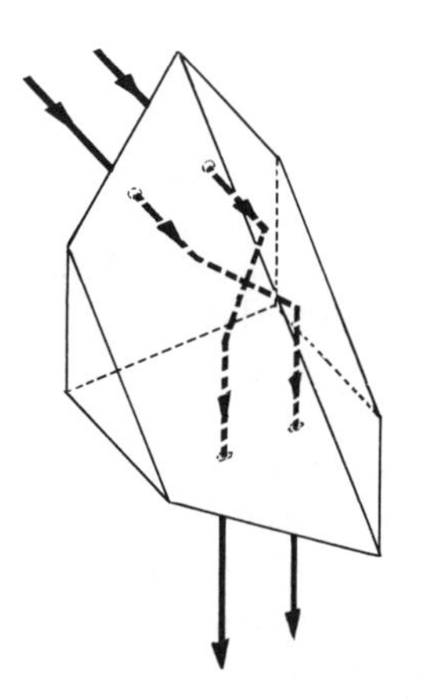

Fig. 2-24. The Amici prism.

This angle, very conveniently, is slightly less than 45°, which makes possible the use in many optical instruments of prisms of angles 45°-45°-90° as totally reflecting surfaces. The advantages of totally reflecting prisms over metallic surfaces as reflectors are, first, that the light is *totally* reflected, while no metallic surface reflects 100% of the light incident on it, and second, the reflecting properties are permanent and not affected by tarnishing. Offsetting these is the fact that there is some loss of light by reflection at the surfaces where light enters and leaves the prism, although recently discovered methods of coating the surfaces with so-called "nonreflecting" films can reduce this loss considerably.

The simplest type of reflecting prism is shown in Fig. 2-20. Its angles are 45°-45°-90°. Light incident normally on one of the shorter faces strikes the inclined face at an angle of incidence of 45°. This is greater than the critical angle, so the light is totally internally reflected and emerges from the second of the shorter faces after undergoing a deviation of 90°.

A 45°-45°-90° prism, used as in Fig. 2-21 is called a *Porro* prism. Light enters and leaves at right angles to the hypotenuse and is reflected at each of the shorter faces. The deviation is 180°. Two Porro prisms are often combined, as in Fig. 2-22.

While a ray incident normally on one of the shorter faces of a 45°-45°-90° prism is deviated through 90°, if the ray is not incident exactly at right angles it is refracted both on entering and emerging from the prism and the deviation is no longer 90°. The prism shown in Fig. 2-23, known as a *penta* prism because of its pentagonal shape, has the important property of producing a deviation of exactly 90° even in rays that are not incident exactly at right angles to its faces, provided only that the rays are parallel to the plane of the diagram. Analysis will show that the rays are incident on the inclined faces at angles smaller than the critical angle, so these faces must be silvered. Penta prisms are used in rangefinders (see Sec. 6-13) where the deviation must be exactly 90°.

Fig. 2-24 illustrates another useful type, the *Amici* prism, in which the hypotenuse of a 45°-45°-90° prism is ground away to form two intersecting surfaces at right angles to one another. Two rays are traced through the prism. The image is inverted and simultaneously deviated through 90°.

2-10 Refraction by a plane parallel plate. Suppose light is incident at an angle ϕ_1 (as in Fig. 2-25) on the upper surface of a transparent plate, the surfaces of the plate being plane and parallel to one another. Let ϕ_1' be the angle of refraction at the upper surface, and ϕ_2 and ϕ_2' the angles of incidence and refraction at the lower surface. Let n be the index of the medium on either side of the plate, and let the index of the plate be n'.

The figure is drawn with $n' > n$. From Snell's law,

$$n \sin \phi_1 = n' \sin \phi_1',$$

$$n' \sin \phi_2 = n \sin \phi_2'.$$

But as is evident from the diagram,

$$\phi_1' = \phi_2.$$

Combining these relations, we find

$$\phi_1 = \phi_2'.$$

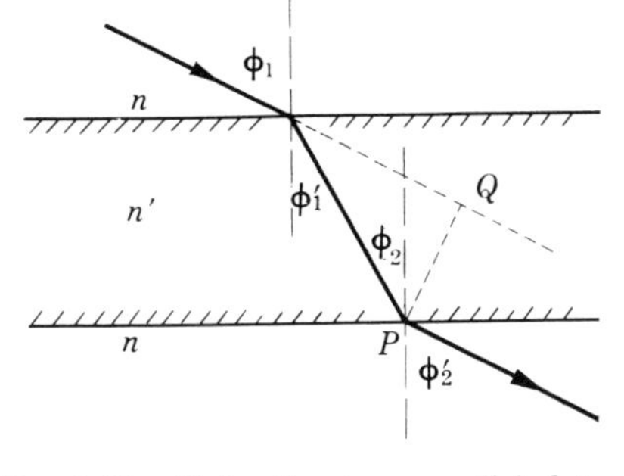

FIG. 2-25. Refraction by a parallel plate.

That is, the emergent ray is parallel to the incident ray. It is not deviated in passing through the plate but is *displaced* by the distance PQ. It will be left as an exercise to show that a ray of light passing through any number of plane parallel plates of different index is not deviated but only displaced from its original path, provided it emerges into a medium of the same index as that in which it was originally traveling.

2-11 Refraction by a prism. The prism, in one or another of its many forms, is second only to the lens as the most useful single piece of optical apparatus. Totally reflecting prisms have been mentioned briefly. We consider now the *deviation* and the *dispersion* produced by a prism.

Consider a light ray incident at an angle ϕ on one face of a prism as in Fig. 2-26 (a). Let the index of the prism be n, the included angle at the apex be A, and let the medium on either side of the prism be air. It is desired to find the *angle of deviation*, δ. This is a straightforward problem in surveying. One has only to apply Snell's law at the first surface, compute the angle of refraction, then by geometry find the angle of inci-

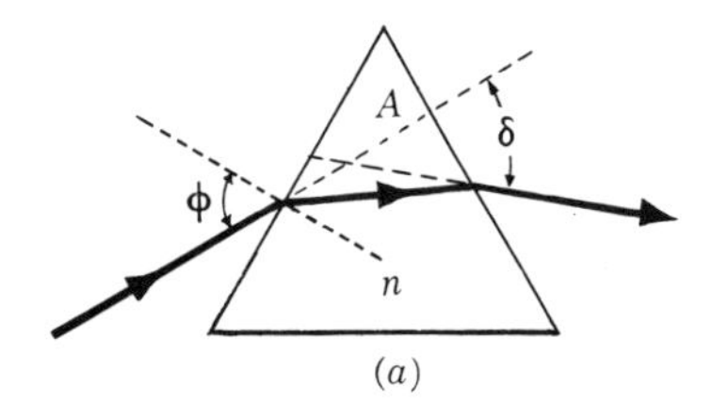

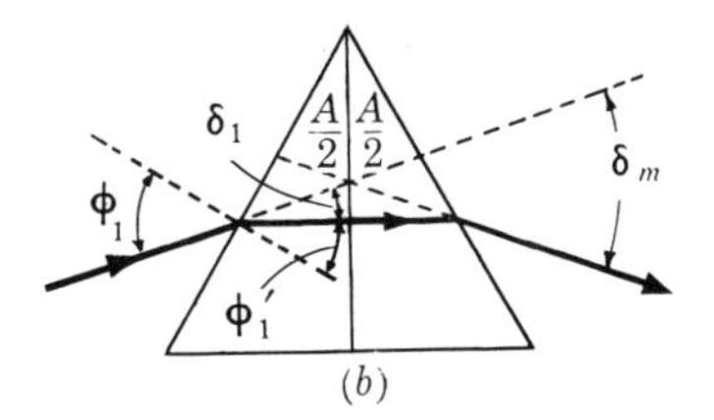

FIG. 2-26. (a) Deviation by a prism. (b) The deviation is a minimum when the ray passes through the prism symmetrically.

dence at the second surface and from a second application of Snell's law find the angle of refraction at the second surface. The direction of the emergent ray is then known and the angle of deviation may be found.

While the method is simple enough, the expression for the angle δ turns out in the general case to be rather complicated. However, as the angle of incidence is, say, decreased from a large value, the angle of deviation decreases at first and then increases, and is a minimum when the ray passes through the prism symmetrically as in Fig. 2-26 (b). The angle δ_m is then called the angle of *minimum deviation* and in this special case it is related to the angle of the prism and its index by the equation

$$n = \frac{\sin \dfrac{A + \delta_m}{2}}{\sin \dfrac{A}{2}}. \tag{2-11}$$

To derive Eq. (2-11), we have from Fig. 2-26 (b),

$$\phi_1' = \frac{A}{2} \text{ (sides mutually perpendicular)},$$

$$\delta_1 = \frac{\delta_m}{2} \text{ (half the deviation takes place at each surface)},$$

$$\phi_1 = \phi_1' + \delta_1 = \frac{A}{2} + \frac{\delta_m}{2} = \frac{A + \delta_m}{2},$$

$$\sin \phi_1 = n \sin \phi_1',$$

$$\therefore \sin \frac{A + \delta_m}{2} = n \sin \frac{A}{2}, \text{ which is Eq. (2-11).}$$

The index of refraction of a transparent solid may be measured, making use of the equation derived above. The specimen whose index is desired is ground into the form of a prism. The angle of the prism A and the angle of minimum deviation δ_m are measured with the aid of a spectrometer (see Sec. 6-15). Since these angles may be determined with a high degree of precision, this method is an extremely accurate one and by means of it indices of refraction may be measured to the sixth place of decimals.

If the angle of the prism is small, the angle of minimum deviation is small also and we may replace the sines of the angles by the angles.

One then obtains

$$n = \frac{A + \delta_m}{A}, \quad \text{or}$$

$$\delta_m = (n - 1) A, \tag{2-12}$$

a useful approximate relation.

2-12 Dispersion. In discussing the phenomena of refraction we have thus far tacitly assumed that the light with which we were dealing was all of a single wave length or, as it is called, *monochromatic* (of one color). Most light beams, however, are *polychromatic,* that is, they are a mixture of waves whose wave lengths extend through (and beyond) the visible spectrum. While the velocity of light waves in a vacuum is the same for all wave lengths, the velocity in a material substance depends on the wave length and hence the index of refraction of a substance is a function of wave length also. A substance in which the velocity of a wave varies with wave length is said to exhibit *dispersion.*

Fig. 2-27 is a diagram showing the variation of index with wave length for a number of the more common optical materials. Notice that the index is in all cases larger for the shorter wave lengths and that the *change* in index with wave length is much greater for some materials than for others.

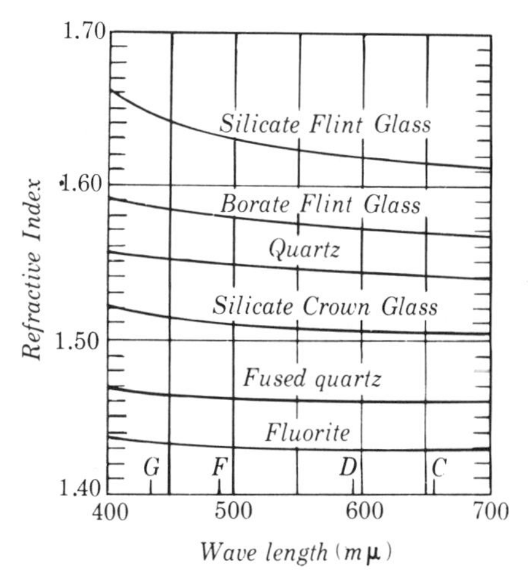

Fig. 2-27. Variation of index with wave length.

Consider a ray of polychromatic light incident on a prism as in Fig. 2-28. Since the deviation produced by the prism increases with increasing index of refraction, violet light is deviated most and red least, with other colors occupying intermediate positions. On emerging from the prism the light is spread out into a

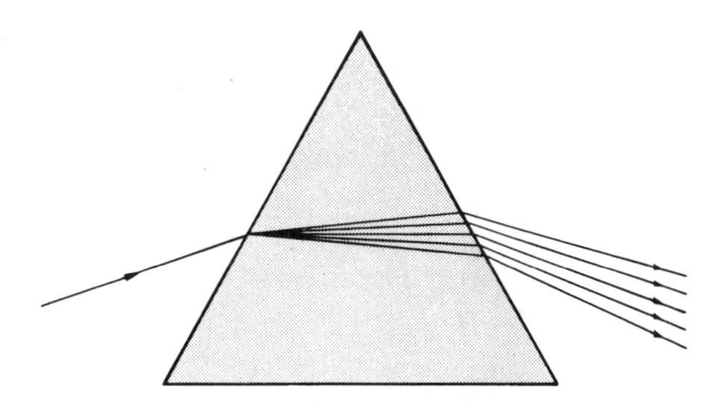

Fig. 2-28. Dispersion by a prism.

fan-shaped beam as shown. The light is said to be *dispersed* into a spectrum and the difference between the angles of deviation of any two rays is called the angular dispersion of those particular rays.

The *dispersive power*[1] of a substance, represented by ω, is defined as follows. Three wave lengths, λ_F, λ_D, and λ_C (see Fig. 2-27), are arbitrarily selected for reference. These wave lengths lie in the blue, yellow, and red portions of the spectrum. The corresponding indices of the substance are n_F, n_D, and n_C. The dispersive power ω is defined by the relation

$$\omega = \frac{n_F - n_C}{n_D - 1}. \tag{2-13}$$

For example, the dispersive power of a silicate flint glass for which $n_F = 1.632$, $n_D = 1.620$, and $n_C = 1.613$, is

$$\omega_{\text{Flint}} = \frac{1.632 - 1.613}{1.620 - 1} = 0.031,$$

while that of a silicate crown glass for which $n_F = 1.513$, $n_D = 1.508$, and $n_C = 1.504$, is

$$\omega_{\text{Crown}} = \frac{1.513 - 1.504}{1.508 - 1} = 0.018.$$

The dispersive power may be interpreted physically as follows. Suppose a narrow beam of polychromatic light is incident on a prism of small angle as in Fig. 2-29. Let δ_F, δ_D, and δ_C represent the angles of deviation of the rays whose wave lengths are λ_F, λ_D, and λ_C. The angle δ_D may be considered the *mean deviation* of the spectrum as a whole, while the angle $\delta_F - \delta_C$ is a measure of the *dispersion* of the spectrum. The dispersion is small if the angle of the prism is small and it may be assumed that each ray passes through the prism at the angle of minimum deviation so that Eq. (2-12) may be applied. Then

$$\delta_F = (n_F - 1)\,A,$$

$$\delta_D = (n_D - 1)\,A,$$

$$\delta_C = (n_C - 1)\,A.$$

[1] The term "power" is greatly overworked in the field of optics. It is used as a synonym for "ability," and one speaks of "dispersive power," "magnifying power," etc. An improved terminology is greatly to be desired.

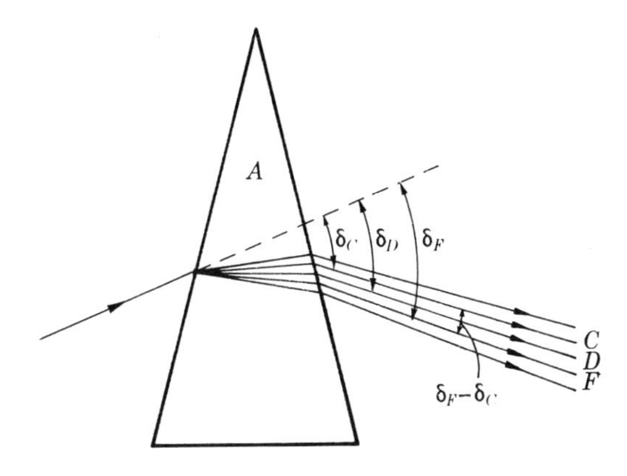

FIG. 2-29.

The dispersion, $\delta_F - \delta_C$, is therefore

$$\delta_F - \delta_C = (n_F - 1)\,A - (n_C - 1)\,A = (n_F - n_C)\,A,$$

and the ratio of the dispersion to the mean deviation is

$$\frac{\delta_F - \delta_C}{\delta_D} = \frac{(n_F - n_C)\,A}{(n_D - 1)\,A} = \frac{n_F - n_C}{n_D - 1} = \omega.$$

The dispersive power is therefore equal to the ratio of the dispersion $\delta_F - \delta_C$, to the mean deviation δ_D, when light is dispersed at minimum deviation by a prism of small angle.

Example. Compute the mean deviations and dispersions produced by the flint and crown glasses above, if the prism angle is 10.0°.

For the flint glass,

$$\text{Mean deviation} = \delta_D = (n_D - 1)\,A = (1.620 - 1)\,10 = 6.20°,$$

$$\text{Dispersion} = \omega\delta_D = .031 \times 6.20 = 0.192°.$$

For the crown glass,

$$\text{Mean deviation} = (1.508 - 1)\,10 = 5.08°,$$

$$\text{Dispersion} = .018 \times 5.08 = 0.091°.$$

2-13 Direct-vision and achromatic prisms. The dispersive powers of various optical materials are not proportional to their mean indices of refraction and consequently the dispersions produced by two prisms of equal angle are not proportional to the mean deviations. For example,

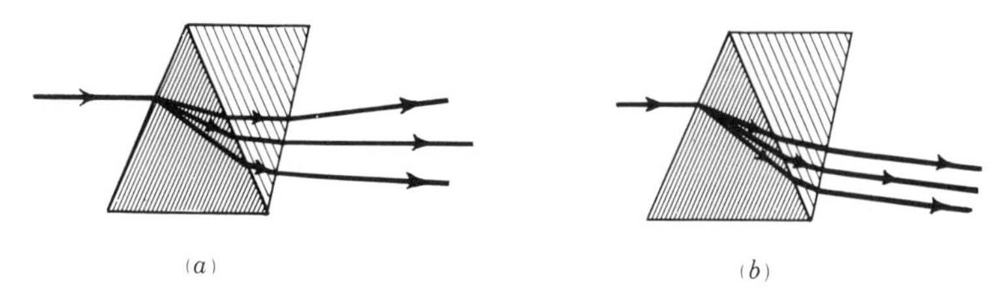

FIG. 2-30. (a) A direct-vision prism. (b) An achromatic prism.

the ratio of the mean deviations produced by the flint and crown prisms in the preceding example is

$$\frac{6.20^\circ}{5.08^\circ} = 1.22,$$

while the ratio of the dispersions is

$$\frac{0.192^\circ}{0.091^\circ} = 2.11.$$

It is therefore possible to combine two (or more) prisms of different materials in such a way that there is no net deviation of a ray of some chosen wave length, while there remains an outstanding dispersion of the spectrum as a whole. Such a device, illustrated in Fig. 2-30 (a), is known as a *direct-vision prism.* (See also Fig. 6-33.) Two prisms may also be designed so that the dispersion of one is offset by the dispersion of the other, although the deviation is not. A compound prism of this sort is called *achromatic* (Fig. 2-30 (b)).

Example. We wish to design both (a) a direct-vision and (b) an achromatic prism, using the flint and crown glasses whose dispersive powers were computed in the preceding section.

(a) Let quantities referring to the flint glass be designated by primed letters and to the crown glass by unprimed letters. We shall assume the prisms to be of small angle so that Eq. (2-11) may be used. The mean angles of deviation, δ_D and δ'_D (see Fig. 2-29) are

$$\delta_D = (n_D - 1)\,A,$$
$$\delta'_D = (n'_D - 1)\,A'.$$

If now the two prisms are combined apex to base as in Fig. 2-30 and the angles A and A' are chosen so that $\delta_D = \delta'_D$, the mean deviation of the first prism is offset by that of the second. The required ratio of the angles is

$$\frac{A}{A'} = \frac{n'_D - 1}{n_D - 1} \tag{2-14}$$

$$= \frac{1.620 - 1}{1.508 - 1}$$

$$= 1.22.$$

If the angle A' of the flint prism is 10.0°, the angle A of the crown prism should be 12.2°.

The dispersion produced by the flint prism is

$$\delta'_F - \delta'_C = (n'_F - n'_C)\, A',$$

and that produced by the crown is

$$\delta_F - \delta_C = (n_F - n_C)\, A.$$

The outstanding dispersion is the difference between these, or

$$\text{Net dispersion} = (n'_F - n'_C)\, A' - (n_F - n_C)\, A. \tag{2-15}$$

In the first term, express A' in terms of A from Eq. (2-14) and multiply numerator and denominator of the second term by $(n_D - 1)$. Then Eq. (2-15) becomes

$$\text{Net dispersion} = A\, (n_D - 1)\, (\omega' - \omega). \tag{2-16}$$

The net dispersion is proportional to the difference between the dispersive powers of the two glasses. When numerical values are inserted in Eq. (2-16) we obtain

$$\begin{aligned}\text{Net dispersion} &= 12.2 \times (1.508 - 1)\,(.031 - .018)\\ &= 0.079°.\end{aligned}$$

(b) If the compound prism is to be achromatic, the dispersions of the two prisms must be equal. That is,

$$(n'_F - n'_C)\, A' = (n_F - n_C)\, A,$$

or

$$\frac{A}{A'} = \frac{n'_F - n'_C}{n_F - n_C}.$$

If the angle A' of the flint prism is 10.0°, we find on inserting numerical values,

$$A = 21.1°.$$

Although the prism is commonly called "achromatic," it is, more precisely, achromatized only for light of wave lengths λ_F and λ_C, or for any pair of wave lengths having the same $(n_F - n_C)$ ratio. That is, the deviations of rays of wave lengths λ_F, λ_D, and λ_C by the prism above are:

$$\begin{aligned}\delta_F &= (n'_F - 1)\, A' - (n_F - 1)\, A = .632 \times 10 - .531 \times 21.1 = -4.50°,\\ \delta_D &= (n'_D - 1)\, A' - (n_D - 1)\, A = .620 \times 10 - .508 \times 21.1 = -4.52°,\\ \delta_C &= (n'_C - 1)\, A' - (n_C - 1)\, A = .613 \times 10 - .504 \times 21.1 = -4.50°.\end{aligned}$$

Thus while the deviations δ_F and δ_C are equal, δ_D is slightly different and the combination is not fully achromatic. Rays of wave lengths λ_F and λ_C are slightly displaced from one another after passing through the prism but since their directions are the same they will be imaged at the same point by the eye or any other image-forming instrument.

The principle illustrated in this example for achromatizing two prisms is also used in designing achromatic lenses. See Sec. 5-7.

2-14 The rainbow. The rainbow is produced by the combined effects of refraction, dispersion, and internal reflection of sunlight by drops of rain. When conditions for its observation are favorable, two bows may be seen, the inner being called the primary bow and the outer the secondary bow. The inner bow, which is the brighter, is red on the outside and violet on the inside, while in the more faint outer bow the colors are reversed. The primary bow is produced in the following manner. Assume that the sun's rays are horizontal, and consider a ray striking a raindrop as in Fig. 2-31 (a). This ray is refracted at the first surface and is in part reflected at the second surface, passing out again at the front surface as shown. An exact computation of the course of such a ray is exceedingly laborious but the French scientist Descartes computed the paths of some thousands of rays incident at different points on the surface of a raindrop and showed that if a ray of any given color were incident at such a point that its deviation was a maximum, all other rays of the same color which struck the surface of the drop in the immediate neighborhood of this point would be reflected in a direction very close to that of the first. The angle of maximum deviation of red light is 138°, or the angle δ in Fig. 2-31 is $180° - 138° = 42°$. The corresponding angle for violet light is 40°, while that for other colors lies intermediate between these.

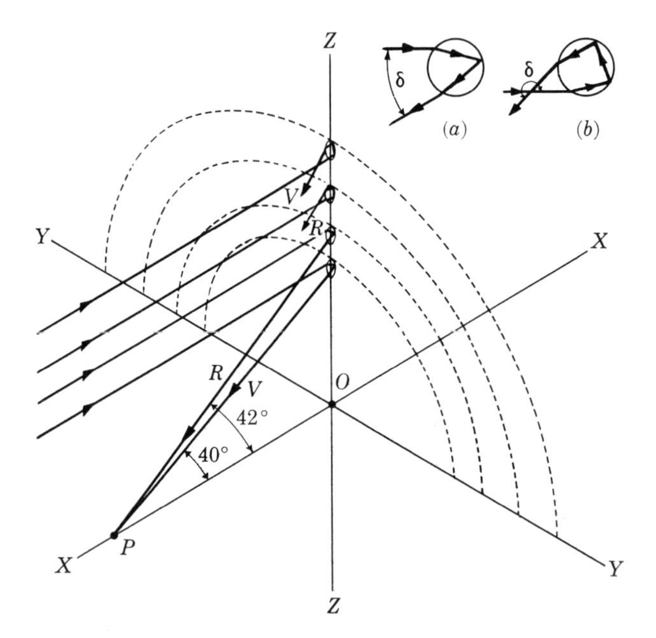

FIG. 2-31. The rainbow.

Consider now an observer at P, Fig. 2-31. The X–Y plane is horizontal and sunlight is coming from the left parallel to the X-axis. All drops which lie on a circle subtending an angle of 42° at P and with the center at O, will reflect red light strongly to P. All those on a circle subtending 40° at P will reflect violet light strongly, while those occupying intermediate positions will reflect the intermediate colors of the spectrum.

The point O, the center of the circular arc of the bow, may be considered the shadow of P on the Y–Z plane. As the sun rises above the horizon the point O moves down, and hence with increasing elevation of the sun a smaller and smaller part of the bow is visible. Evidently an observer at ground level cannot see the primary bow when the sun is more than 42° above the horizon. If the observer is in an elevated position, however, the point O moves up and more and more of the bow may be seen. In fact, it is not uncommon for a complete circular rainbow to be seen from an airplane.

The secondary bow is produced by two internal reflections, as shown in Fig. 2-31 (b). As before, the light which is reflected in any particular direction consists largely of the color for which that direction is the angle of maximum deviation. Since the angle of deviation is here the angle δ and since the violet is deviated more than the red, the violet rays in the secondary bow are deflected down at a steeper angle than the red and the secondary bow is red on the inside and violet on the outside edge. The corresponding angles are 50.5° for red and 54° for violet.

Problems—Chapter 2

(1) (a) Prove that a ray of light reflected from a plane mirror rotates through an angle 2θ when the mirror rotates through an angle θ about an axis perpendicular to the plane of incidence. (b) In a Foucault apparatus for measuring the velocity of light, light is reflected from a rotating mirror to a stationary mirror 1000 m distant, and returned to the rotating mirror. Find the angular velocity of the rotating mirror, if the return beam after reflection from it makes an angle of 1 minute with the original incident beam.

(2) A ray of light is incident on a plane surface separating two transparent media of indices 1.60 and 1.40. The angle of incidence is 30° and the ray originates in the medium of higher index. Compute (a) the angle of refraction, (b) the angle of deviation.

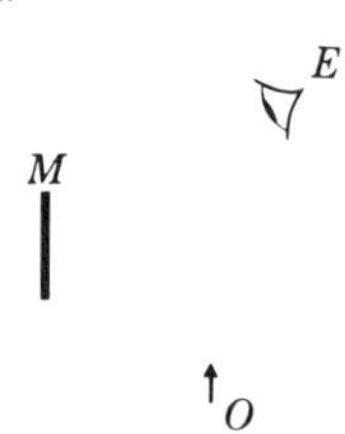

FIG. 2-32.

(3) (a) Show in a diagram the position of the image of the arrow O, formed by the plane mirror M in Fig. 2-32. (b) Show also the pencil of rays by which an observer E sees the image of the head of the arrow.

(4) What is the height of the smallest vertical plane mirror in which an observer, standing erect, can see his full length image?

(5) Two identical beakers, one filled with carbon disulphide and the other with water, are viewed from above. (a) Which beaker appears to contain the greater depth of liquid? (b) What is the ratio of the apparent depths?

(6) A layer of ether ($n = 1.36$) 2 cm deep floats on water ($n = 1.33$) 4 cm deep. What is the apparent distance from the ether surface to the bottom of the water layer, when viewed at normal incidence?

(7) A glass plate 1 inch thick, of index 1.50, having plane parallel faces, is held with its faces horizontal and its lower face 4 inches above a printed page. Find the position of the image of the page, formed by rays making a small angle with the normal to the plate.

(8) A plane mirror M is at a height h above the bottom of an empty beaker as shown in Fig. 2-33. (a) Where is the image, formed by the mirror, of a scratch on the bottom of the beaker? (b) When the beaker is filled with water, does the image of the scratch move up or down? (c) Deduce a general expression for the distance from the scratch to its image in the mirror, in terms of h and the depth d of the water in the beaker.

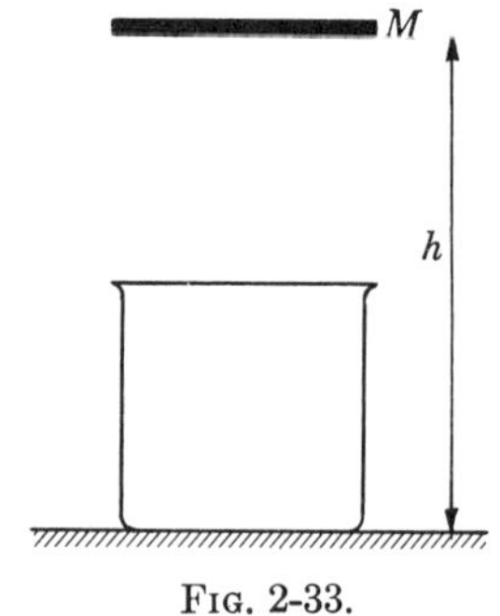

FIG. 2-33.

(9) A cork floats on the surface of a quiet pond 5.3 feet deep. Where is the shadow of the cork on the bottom of the pond just as the sun is setting?

(10) Find the distance of the image point P' in Fig. 2-16 below the refracting surface, if point P is 100 mm below the surface, and the angle of incidence of rays 2, 3, 4, and 5 is (a) 2°, (b) 30°, (c) 45°. Let $n = 1.33$, $n' = 1.00$.

(11) A point light source is 2 inches below a water-air surface. (a) Compute the angles of refraction of rays from the source making angles with the normal of 10°, 20°, 30°, and 40°, and show these rays in a carefully drawn full size diagram. (b) Find the coordinates of the astigmatic line images of the source, formed by a pencil whose central ray is incident at an angle of 25°.

(12) Let the point P in Fig. 2-17 represent a submarine which is being spotted by a plane. The submarine is 50 feet below the surface. What is the depth below the surface of the astigmatic image P'', if the angle of refraction of the central ray is (a) 60°, (b) 45°, (c) 30°, (d) 0°?

(13) A ray of light is incident at an angle of 60° on one surface of a glass plate 2 cm thick, of index 1.50. The medium on either side of the plate is air. Find the transverse displacement between the incident and emergent rays.

(14) A body of water is covered with a layer 1 cm thick of oil of index 1.63. A light ray originating in the water strikes the water-oil boundary at an angle of incidence of 30°. If the medium above the oil is air, will the ray be totally reflected or not?

(15) A ray of light is incident on the left vertical face of a glass cube of index 1.5, as shown in Fig. 2-34. The plane of incidence is in the plane of the paper, and the cube is surrounded by water. At what maximum angle must the ray be incident on the left vertical surface of the cube if total internal reflection is to occur at the top surface?

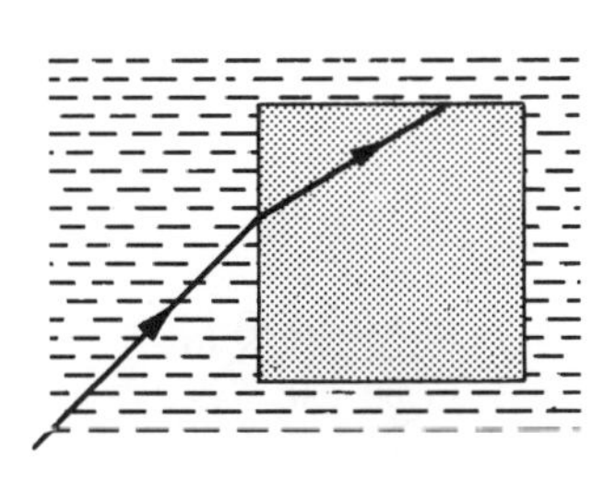

FIG. 2-34.

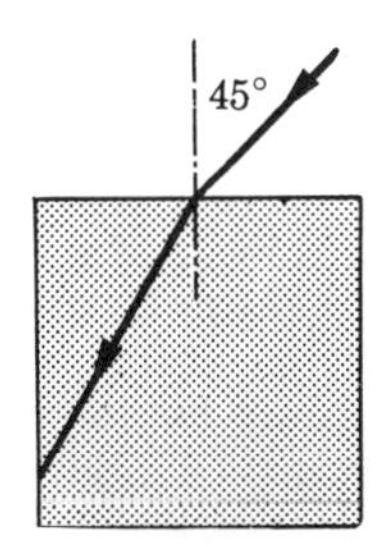

FIG. 2-35.

(16) Light is incident at an angle of 45° on the upper surface of a glass cube as in Fig. 2-35. The index of the glass is 1.414. Will the ray be totally reflected at the vertical face?

(17) A 45°-45°-90° prism is immersed in water. What is the minimum index of refraction the prism may have if it is to reflect totally a ray incident normally on one of its shorter faces?

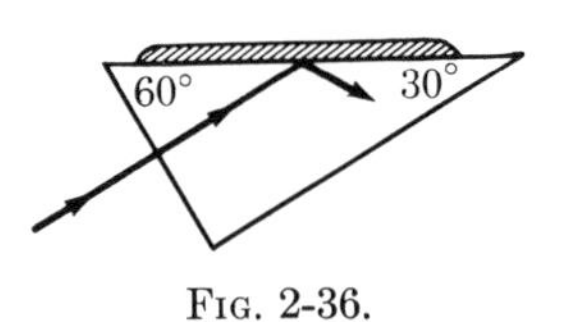

FIG. 2-36.

(18) Light is incident normally on the short face of a 30°-60°-90° prism as in Fig. 2-36. A drop of liquid is placed on the hypotenuse of the prism. If the index of the prism is 1.50, find the maximum index the liquid may have if the light is to be totally reflected.

(19) A fish looks upward at an unobstructed overcast sky. What total angle does the sky appear to subtend?

(20) A point source of light is 8 inches below the surface of a body of water. Find the diameter of the largest circle at the surface through which light can emerge from the water.

(21) A cylindrical tin can open at the top has a diameter of 8 inches and a height of 3 inches. In the center of its inside bottom surface is a tiny black dot, and the can is completely filled with water. Compute the radius of the smallest opaque circular disk that would prevent the dot from being seen, if the disk were floated centrally on the surface of the water.

(22) If the allowable deviation of light by a sheet of window glass is 1 minute of arc, how nearly parallel must the two faces be?

(23) What is the angle of minimum deviation of an equiangular prism whose index of refraction is 2? Illustrate by a diagram the path of a ray that traverses such a prism at minimum deviation.

(24) An equiangular prism is constructed of the silicate flint glass whose index of refraction is given in Fig. 2-27. Find the angles of minimum deviation for light of wave length 400 mμ and 700 mμ.

(25) The indices of refraction of an equiangular prism, for light of wave length λ_D and λ_C, are $n_D = 1.620$, $n_C = 1.613$. (a) For what angle of incidence will rays from a sodium flame (wave length $= \lambda_D$) pass through the prism at minimum deviation? (b) What is their angle of deviation? (c) If a ray of wave length λ_C is incident at the angle computed in part (a), what angle does the refracted ray within the prism make with the prism base? Draw a diagram.

(26) The index of refraction of the prism shown in Fig. 2-37 is 1.56. A ray of light enters the prism at point a and follows in the prism the path ab which is parallel to the line cd. (a) Sketch carefully the path of the ray from a point outside the prism at the left, through the glass, and out some distance into the air again. (b) Compute the angle between the original and final directions in air. (Dotted lines are construction lines only.)

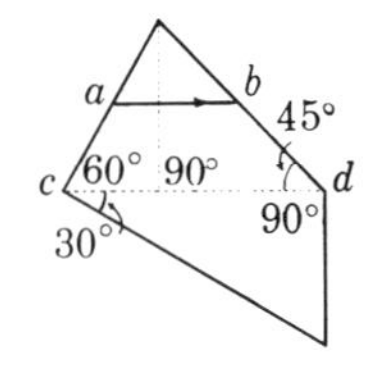

FIG. 2-37.

(27) A ray of light strikes a plane mirror M at an angle of incidence of 45° as in Fig. 2-38. After reflection, the ray passes through a prism of index 1.50, whose apex angle is 4°. Through what angle must the mirror be rotated if the total deviation of the ray is to be 90°?

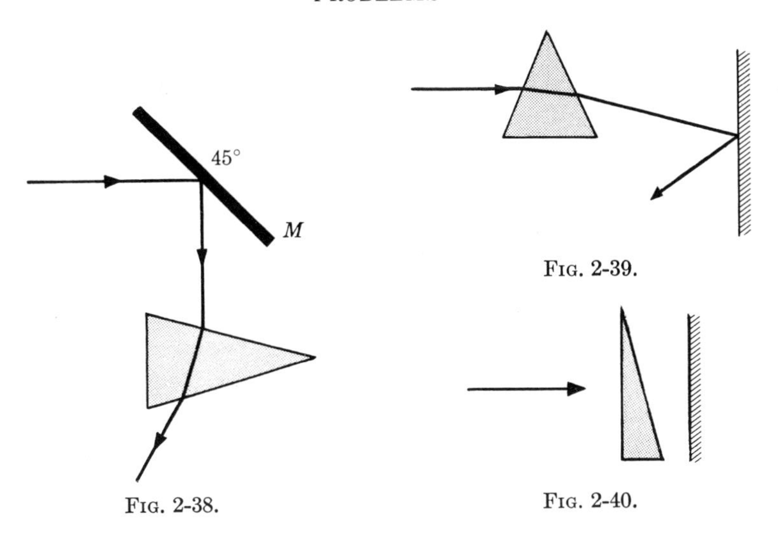

FIG. 2-38.

FIG. 2-39.

FIG. 2-40.

(28) A horizontal ray of light passes through a prism of index 1.50 whose apex angle is 4° and then strikes a vertical mirror, as shown in Fig. 2-39. Through what angle must the mirror be rotated if after reflection the ray is to be horizontal?

(29) A prism having an apex angle of 4° and refractive index of 1.50 is located in front of a vertical plane mirror, as shown in Fig. 2-40. A horizontal ray of light is incident on the prism. (a) Make a careful sketch showing the complete path of the ray. (b) What is the angle of incidence at the mirror? (c) Through what total angle is the ray deviated?

(30) A silicate crown prism of apex angle 15° is to be combined with a prism of silicate flint so as to result in no net deviation of light of wave length 550 mμ. (See Fig. 2-27 for indices of refraction.) (a) Find the angle of the flint prism. (b) Find the angular dispersion of two rays of wave length 400 mμ and 700 mμ.

(31) A silicate crown prism of apex angle 15° is to be combined with a prism of silicate flint so as to be achromatized for rays of wave length 400 mμ and 700 mμ. (See Fig. 2-27 for indices of refraction.) Find the angle of the flint prism.

CHAPTER 3

REFLECTION AND REFRACTION AT SPHERICAL SURFACES

3-1 Refraction at a spherical surface. With few exceptions, the surfaces of all lenses and mirrors are spherical or plane, because only spherical and plane surfaces can be produced by machine methods at reasonable cost. When a train of light waves passes through an optical instrument, the curvature of the wave fronts is altered at each boundary surface. However, a wave front that is originally spherical or plane does not have a geometrically simple form after refraction at a spherical surface. This makes it practically impossible to analyze the passage of light through an optical instrument in terms of wave surfaces, and it is here that the simplifications of the ray method are most useful. A ray, in its passage through an optical instrument, is made up of a number of segments of straight lines, deviated at reflecting or refracting surfaces by angles which can be computed from the law of reflection or from Snell's law. Hence the problem of tracing the path of a ray reduces to a problem in geometry and this branch of optics is called *geometrical optics* (*trigonometrical* optics would be more appropriate).

The line AA in Fig. 3-1 represents a spherical surface of radius R separating two transparent substances having different indices of refraction. The center C of the spherical surface is called the *center of curvature.* A line such as PVC through the center of curvature is called an *axis.* The point V, where the axis intersects the surface, is called the *vertex.* Let the index of the substance at the left of the surface be n and that at the right be n'. The diagram is drawn with n' greater than n.

Fig. 3-1 (a) shows a pencil of rays diverging from a point P, called the *object.* The distance from the object to the vertex is called the *object distance* and is represented by s. Let us select an arbitrary ray, such as the one indicated by a heavy line, and trace it through the surface. This ray is shown separately in Fig. 3-1 (b). It makes an angle u with the axis and is incident on the refracting surface at point B. Draw the radius CB and extend it beyond point B as shown. Since the radius is normal to the surface at point B, the angle ϕ is the angle of incidence of the chosen ray, and ϕ' is the angle of refraction. The refracted ray crosses the axis at point P' at a distance s' to the right of the vertex, and makes an angle u' with the axis. We shall show that for small values of the angle u, all rays from P intersect at P', which is called the *image* of point P. The distance s' is the *image distance.*

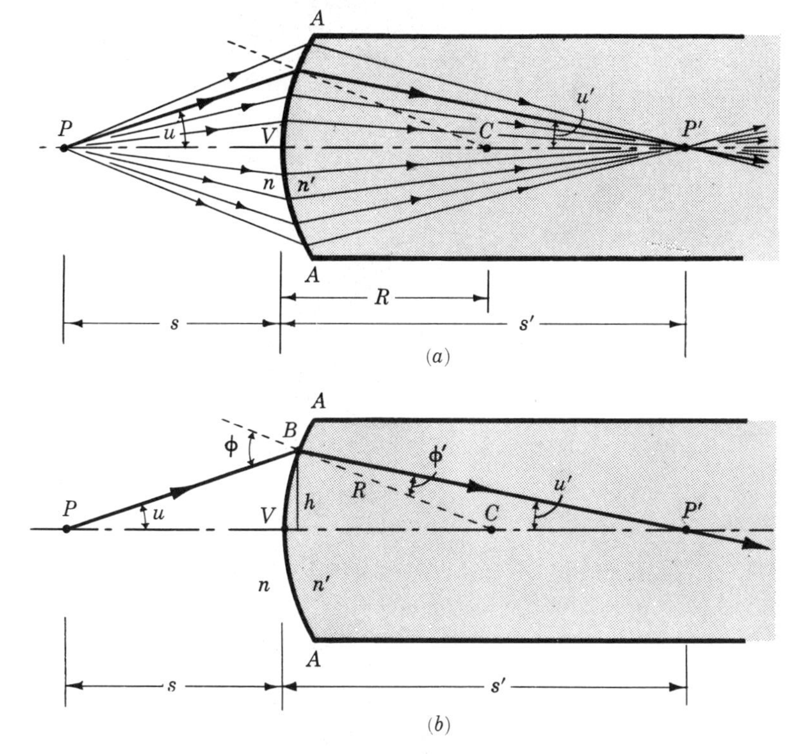

FIG. 3-1. Refraction of rays at a spherical surface.

It is necessary at this point to adopt a convention of signs for distances and angles. Many such conventions are in use and all have some points in their favor. We shall use the following set of conventions.

1. *Draw all figures with light incident on the refracting or reflecting surface from the left.*

2. *Consider object distances* (s) *positive when the object lies at the left of the vertex of the refracting or reflecting surface.*

3. *Consider image distances* (s') *positive when the image lies at the right of the vertex.*

4. *Consider radii of curvature* (R) *positive when the center of curvature lies at the right of the vertex.*

5. *Consider angles positive when the slope of the ray with respect to the axis* (*or with respect to a radius of curvature*) *is positive.*

6. *Consider transverse dimensions positive when measured upward from the axis.*

It is advisable that these conventions of sign be memorized, since many of the difficulties that arise in this particular branch of optics have been found to result from a lack of familiarity with them, rather than from a failure to grasp the physical principles involved.

Let us now return to the problem of tracing an arbitrary ray from P through the refracting surface in Fig. 3-1 (b). From the triangle PBC, by the law of sines,

$$\frac{\sin(\pi - \phi)}{\sin u} = \frac{R + s}{R},$$

and since $\sin(\pi - \phi) = \sin \phi$,

$$\sin \phi = \frac{R + s}{R} \sin u. \tag{3-1}$$

The angle of refraction, ϕ', may now be found from Snell's law.

$$\sin \phi' = \frac{n}{n'} \sin \phi. \tag{3-2}$$

The slope angle u' of the refracted ray can be computed from the triangle PBP', making use of the fact that the sum of the interior angles of a triangle is 180° or π radians. Note that according to our convention of signs the angle u' is negative, since the slope of the ray BP' is negative with respect to the axis.

$$u + (\pi - \phi) + \phi' - u' = \pi,$$

$$u' = \phi' + u - \phi. \tag{3-3}$$

Finally, the image distance s' is found from the triangle CBP', by the law of sines.

$$\frac{s' - R}{R} = \frac{\sin \phi'}{-\sin u'},$$

$$s' = R - R\frac{\sin \phi'}{\sin u'}. \tag{3-4}$$

Eqs. (3-1), (3-2), (3-3), and (3-4) suffice to determine the distance s' and the slope angle u' of any ray, in terms of the distance s, the original slope angle u, and the constants of the system, n, n', and R.

On solving the preceding equations for s', it is found that this distance depends on the slope angle u of the incident ray. It is greatest when the angle u is zero, and becomes smaller as u increases. That is, the rays originating at point P do not all intersect at a common point after refraction, which is another way of stating that the refracted wave fronts are not spherical. This characteristic of spherical refracting surfaces is called *spherical aberration*, and is discussed further in Chap. 5.

An important simplification results if the slope angles of the incident rays are small. Such rays, nearly parallel to the axis, are called *paraxial rays*. If the angle u is small, the angles ϕ, ϕ', and u' will be small also, and the sines of the angles, in Eqs. (3-1) to (3-4) can be replaced by the angles. The *trigonometric* equations then become *algebraic* equations, and may be combined by straightforward algebraic operations to give the following relation between s and s' and the constants of the system:

$$\boxed{\frac{n}{s} + \frac{n'}{s'} = \frac{n' - n}{R}.} \tag{3-5}$$

The series expansion of $\sin \phi$ is

$$\sin \phi = \phi - \frac{\phi^3}{3!} + \frac{\phi^5}{5!} - \cdots\cdot\cdot$$

To assume that $\sin \phi = \phi$ is to neglect all terms in the series beyond the first, and the equations that result from this approximation are known as those of *first order theory*. A better approximation is obtained by including the next term, leading to *third order theory*, but the resulting cubic equations are not easy to manipulate.

Since the angle u does not appear in Eq. (3-5), it follows that *all* rays from an axial point P, making small angles with the axis, intersect at a common point P' after refraction at a spherical surface. In other words, a small portion of the refracted wave front near the axis is approximately a spherical surface with center at P'. Point P' is the *image* of the point P.

The radius R of the refracting surface in Fig. 3-1, the indices of refraction n and n', and the object distance s, have values such that the rays diverging from P are deviated sufficiently to become converging after refraction. For other values of these quantities the deviation may not be sufficient to cause the rays to converge. They then appear to diverge, as in Fig. 3-2, from a point P' at the left of the vertex. (The rays may, of course, be parallel to the axis after refraction, in which case one may consider either that they converge to a point at $+\infty$, or diverge from a point at $-\infty$.)

Point P' in Fig. 3-2 is a *virtual* image of P, point P' in Fig. 3-1 is a *real* image. Thus we see that a *positive* value of the image distance s' indicates a *real* image, while a *negative* value indicates a *virtual* image.

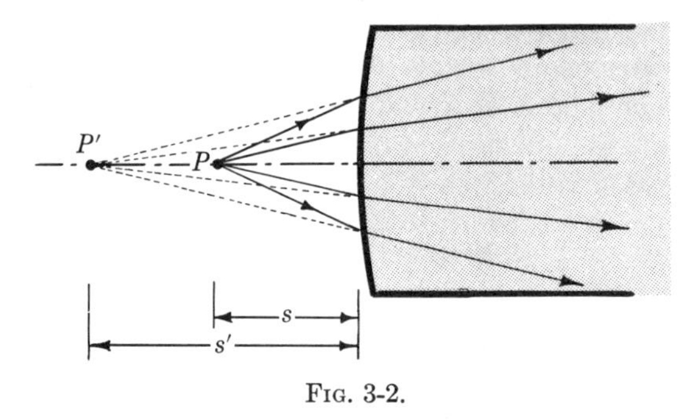

FIG. 3-2.

Notice carefully that the rays from P in Fig. 3-1, or the waves they represent, do not stop at the point P' but continue beyond it. That is, within the cone bounded by the outermost rays striking the refracting surface, light waves spread out toward the right from P' just as they would from a material luminous object at this point. Notice also that the image is something that exists in space, and that an image is formed at P' whether or not there is a screen at this point.

Although Eq. (3-5) does not apply, in general, to rays making large angles with the axis, it does apply to such rays in optical systems which have been corrected for spherical aberration. It is therefore one of the most important equations in geometrical optics. For reasons of clarity in the illustrations in this and following chapters, rays making large angles with the axis have been drawn as if they obeyed the equations of first order theory.

Another viewpoint regarding the formation of an image by a refracting surface should be considered at this time, since we shall need to make use of it later in Chap. 9. All paraxial rays diverging from point P in Fig. 3-1 intersect at P'. Let us compare the number of waves in the ray PVP' with the number in the ray PBP'. The former travels a smaller distance than the latter in the first medium of index n, but a greater distance in the medium of index n'. If $n' > n$, the wave length in the first medium is greater than that in the second. We shall show that these differences just compensate one another, and that the number of waves is the same in all paraxial rays which diverge from P and intersect at P'. Since the waves start from P in phase, they reach P' in phase and their amplitudes add. This is the true significance of an *image*. However, the fact that the waves are all in phase at the geometrical image point P', does not imply that they are exactly out of phase at all other points. We shall show in Chap. 9 that in the immediate vicinity of the geometrical image point the waves, while not exactly in phase, are nearly enough so

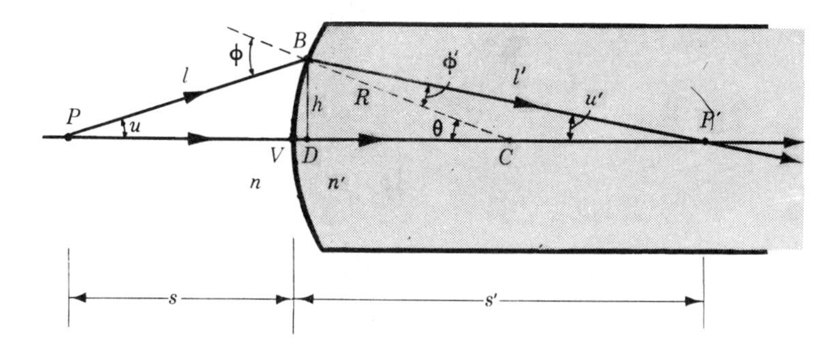

FIG. 3-3. The time is the same along all paths between an object and its image.

as to cause the "image" to spread out over an appreciable area. This phenomenon is called *diffraction.*

The proof of the proposition above is as follows. Let λ_0 be the wave length of the light in vacuum. The wave lengths in the first and second media are then

$$\lambda = \frac{\lambda_0}{n}, \quad \lambda' = \frac{\lambda_0}{n'}.$$

The number of waves in the ray PVP' is

$$\frac{s}{\lambda} + \frac{s'}{\lambda'}.$$

Let $PB = l$, $BP' = l'$. The number of waves in the ray PBP' is

$$\frac{l}{\lambda} + \frac{l'}{\lambda'},$$

and we wish to show that

$$\frac{s}{\lambda} + \frac{s'}{\lambda'} = \frac{l}{\lambda} + \frac{l'}{\lambda'},$$

or, in terms of indices of refraction,

$$ns + n's' = nl + n'l'.$$

From Fig. 3-3, we see that

$$R = VC = VD + DC,$$
$$DC = R \cos \theta,$$
$$R = VD + R \cos \theta,$$
$$VD = R\,(1 - \cos \theta).$$

Also,

$$s = PD - VD,$$
$$PD = l \cos u.$$

Hence

$$s = l \cos u - R(1 - \cos \theta).$$

For paraxial rays, the angles u, u', and θ are small and we can use the approximation

$$\cos\theta = 1 - \frac{\theta^2}{2}, \quad \cos u = 1 - \frac{u^2}{2}.$$

Then

$$\begin{aligned} s &= l\left(1 - \frac{u^2}{2}\right) - R\frac{\theta^2}{2} \\ &= l - \frac{1}{2}(lu^2 + R\theta^2). \end{aligned}$$

But from Fig. 3-3,

$$h = l\sin u = l'\sin u' = R\sin\theta,$$

and since the angles are small,

$$lu = l'u' = R\theta.$$

Hence

$$s = l - \frac{lu}{2}(u + \theta).$$

But

$$u + \theta = \phi,$$

so

$$s = l - \frac{lu}{2}\phi.$$

In the same way, we find that

$$s' = l' + \frac{lu}{2}\phi'.$$

Now multiply the expressions for s and s' by n and n' respectively, and add.

$$ns + n's' = nl + n'l' - \frac{lu}{2}(n\phi - n'\phi').$$

But if ϕ and ϕ' are small, Snell's law becomes

$$n\phi = n'\phi'.$$

Hence the term in parentheses is zero and for paraxial rays,

$$ns + n's' = nl + n'l',$$

which was to be proved.

The fact that the number of waves is the same in all paraxial paths from P to P' is equivalent to stating that the time is the same along these paths, so that this is another example of Fermat's principle.

3-2 Reflection at a spherical surface. As explained in Sec. 2-2, when light is incident on a surface bounding two transparent media of different indices of refraction, some light is reflected at the surface, with the angle of reflection of any ray equal to the angle of incidence. Hence a ray such as PB in Fig. 3-1 (a) gives rise to a reflected ray as well as a refracted ray. If the surface is convex toward the left, as in Fig. 3-4, the reflected rays appear to diverge from a virtual image of P at P_1. The fraction of the incident light reflected from an air-glass boundary surface is only of the order of a few percent; nevertheless, interreflections at the surfaces of camera lenses often result in enough stray light reaching the film to cause a serious reduction in contrast and, if the instrument is directed toward the sun or any intense source, may produce out-of-focus images called "ghosts." The fraction of light reflected can be reduced to negligible amounts by so-called "nonreflecting" coatings (see Sec. 8-2) on the lens surfaces. On the other hand, one often wishes to utilize the reflected light only, as in a telescope mirror. The surface is then coated with silver or aluminum to make it highly reflecting. High quality mirrors are almost invariably made by grinding and polishing a glass surface to the curvature desired and then metallizing it. Where high quality is not essential, the mirror may be stamped from a metal sheet.

The angle of incidence ϕ of a ray from P in Fig. 3-4 is of course given by Eq. (3-1). The angle between the *reflected* ray and the normal is found from the law of reflection instead of from Snell's law as in Eq. (3-2). That is, if r is the angle of reflection, we have instead of Eq. (3-2),

$$r = -\phi.$$

The negative sign enters since the reflected ray has a negative slope with respect to the normal. By introducing a mathematical artifice at

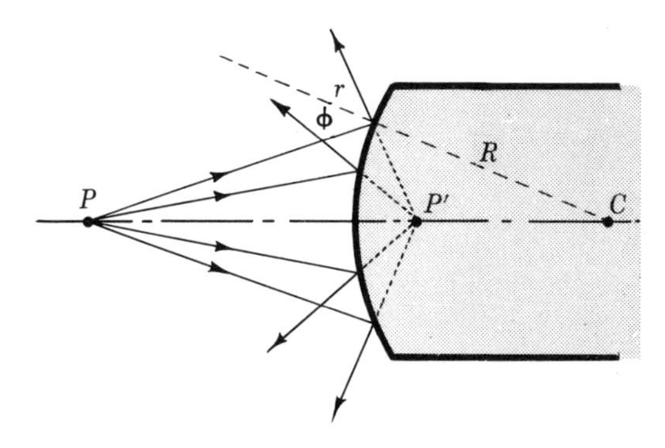

FIG. 3-4. Reflection at a spherical surface.

this point, the same formulas that were derived for refraction can be used for reflection also. If we assume in Snell's law that $n' = -n$, this law reduces to

$$\sin \phi' = -\sin \phi,$$

or

$$\phi' = -\phi.$$

Hence if we let $n' = -n$, and let the primed quantities in Eqs. (3-1) to (3-5) apply to the *reflected* as well as the refracted rays, these equations can be used for reflecting as well as for refracting surfaces. It should be noted, however, that a virtual image formed by a mirror lies at the *right* of the reflecting surface, so that a positive value of s' corresponds to a virtual image and a negative value of s' to a real image. As an illustration, consider the formation of an image by a plane mirror. The radius of curvature of such a mirror is infinite, and setting $R = \infty$ and $n' = -n$ in Eq. (3-5), we find

$$s = s'.$$

Since the object distance s is positive if the object is real, the image distance s' is positive also. Therefore a plane mirror forms a virtual image of a real object, at the same distance behind the mirror that the object is in front of it.

If the radius of curvature of the mirror is finite, Eq. (3-5) becomes

$$\frac{1}{s} - \frac{1}{s'} = -\frac{2}{R}. \tag{3-9}$$

3-3 Lateral magnification. Ordinarily, the rays incident on a refracting surface originate not merely at a single point, but at all points of the surface of an object of finite size. The arrow PQ in Fig. 3-5 is a conventional method of indicating an object of finite size in a plane at right angles to the axis of a refracting surface. Paraxial rays diverging from point P are imaged at P' as in Fig. 3-1, and, to avoid confusion, are not shown. The cone of rays diverging from Q is imaged at Q'. An axis drawn from Q through the center of curvature C has its vertex at V'. The distance QV' is greater than PV, and it will be seen from Eq. (3-5) that the distance from the vertex V' to the image Q' is less than that from the vertex V to the image P'. As a consequence, the image $P'Q'$ does not lie in a plane at right angles to the axis PVC, but on a surface concave toward the left. This effect is known as *curvature of field.* However, if the angle subtended by the object (or its image) at the refracting surface is small, the image will lie approximately in a plane at right angles to the axis, and the distances s and s' can be considered to apply to the entire object and image, not merely to the axial points P and P'.

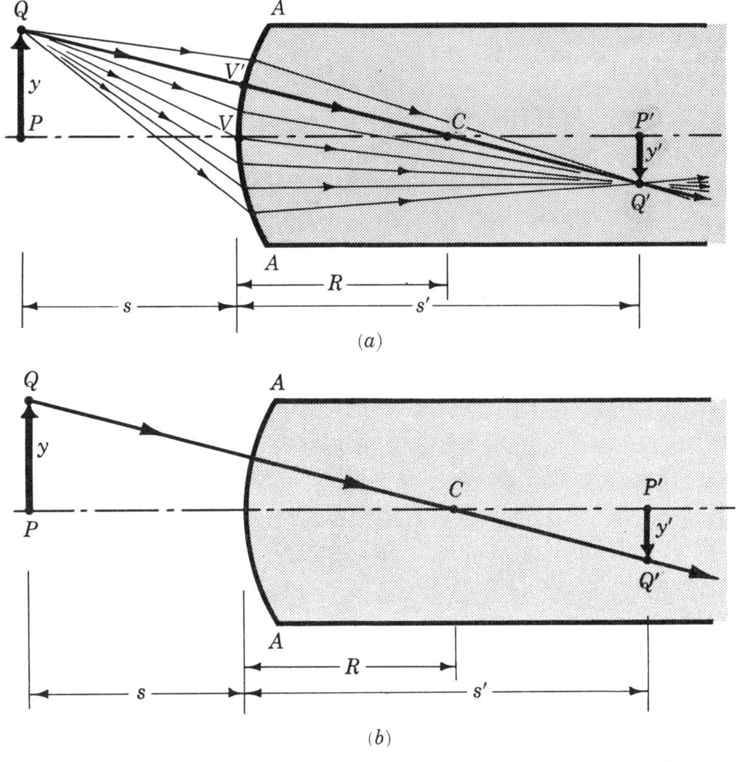

FIG. 3-5. Rays diverging from Q are imaged at Q' after refraction.

The ratio of the height y' of the image $P'Q'$ to the height y of the object PQ is called the *lateral* or *transverse magnification* and is represented by m.

$$m = \frac{y'}{y}.$$

To derive the expression for the magnification, let us select from the cone of rays diverging from point Q in Fig. 3-5 (a) that particular ray passing through the center of curvature C, shown by a heavy line. This ray is drawn separately in Fig. 3-5 (b). Since it is incident normally on the surface AA, it is not deviated and proceeds in a straight line from Q to Q'. From the similar triangles PQC and $P'Q'C$,

$$\frac{-y'}{y} = \frac{CP'}{PC} = \frac{s' - R}{s + R}. \tag{3-10}$$

Now refer to Fig. 3-1 and the equations derived from it. Eqs. (3-1) and (3-4) may be put in the form

$$s' - R = -R\frac{\sin \phi'}{\sin u'}, \quad s + R = R\frac{\sin \phi}{\sin u}.$$

When these expressions for $s' - R$ and $s + R$ are inserted in Eq. (3-10) we get

$$m = \frac{y'}{y} = \frac{\sin \phi'}{\sin \phi}\frac{\sin u}{\sin u'},$$

and since by Snell's law $\sin \phi'/\sin \phi = n/n'$, it follows that

$$m = \frac{y'}{y} = \frac{n \sin u}{n' \sin u'}, \tag{3-11}$$

or

$$ny \sin u = n'y' \sin u'. \tag{3-12}$$

Either of the preceding equations is known as *Abbe's sine condition.*

In the special case of paraxial rays, where u and u' are small, the sines of these angles can be approximated (see Fig. 3-1 (b)) by

$$\sin u = \frac{h}{s}, \quad \sin u' = \frac{h}{s'}.$$

Hence $\sin u/\sin u' = -s'/s$, approximately, and for paraxial rays

$$m = \frac{y'}{y} = -\frac{ns'}{n's}. \tag{3-13}$$

If m is positive, y and y' have the same sign, and object and image lie on the same side of the axis. In this case the image is called *erect.* If m is negative, y and y' are of opposite sign and the image is *inverted.*

The expression for the lateral magnification produced by a reflecting surface is obtained by setting $n = -n'$ in Eq. (3-13). This gives

$$m = \frac{s'}{s}.$$

That is, the magnification equals the ratio of image distance to object distance.

Examples. (1) One end of a cylindrical glass rod of index 1.50 is ground and polished to a hemispherical surface of radius $R = 20$ mm. (See Fig. 3-6(a).) An object in the form of an arrow 1 mm high, at right angles to the axis of the rod, is

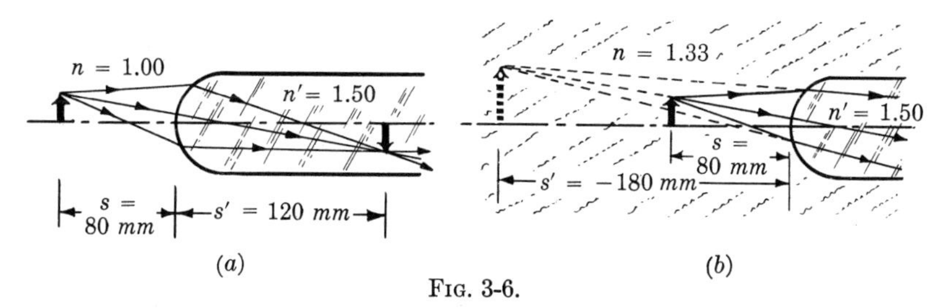

FIG. 3-6.

located 80 mm to the left of the vertex of the surface. Find the position and magnification of the image. The rod is in air.

From the given data,

$$n = 1, \quad n' = 1.5, \quad R = +20 \text{ mm}, \quad s = +80 \text{ mm}.$$

$$\frac{1}{80} + \frac{1.5}{s'} = \frac{1.5 - 1}{20},$$

$$s' = +120 \text{ mm}.$$

The image is therefore at the right of the vertex (s' is positive) and at a distance of 120 mm from it. It is a *real* image; the refracted rays converge toward it, and diverge from it after intersecting.

The lateral magnification is

$$m = \frac{y'}{y} = -\frac{1 \times 120}{1.5 \times 80} = -1.$$

Therefore y' is numerically equal to y but is of opposite sign. That is, the image is the same height as the object but is inverted.

(2) Let the same rod be immersed in water of index 1.33. Find the position and magnification of an object 80 mm to the left of the vertex. (Fig. 3-6(b).)

We now have

$$\frac{1.33}{80} + \frac{1.5}{s'} = \frac{1.5 - 1.33}{20},$$

$$s' = -180 \text{ mm}.$$

The fact that s' is negative means that the rays after refraction are not converging, but appear to diverge from a point 180 mm to the left of the vertex. In other words, the image is *virtual.*

The lateral magnification is

$$m = -\frac{1.33 \times (-180)}{1.5 \times 80} = +2.$$

The image is erect (m is positive) and twice the height of the object.

(3) (a) What type of mirror is required to form an image, on a wall 3 m from the mirror, of the filament of a headlight lamp 10 cm in front of the mirror? (b) What is the height of the image, if the height of the object is 5 mm?

(a) From the given data,

$$s = 10 \text{ cm}, \quad s' = -300 \text{ cm}.$$

$$\frac{1}{10} - \frac{1}{-300} = -\frac{2}{R},$$

$$R = -19.4 \text{ cm}.$$

Since the radius is negative, a concave mirror is required.

(b) The lateral magnification is

$$m = \frac{s'}{s} = \frac{-300}{10} = -30.$$

The image is therefore inverted (m is negative) and is 30 times the height of the object, or 150 mm in height.

(4) An object is 4 in. to the left of the vertex of a concave mirror of radius of curvature 12 in. Find the position and magnification of the image.

$$s = 4 \text{ in}, \quad R = -12 \text{ in}.$$

$$\frac{1}{4} - \frac{1}{s'} = -\frac{2}{-12},$$

$$s' = 12 \text{ in}.$$

$$m = \frac{12}{4} = +3.$$

The image is therefore 12 in. to the right of the vertex (s' is positive), is virtual (s' is positive), erect (m is positive), and 3 times the height of the object. See Fig. 3-7.

(5) As an illustration of the general applicability of Eq. (3-5), let us apply it to a problem that has been already solved by other methods, namely, to find the position of the image of an object at a depth h below the surface of a body of water, as seen by an observer looking vertically downward.

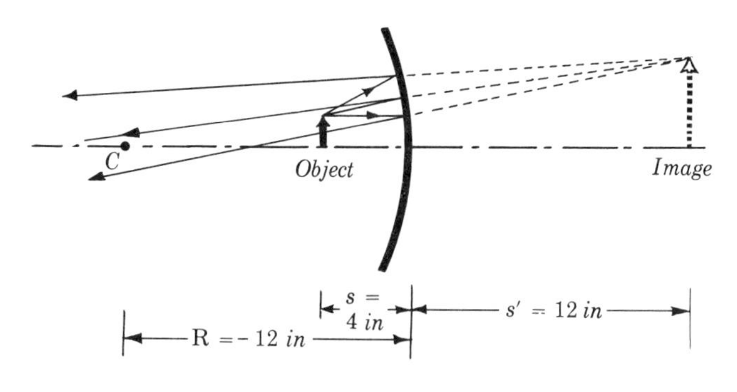

FIG. 3-7.

In this case the refracting surface is plane, so $R = \infty$. Also, $n = 1.33$, $n' = 1$, $s = h$ and is positive. Then

$$\frac{1.33}{h} + \frac{1}{s'} = \frac{1 - 1.33}{\infty},$$

$$s' = -\frac{3}{4}h,$$

in agreement with the result derived in Sec. 2-7.

3-4 Focal points and focal lengths. The *first focal point* of a refracting or reflecting surface is defined as that axial object point which is imaged by the surface at infinity. That is, rays diverging from the first focal point are parallel to the axis of the surface after reflection or refraction. Fig. 3-8 (a) shows the first focal point F of a refracting surface and Fig. 3-8 (b) that of a concave mirror.

The distance from the first focal point to the vertex of the surface is called the *first focal length*, f. The first focal length can be computed from Eq. (3-5) by setting $s' = \infty$. This gives

$$\frac{n}{s} + \frac{n'}{\infty} = \frac{n' - n}{R},$$

and since in this case, by definition, the object distance s equals the first focal length f,

$$f = \frac{n}{n' - n} R. \tag{3-14}$$

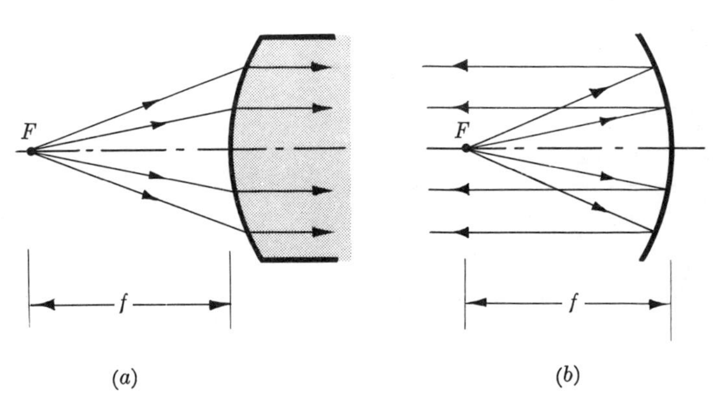

FIG. 3-8. (a) First focal point of a refracting surface. (b) First focal point of a concave mirror.

For a mirror, we set $n' = -n$, and obtain

$$f = -\frac{R}{2}.$$

If the mirror is concave, as in Fig. 3-8, R is a negative quantity, f is positive, and the first focal point lies at the left of the vertex, halfway between it and the center of curvature.

The *second focal point* of a surface, designated F', is defined as the image of an infinitely distant object point on the axis. That is, it is the point at which incident rays originally parallel to the axis intersect after reflection or refraction. (See Fig. 3-9 (a).) The distance from the vertex to the second focal point is the *second focal length*, f'. If we set $s = \infty$ in Eq. (3-5) we find

$$\frac{n}{\infty} + \frac{n'}{s'} = \frac{n' - n}{R},$$

and since $s' = f'$, by definition,

$$f' = \frac{n'}{n' - n} R. \tag{3-15}$$

Setting $n' = -n$, we find for the second focal length of a mirror,

$$f' = \frac{R}{2}.$$

If the mirror is concave, R is negative, s' and f' are negative also and, since they refer to *image* distances, the second focal point lies at a distance $R/2$ to the left of the vertex as in Fig. 3-9 (b). In other words, the first and second focal lengths of a mirror are equal, and the first and second

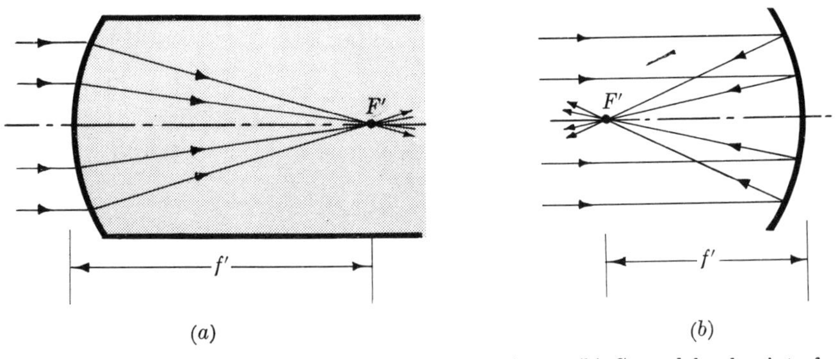

FIG. 3-9. (a) Second focal point of a refracting surface. (b) Second focal point of a concave mirror.

focal points of a mirror coincide. It is therefore unnecessary to distinguish between them and ordinarily one simply speaks of *the* focal point of a mirror. The focal length of a mirror is considered to be the first focal length, $f = -R/2$.

The focal lengths of a refracting surface, however, are unequal, as is seen by a comparison of Eqs. (3-14) and (3-15).

Example. Find the first and second focal lengths of the spherical surface in Example (1) at the end of Sec. 3-3.

The first focal length f is

$$f = \frac{1}{1.5 - 1} \times 20 = 40 \text{ mm}.$$

The second focal length f' is

$$f' = \frac{1.5}{1.5 - 1} \times 20 = 60 \text{ mm}.$$

Hence the first focal point F lies 40 mm to the left of the vertex, and the second focal point F' lies 60 mm to the right of the vertex.

When the positions of the focal points of a refracting or reflecting surface are known, the position and size of the image of a given object, formed by the surface, can be found by a simple graphical construction. The method is not of high precision but it is often useful in obtaining approximate values of the position and size of an image.

Fig. 3-10 shows a number of rays diverging from point Q of an object PQ. Ray 1 is parallel to the axis and therefore passes through the second focal point F' after refraction. Ray 2, directed toward the center of curvature of the surface, strikes the surface normally and is undeviated. Ray 3 strikes the surface after passing through the first focal point F, and is parallel to the axis after refraction.

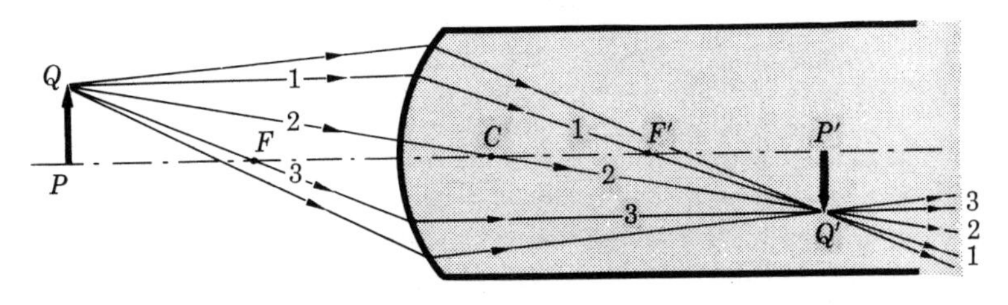

Fig. 3-10. Graphical method for locating an image formed by refraction at a single surface.

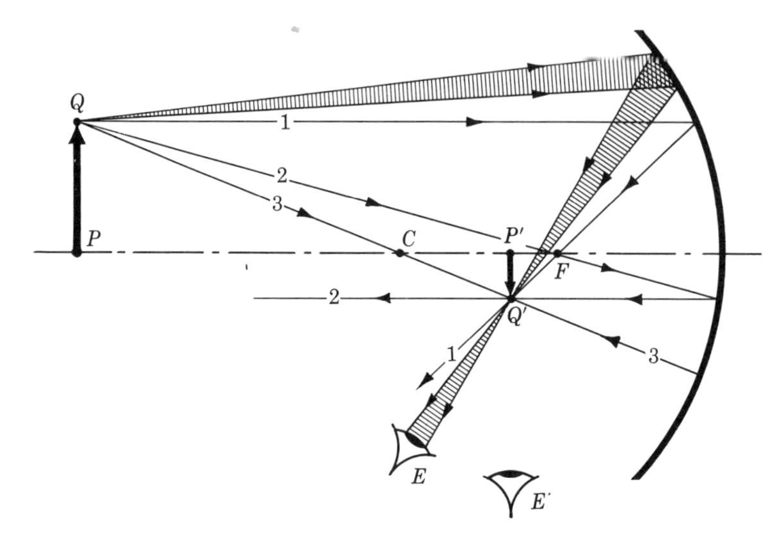

FIG. 3-11. Graphical method for locating the image formed by a spherical mirror.

The point of intersection of any two of these rays suffices to locate the image Q' of point Q, and since (in the absence of aberrations) *all* rays from the object point pass through the same image point, the paths of any other rays, such as the two unnumbered ones, can be determined after the image has been located. Note that the construction must be made for an object point *not* on the axis of the surface.

Fig. 3-11 illustrates the graphical method applied to the formation of an image by a concave mirror. The arrow PQ represents the object, and the points C and F are respectively the center of curvature and the focal point of the mirror. Ray 1, parallel to the axis before striking the mirror, passes through the focal point F after reflection. Ray 2, incident on the mirror after passing through the focal point, is parallel to the axis after reflection. Ray 3, incident on the mirror after passing through the center of curvature C, strikes the mirror at normal incidence and hence retraces its original path. The point of intersection of any two of these rays suffices to locate the image of the head of the arrow. Once this point has been found, the paths of any other rays may be traced.

The image $P'Q'$ is a real image. How may it be seen? In order to see the image one must look at it, just as one looks at a material object in order to see it. Thus an observer E in Fig. 3-11 sees the head of the arrow by means of the narrow shaded pencil of rays. Of course, the eye must be sufficiently far away from the image $P'Q'$ for a distinct image to be formed on the retina, just as the eye must be at or beyond a certain minimum distance from any material object in order to see it clearly. This minimum distance for the average observer is about 10 in. or 25 cm. (See Sec. 6-1.) Note, however, that in contrast to a material object, the image cannot be seen from all directions, but only by an observer within the cone limited by those rays which strike the outer edges of the mirror. Thus the observer at E' in Fig. 3-11 cannot see the image $P'Q'$, since no rays from it travel toward his eye.

It is not necessary to project the image on a screen in order to see it. Strictly speaking, if this is done one sees not the image, but the *screen*, illuminated to a greater or less extent depending upon the amount of light incident at various points of the image.

3-5 Virtual objects. Fig. 3-1 has been drawn for the special case in which a *divergent* cone of rays, incident on a refracting surface, becomes a *convergent* cone after refraction. In general, however, the rays incident on a reflecting or refracting surface may be either diverging or converging and, similarly, the refracted rays may either converge or diverge. As special cases, either set of rays, or both, may be parallel to one another. The various possibilities can be distinguished in at least three different but equivalent ways, (1) by the algebraic signs of object distances s and image distances s', (2) by the terms "real" and "virtual," and (3) by the convergence or divergence of the rays. Four different possibilities are illustrated in Fig. 3-12.

In Fig. 3-12 (a), the incident rays are diverging and the refracted rays are converging. Object and image are both real and s and s' are both positive.

In Fig. 3-12 (b), both the incident and refracted rays are diverging. The object is real and the object distance s is positive, while the image is virtual and the image distance s', by convention (3), is negative.

The situations depicted in Fig. 3-12 (c) and (d) are less familiar than those in parts (a) and (b), but are of common occurrence in optical instruments. A refracting surface or surfaces at the left of the one shown have produced a *converging* cone of rays, which would converge to a real image at P but for the interposition of the refracting surface. The object distance s is the distance from the vertex to the point P at which the real image would have been formed, had the surface not been interposed. Point P is called a *virtual object* and, since it lies at the right of the vertex, the object distance s, by convention (2), is negative. In Fig. 3-12 (c) the rays are converging after refraction, as in Fig. 3-12 (a), so the image is real and the image distance s' is positive.

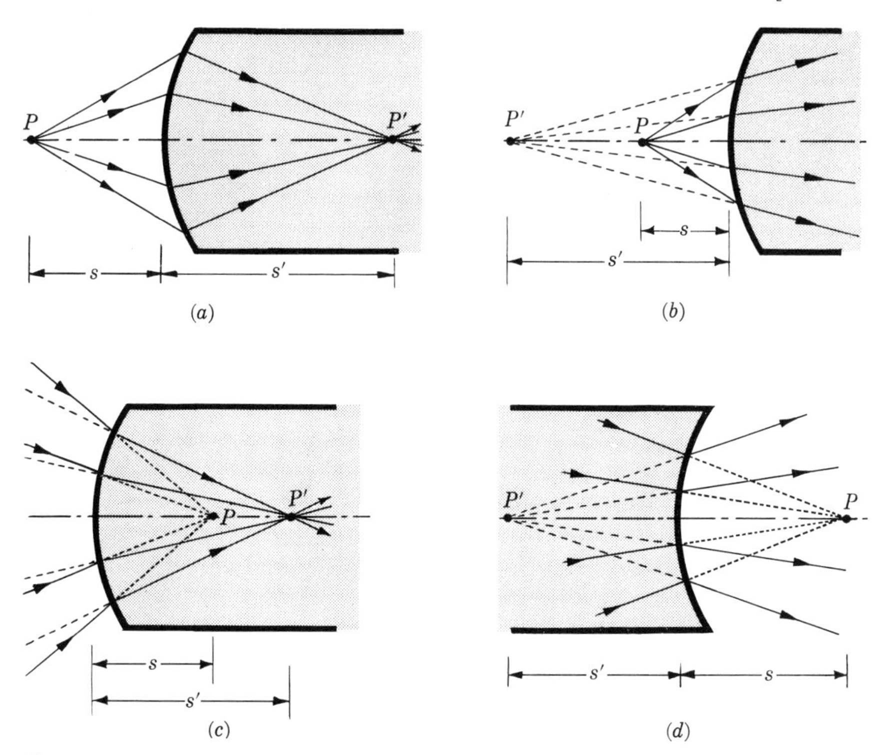

Fig. 3-12. (a) A real image of a real object. (b) A virtual image of a real object. (c) A real image of a virtual object. (d) A virtual image of a virtual object.

In Fig. 3-12 (d) the incident rays are converging and the refracted rays are diverging. Both object and image are virtual, and s and s' are both negative.

To sum up, then, whenever a *converging* cone of rays is incident on a surface, the point toward which the rays are converging serves as the object for the surface. It is called a *virtual object,* and the object distance s is *negative.*

If the incident rays are parallel to one another, one can consider either that they diverge from a point at an infinite distance to the left of the vertex, or that they converge toward a point at an infinite distance to the right of the vertex. In the first case the object is real and $s = +\infty$; in the second, the object is virtual and $s = -\infty$. Similarly, if the refracted rays are parallel, one can consider either that $s' = +\infty$ and the image is real, or that $s' = -\infty$ and the image is virtual.

These relations between convergence and divergence, object and image distances, and real and virtual images, are summarized in Fig. 3-13.

Objects	Incident rays diverging	Real object	s positive
	Incident rays converging	Virtual object	s negative
	Incident rays parallel	Object either real or virtual	$s = \pm\infty$
Images	Refracted rays converging	Real image	s' positive
	Refracted rays diverging	Virtual image	s' negative
	Refracted rays parallel	Image either real or virtual	$s' = \pm\infty$

FIG. 3-13.

Examples. (1) Find the position of the first focal point of a convex mirror. (See Fig. 3-14.)

By definition, the first focal point of a surface is that axial object point which is imaged at infinity. When we set $s' = \infty$ in Eq. (3-9), we get

$$\frac{1}{s} - \frac{1}{\infty} = -\frac{2}{R},$$

$$s = -\frac{R}{2}.$$

The radius of curvature of a convex mirror is positive, so the object distance s in the equation above is negative. That is, a convex mirror forms an image at infinity of a virtual object at a distance to the right of the mirror equal to one-half its radius of curvature, as illustrated in Fig. 3-14. In other words, incident rays converging toward this virtual object point are parallel to the axis of the mirror after reflection.

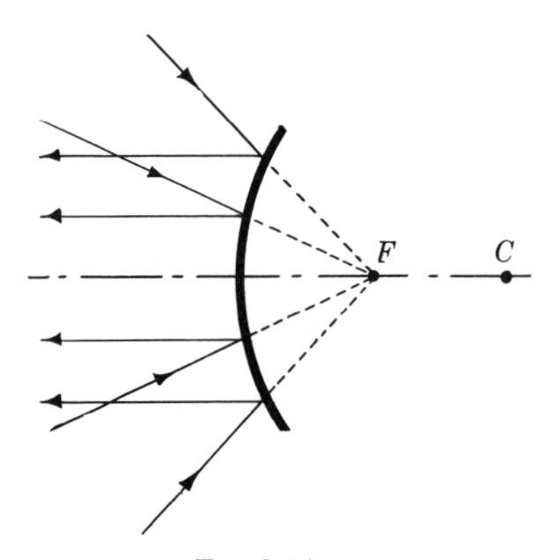

FIG. 3-14.

(2) The radius of curvature of the spherical surface in Fig. 3-12 (c) is 40 mm, the index at the left of the surface is 1.0, that at the right is 1.5, and the incident rays are converging toward a virtual object P, 20 mm to the right of the vertex. Find the position of the image P'.

From the given data,

$$n = 1.0, \quad n' = 1.5, \quad s = -20 \text{ mm}, \quad R = +40 \text{ mm}.$$

$$\frac{1}{-20} + \frac{1.5}{s'} = \frac{1.5 - 1}{40},$$

$$s' = 24 \text{ mm.}$$

The image is real and lies 24 mm to the right of the vertex.

(3) Many readers will probably have wondered why it is necessary to complicate matters by talking about virtual objects, and why a problem such as the preceding one cannot be solved merely by assuming that a real object is located at point P in Fig. 3-12 (c) and by computing the position of its image. To do so is an excellent example of the generally unwise procedure of attempting to solve a given problem by making up a second problem, thought to be equivalent to the first, and solving that instead. Physically, the two problems are *not* equivalent, because in the first a specified cone of *converging* rays is incident on the *convex* side of the refracting surface, while in the presumed equivalent problem a cone of *diverging* rays is incident on the *concave* side of the surface. That is, the presumed equivalent problem assumes that rays diverge from P in the directions of the dotted lines in Fig. 3-12 (c), and computes the directions of these rays after refraction.

As a numerical example to show that the two problems are not equivalent, let us compute the position of the image of a real object at point P in Fig. 3-12 (c). To accord with our sign convention, the diagram must be redrawn as in Fig. 3-15, with P at the left of the refracting surface. We now have

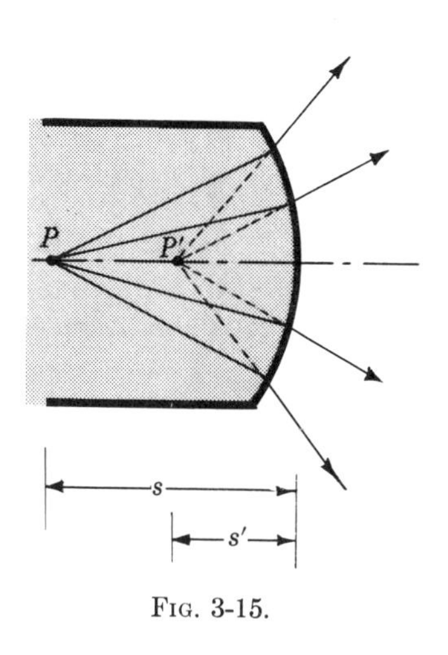

FIG. 3-15.

$$n = 1.5, \qquad n' = 1.0,$$

$$s = +20 \text{ mm},$$

$$R = -40 \text{ mm},$$

$$\frac{1.5}{20} + \frac{1}{s'} = \frac{1 - 1.5}{-40},$$

$$s' = -16 \text{ mm.}$$

Hence a real object at P would be imaged at a point 16 mm to the left of the surface in Fig. 3-15, or 16 mm to the right of the surface in Fig. 3-12 (c), whereas the given cone of converging rays in Fig. 3-12 (c) is imaged 24 mm to the right of the surface.

3-6 Images as objects. Most optical systems include more than one reflecting or refracting surface. The image formed by the first surface serves as the object for the second; the image formed by the second serves as the object for the third; etc. In Fig. 3-16, the arrow at point O represents a small object at right angles to the axis. A narrow cone of rays diverging from the head of the arrow is traced through the system. Surface 1 forms a real image of the arrow at point P. The object distance for the first surface is OV_1 and the image distance is V_1P. Both of these are positive.

The image at P, formed by surface 1, serves as a real object for surface 2. The object distance is PV_2 and is positive. The second surface forms a virtual image at point Q. The image distance is QV_2 and is negative.

The image at Q, formed by surface 2, serves as the object for surface 3. The object distance is QV_3 and is positive, since Q is at the left of V_3. The image at Q, although virtual, constitutes a real object as far as surface 3 is concerned. The rays incident on surface 3 are rendered converging and, except for the interposition of surface 4, would converge to a real image at point R. Even though this image is never formed, distance V_3R is the image distance for surface 3 and is positive.

The rays incident on surface 4 are converging, and the image at R, toward which they converge, is a virtual object for surface 4. The object distance is V_4R and is negative. These rays are rendered converging after refraction at surface 4, which forms a real image at I. The final image distance is V_4I and is positive.

The lateral magnification produced by the system is the product of the lateral magnifications at each surface.

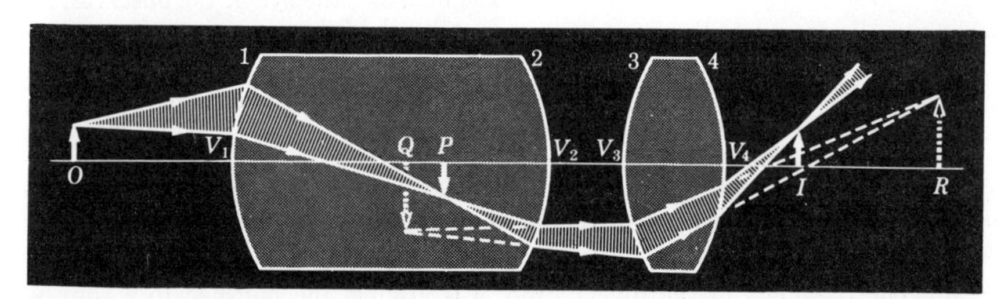

FIG. 3-16. The object for each surface, after the first, is the image formed by the preceding surface.

Problems—Chapter 3

(1) A small particle is embedded in a clear glass marble 1 inch in diameter, at a radial distance of ¼ inch beneath the surface. The index of the glass is 1.50. Find the position and lateral magnification of the image of the particle, formed by paraxial rays from the particle directed (a) toward the center, (b) radially outward from the center.

(2) A paperweight in the form of a glass hemisphere is placed on a printed page with its flat surface down, and an observer looks vertically down toward the center of the hemisphere. Find (a) the position, and (b) the lateral magnification of the image of the type at the center of the flat surface of the hemisphere. (c) Draw a diagram, represent an object at the center of the flat surface by an arrow, and construct approximately to scale a few rays from a point of the object, not on the axis. Take the index of the glass as 1.50.

(3) A small tropical fish is at the center of a spherical fish bowl 1 ft in diameter. Find (a) the position and (b) the lateral magnification of the image of the fish, seen by an observer outside the bowl. The effect of the thin glass walls of the bowl may be neglected.

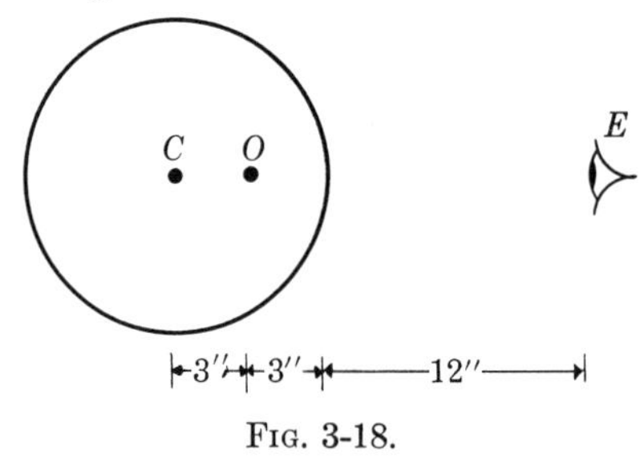

FIG. 3-18.

(4) A small tropical fish is at point O in Fig. 3-18, 3 inches from the center C of a thin-walled spherical fish bowl 1 ft in diameter. (a) Find the position and lateral magnification of the image of the fish seen by an observer at E. (b) Find the position and lateral magnification of the image of the eye of the observer, as seen by the fish.

(5) The left end of a long glass rod 10 cm in diameter, of index 1.50, is ground and polished to a convex hemispherical surface of radius 5 cm. An object in the form of an arrow 1 mm long, at right angles to the axis of the rod, is located on the axis 20 cm to the left of the vertex of the convex surface. Compute (a) the position and (b) the magnification of the image of the arrow formed by paraxial rays incident on the convex surface.

(6) The right end of the rod in Prob. 5 is ground and polished to a convex spherical surface of radius 10 cm. The length of the rod between vertices is 60 cm. An arrow 1 mm long, at right angles to the axis and 20 cm to the left of the first vertex, constitutes the object for the first surface. (a) What constitutes the object for the second surface? (b) What is the object distance for the second surface? (c) Is the object real or virtual? (d) What is the position of the image formed by the second surface? (e) Is the image real or virtual? Erect or inverted, with respect to the original object? (f) What is the height of the final image?

(7) The same rod as in Probs. 5 and 6 is now shortened to a distance of 10 cm between its vertices, the curvatures of its ends remaining the same as in Prob. 6. (a) What is the object distance for the second surface? (b) Is the object real or

virtual? (c) What is the position of the image formed by the second surface? (d) Is the image real or virtual? Erect or inverted, with respect to the original object? (e) What is the height of the final image?

(8) The rod in Prob. 5 is immersed in water of index 1.33. An arrow of the same size and in the same position serves as the object. (a) Find the position of the image formed by refraction of paraxial rays at the convex surface. (b) Is the image real or virtual? Erect or inverted? What is its height?

(9) The left end of a glass rod of index 1.50 is ground and polished to a convex spherical surface of radius 2 cm. The rod is in air. (a) Find the first and second focal lengths of the surface, and show the first and second focal points in a diagram. (b) An arrow 1 cm long, at right angles to the axis, lies 8 cm to the left of the vertex. Find *graphically* the position of the image of the arrow. (c) Check your graphical construction by computing the position and magnification of the image.

(10) A narrow beam of parallel rays enters a solid glass sphere in a radial direction. The radius of the sphere is 3 cm and its index is 1.50. (a) At what point outside the sphere are these rays brought to a focus? (b) What should be the index of refraction of the sphere in order that the rays will be brought to a focus at the vertex of the second surface?

(11) A glass plate 1 inch thick, of index 1.50, having plane parallel faces, is held with its faces horizontal and its lower face 4 inches above a printed page. Find (a) the position and (b) the magnification of the image of the page, formed by rays making a small angle with the normal to the plate.

(12) A parallel beam of light is incident as shown in Fig. 3-19 on the surface of a transparent hemisphere of index 2.0. (a) Show whether or not the central ray in the beam is totally reflected at the plane surface. (b) Find the position of the image formed by refraction at the first surface. (c) Find the position of the image formed by reflection at the plane surface. (d) Show clearly in a diagram the subsequent paths of the three incident rays in Fig. 3-19, after reflection at the plane surface.

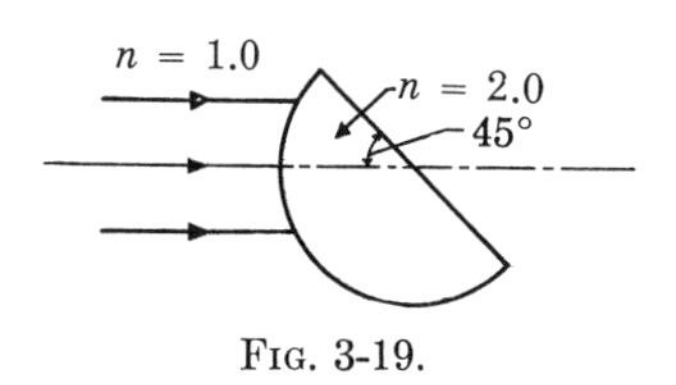

FIG. 3-19.

(13) Find graphically by a carefully drawn full size diagram the position and magnification of the image formed by a concave spherical mirror of radius of curvature 4 inches, of an arrow ¼ inch high, 3 inches to the left of the vertex of the mirror. Check your results by calculation. Indicate by shading, the cone of rays that form the image of the head of the arrow, if the mirror is 1 inch in diameter.

(14) A concave mirror is to form an image of the filament of a headlight lamp on a screen 4 m from the mirror. The filament is 5 mm high, and the image is to be 40 cm high. What should be the focal length of the mirror, and how far in front of the vertex of the mirror should the filament be placed?

(15) The diameter of the moon is 2160 mi and its distance from the earth is 240,000 mi. Find the diameter of the image of the moon formed by a spherical concave telescope mirror of focal length 12 ft.

(16) An object 1 cm high is 20 cm from the vertex of a concave spherical mirror whose radius of curvature is 50 cm. Compute the position and size of the image. Is it real or virtual? Erect or inverted?

(17) A concave spherical mirror has a radius of curvature of 50 cm. (a) Find two positions of an object for which the image is four times as large as the object. (b) What is the position of the image in each case? (c) Is it real or virtual?

(18) A man looks at himself in a concave shaving mirror whose radius of curvature is 48 inches. One of his eyes is on the axis of the mirror, 12 inches in front of the vertex. (a) Find the position and magnification of the image of the eye. (b) Is the image real or virtual? Erect or inverted?

(19) The nose of an airplane is a polished spherical surface whose radius of curvature is 1 m. (a) Where is the image of the sun, formed by reflection in the polished surface? (b) How close to the sun would the airplane need to be in order to displace the image from this position by 1 mm, in a radial direction?

(20) An image of the sun is formed by reflection in a silvered glass globe 20 inches in diameter. (a) Where is the image? (b) What is its magnification? (c) Where must an object be placed if its image by reflection in the globe is to be of the same size as the object, but inverted?

(21) A spherical lamp bulb 2 inches in diameter is imaged by a steel ball bearing located 20 inches away from the bulb. If the ball bearing is 1 inch in diameter, (a) where is the image of the lamp bulb, and (b) what is its diameter?

(22) A convex mirror is sometimes used as the rear view mirror of an automobile. Prove that the images formed by such a mirror are always virtual, always erect, and always smaller than the object.

(23) A solid glass sphere of radius R and index 1.50 is silvered over one hemisphere as in Fig. 3-20. A small object is located on the axis of the sphere at a distance $2R$ from the pole of the unsilvered hemisphere. Find the position of the final image formed by the refracting and reflecting surfaces.

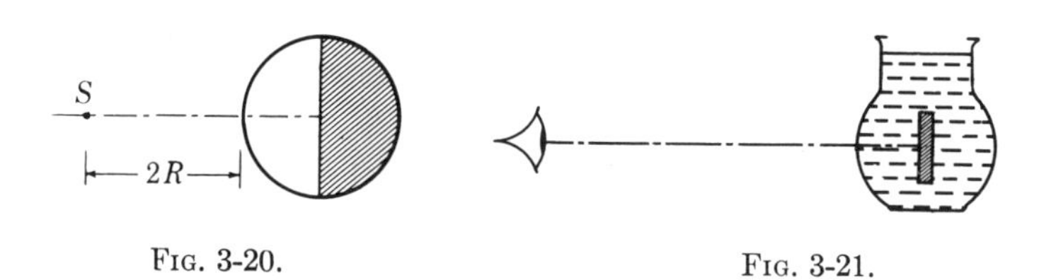

FIG. 3-20. FIG. 3-21.

(24) An observer places his eye at a distance of $10R$ to the left of the vertex of the sphere in Fig. 3-20 and looks toward the sphere. (a) Does he see an image of his eye? (b) Where is this image? (c) Is it real or virtual?

(25) A plane mirror is suspended vertically at the center of a large thin-walled spherical flask filled with water. The diameter of the flask is 10 inches. An observer whose eye is 35 inches from the mirror as shown in Fig. 3-21 sees an image of his own eye. Where is the image which he sees? The effect of the thin glass walls of the flask may be neglected.

(26) A glass hemisphere of index 1.5 and radius R is silvered on its flat face. A small object of height h is located on the axis of the hemisphere at a distance of $2R$ from the vertex, as shown in Fig. 3-22. (a) Find the position of the first image formed by the spherical glass surface. Consider only paraxial rays. (b) Where is the image formed by the mirror? (c) Locate the final image formed by the system. (d) What is the height of the final image? (e) Is it erect or inverted?

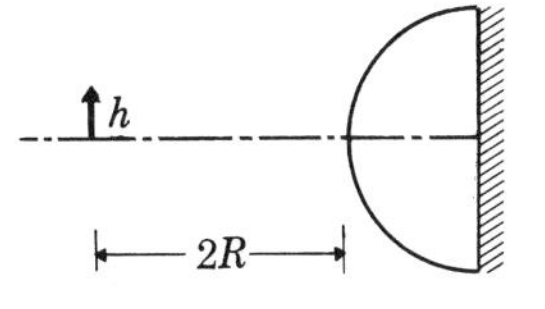

FIG. 3-22.

CHAPTER 4

LENSES

4-1 Lenses. A lens is an optical system bounded by two or more refracting surfaces having a common axis. If the lens has two surfaces only, it is called a *simple* lens; if there are more than two surfaces, a *compound* lens. All high quality lenses are compound lenses, that is, they consist of a number of simple lenses having a common axis. The surfaces of the lenses may be in contact, or there may be air spaces between them.

Except when it is incident normally on one of the surfaces, a ray passing through a simple lens undergoes deviation at both surfaces. The axial thickness of many simple lenses is sufficiently small so that the entire deviation of a ray can be considered to take place in a single plane through the center of the lens. When this approximation can be made, the lens is called a *thin* lens. Any lens, simple or compound, that is not a thin lens, is called for brevity a *thick* lens.

4-2 The simple lens in air. The general problem of refraction by a lens is solved by applying the methods of Sec. 3-1 to each surface in turn, the object for each surface after the first being the image formed by the preceding surface.

Fig. 4-1 shows a pencil of rays[1] diverging from point Q of an object PQ and incident on a simple lens L. Let n represent the index of refraction of the material of which the lens is constructed, and R_1 and R_2 the radii of curvature of its first and second surfaces. The lens is in air. The first surface of the lens forms a virtual image of Q at Q'. This virtual image serves as a real object for the second surface, which forms a real image of Q' at Q''. The distance of the object from the first vertex is s_1. The corresponding image distance, s_1', is found from Eq. (3-5).

$$\frac{1}{s_1} + \frac{n}{s_1'} = \frac{n-1}{R_1}. \tag{4-1}$$

The distance of the object $P'Q'$ from the second vertex is

$$s_2 = t - s_1', \tag{4-2}$$

[1] The formulas in this chapter are based on first-order theory and apply to paraxial rays. For clarity, all diagrams have been exaggerated laterally and aberrations have been ignored.

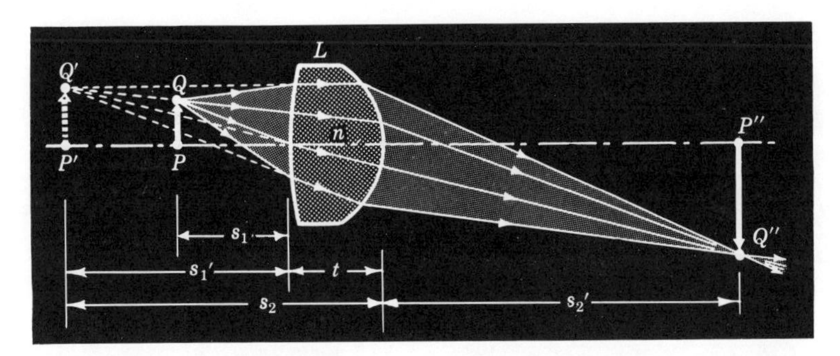

FIG. 4-1. Formation of an image by a simple lens. The image formed by the first surface serves as the object for the second.

where t is the axial thickness of the lens. (We must write $t - s_1'$ and not $t + s_1'$, since the image distance s_1' is a negative quantity.)

When Eq. (3-5) is applied to the second surface we get

$$\frac{n}{s_2} + \frac{1}{s_2'} = \frac{1 - n}{R_2}. \tag{4-3}$$

Eqs. (4-1), (4-2), and (4-3) may now be solved for s_2', which gives the distance of the final image $P''Q''$ from the second vertex. The height of the final image is obtained by multiplying the height of the object PQ by the product of the lateral magnifications produced by each surface.

Obviously this method of determining the position and size of an image can be extended to any number of surfaces, as in the example in Sec. 3-6.

4-3 Focal points and focal planes. The *first focal point* of a lens may be defined as that object point on the lens axis which is imaged by the lens at infinity. In other words, rays diverging from the first focal point are parallel to the axis of the lens after refraction, as shown in Fig. 4-2 (a). The first focal point is lettered F.

A plane through the first focal point, at right angles to the axis, is called the *first focal plane.* Fig. 4-2 (b) shows a pencil of rays diverging from a point P in the first focal plane. These rays are *parallel to one another* after refraction, but not parallel to the lens axis. That is, they converge to an infinitely distant image of the point P.

The *second focal point* of a lens may be defined as the image point of an infinitely distant point object on the lens axis. In other words, rays incident on the lens, parallel to the lens axis, pass through the second focal point after refraction as in Fig. 4-2 (c). The second focal point is lettered F'.

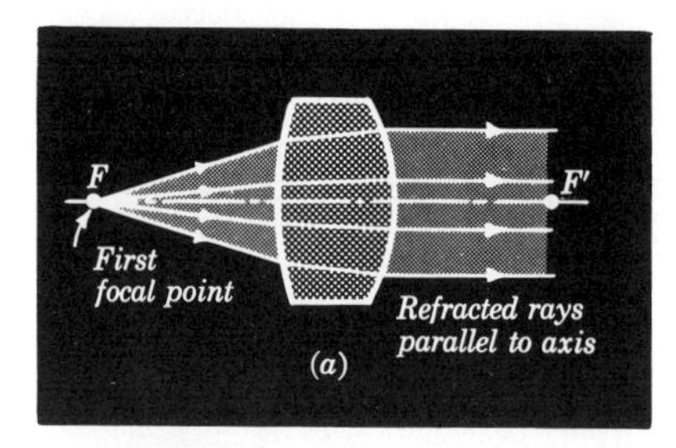

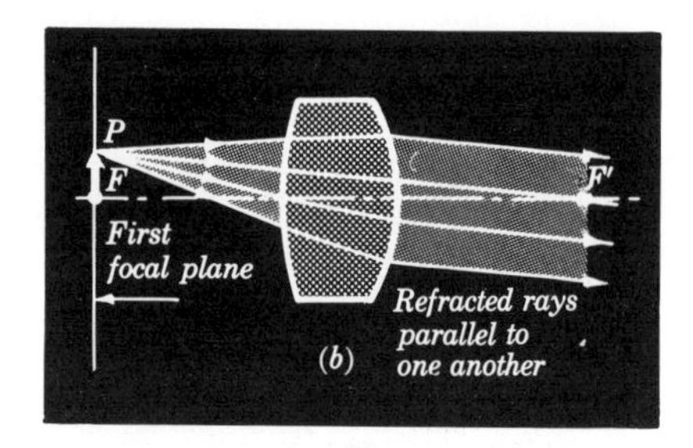

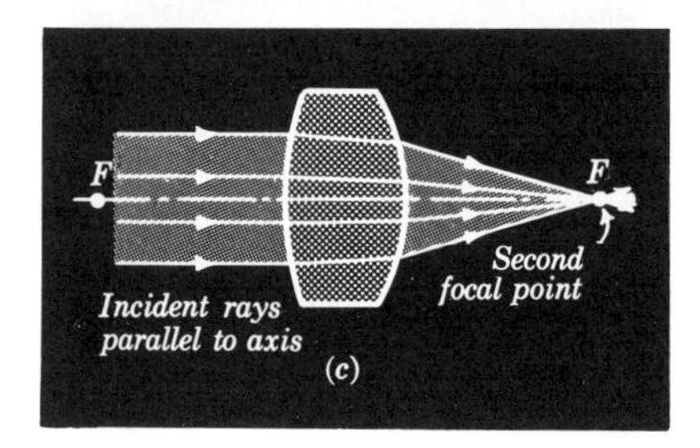

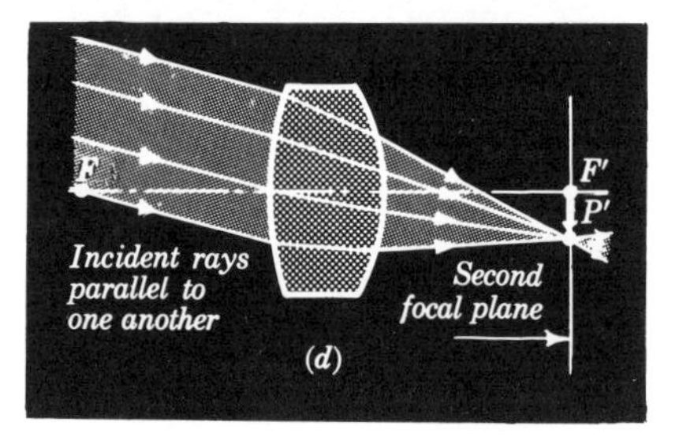

FIG. 4-2. (a) Rays diverging from the first focal point are parallel to the axis after refraction. (b) Rays diverging from a point in the first focal plane are parallel to one another after refraction. (c) Incident rays parallel to the axis pass through the second focal point after refraction. (d) Incident rays parallel to one another converge to an image in the second focal plane.

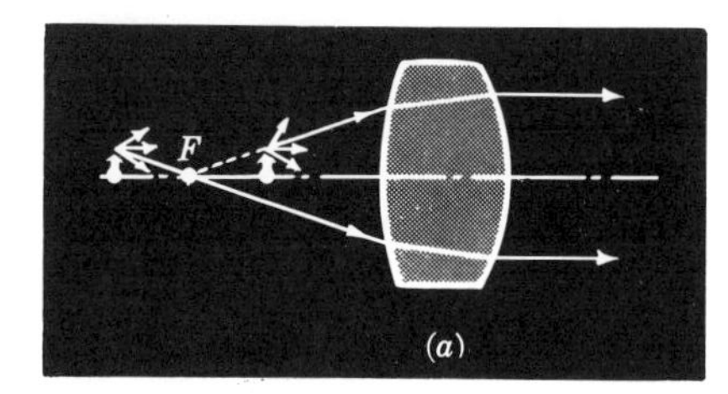

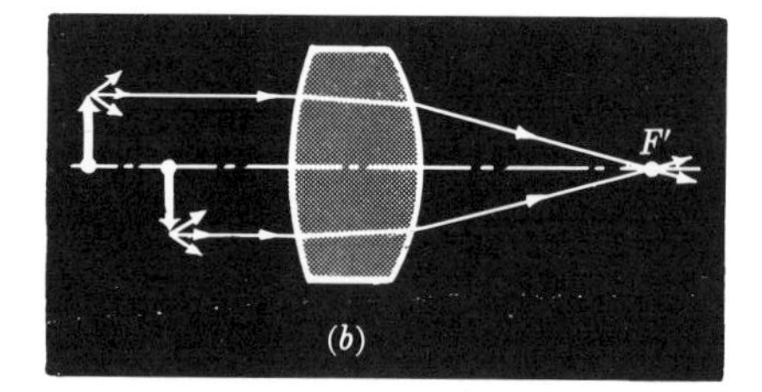

FIG. 4-3.

A plane through the second focal point, at right angles to the axis, is called the *second focal plane*. In Fig. 4-2 (d), a bundle of parallel rays from an infinitely distant point object, not on the lens axis, converges to an image P' in the second focal plane of the lens.

It is not necessary that a ray shall actually *originate* at the first focal point of a lens, in order that its direction shall be parallel to the lens axis after refraction. The deviation of a ray incident at any point of a lens depends only on its *direction* at the point of incidence. This is illustrated in Fig. 4-3 (a), which shows the deviation of two rays, one of which originates at the left of the first focal point while the other originates at the right of this point. Similarly, any ray incident on a lens parallel to the lens axis, passes through the second focal point after refraction as in Fig. 4-3 (b).

To find the position of the first focal point of a simple lens, let s_2' equal infinity in Eq. (4-3) and solve this equation for s_2. Then find s_1' from Eq. (4-2), insert this value in Eq. (4-1), and solve for s_1. This gives the distance from the first focal point to the first vertex. The position of the second focal point is found in a similar manner, letting s_1 equal infinity and solving for s_2', which gives the distance from the second vertex to the second focal point.

If the lens has more than two surfaces, the focal points are located by an extension of the process above. One sets the final image distance equal to infinity and solves for the corresponding object distance, then sets the object distance equal to infinity and solves for the image distance.

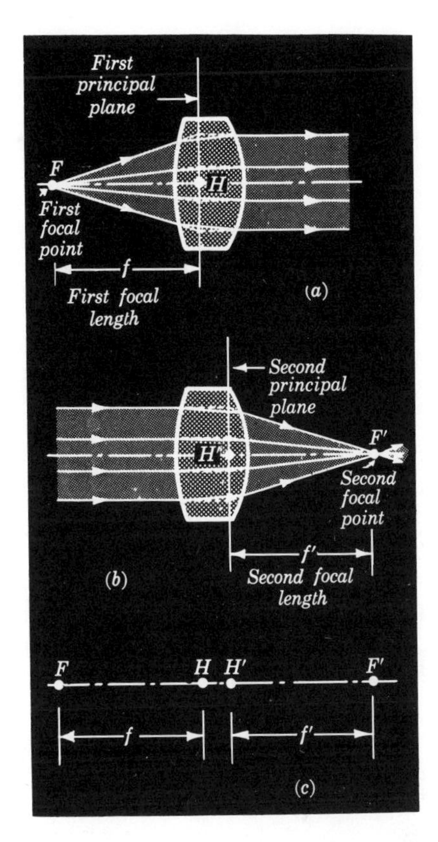

FIG. 4-4. (a) The entire deviation of rays from the first focal point can be assumed to take place in the first principal plane. (b) The entire deviation of incident rays parallel to the axis can be assumed to take place in the second principal plane. (c) The distance between each focal point and the corresponding principal point is the focal length.

4-4 Principal points and focal lengths. Computation of the position and magnification of the image of a given object at a given distance from a lens is facilitated by determining the positions of two points on the lens axis known as the *principal points*. These are defined as follows.

Fig. 4-4 (a) shows a cone of rays diverging from the first focal point of a simple lens and deviated at both of the lens surfaces. When the incident and emergent rays are projected ahead and back, as indicated by dotted lines, the points of intersection lie in a common plane known as the *first principal plane* of the lens. That is, the two deviations at the two lens surfaces are equivalent to a single deviation in the first principal plane. The point of intersection of the first principal plane with the axis is the *first principal point* and is lettered H.

Fig. 4-4 (b) shows a bundle of rays incident on a simple lens and parallel to its axis. When the incident and emergent rays are projected ahead and back, the points of intersection lie in a plane known as the *second principal plane*. The point of intersection of this plane with the axis is the *second principal point* and is lettered H'. Except in corrected lenses, it is only for paraxial rays that the focal "planes" and principal "planes" are actually planes at right angles to the axis. In general, they are curved surfaces and should properly be called the *focal surfaces* and *principal surfaces*.

The distance from the first focal point F to the first principal point H is called the *first focal length* of the lens and is lettered f. The distance from the second principal point H' to the second focal point F' is called the *second focal length* and is lettered f'. However, if the medium on both sides of the lens has the same index, which is the case if the lens is in air, the two focal lengths are equal. This is the only case we shall consider (except in some problems) and we shall represent the common value of the focal length by f.

We shall show in the next section how to compute the focal length of a thin lens and how to locate its focal points and principal points. The more general problem of a thick lens will be discussed in Sec. 4-9.

One type of retrodirective reflector used in highway signs was described in Sec. 2-6. Another common type, the so-called "reflector button," is illustrated in Fig. 4-5. It consists of a short-focus converging thick lens combined with a concave mirror conforming to the second focal surface of the lens. (As mentioned above, the focal "planes" of an uncorrected lens are actually curved surfaces.) A parallel incident beam, whether parallel to the lens axis or not, is converged to a point on the mirror

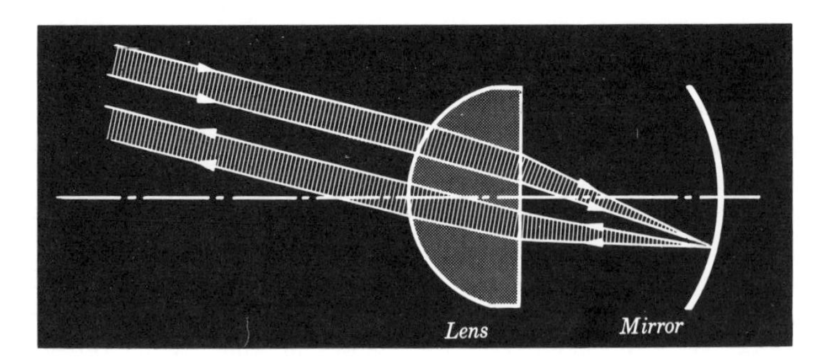

FIG. 4-5. Principle of the retrodirective reflector button.

and returned by the latter to the lens as a beam diverging from the point at which reflection occurred. Since this point is in the second focal surface, the lens renders the emergent beam parallel, with its direction of travel just the reverse of the incident direction. The aberrations of a simple spherical lens introduce the necessary spreading in the reflected beam.

4-5 The thin lens. The entire deviation of any ray passing through a *thin* lens can be considered to take place in a plane through the center of the lens. That is, the first and second principal planes of a thin lens coincide. Hence the first and second principal points coincide also, and both coincide with the center of the lens. The focal length of a thin lens then becomes merely the distance from the center of the lens to either focal point. See Fig. 4-6.

The first focal point of any lens is, by definition, that axial object point which is imaged at infinity. To find the position of the first focal point of a thin lens, we let s_2' equal infinity in Eq. (4-3). This gives

$$\frac{n}{s_2} = \frac{1-n}{R_2}. \tag{4-4}$$

Since the axial thickness t of a thin lens is negligible, we have from Eq. (4-2)

$$s_1' = -s_2,$$

and hence

$$\frac{n}{s_2} = -\frac{n}{s_1'}. \tag{4-5}$$

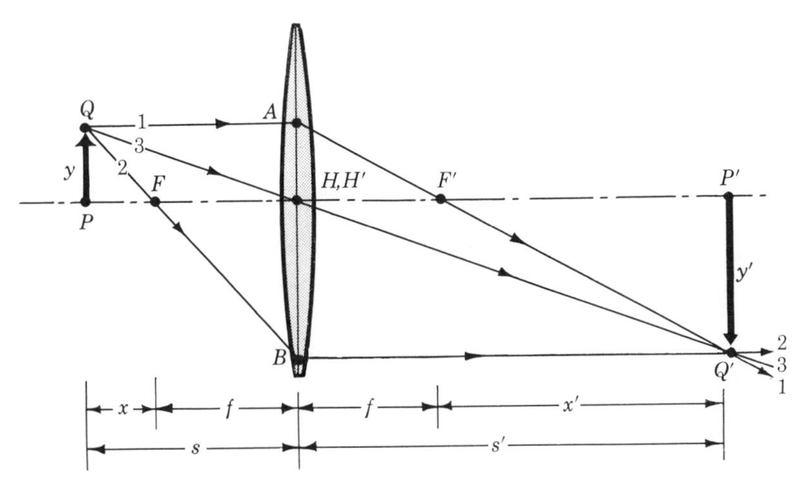

FIG. 4-6. The principal points of a thin lens coincide.

When Eqs. (4-4) and (4-5) are combined with Eq. (4-1) we get

$$\frac{1}{s_1} = (n - 1)\left(\frac{1}{R_1} - \frac{1}{R_2}\right). \tag{4-6}$$

The distance s_1 is, strictly speaking, the distance from the first focal point to the first vertex, but if the lens is thin, the principal points and the lens vertices can be assumed to coincide. Hence s_1 is also the distance from the first focal point F to the first principal point H, and therefore by definition is equal to the focal length f. Then from the preceding equation,

$$\boxed{\frac{1}{f} = (n - 1)\left(\frac{1}{R_1} - \frac{1}{R_2}\right).} \tag{4-7}$$

The focal length of a thin lens in air therefore depends only on the index of the lens and the radii of curvature of its surfaces. Eq. (4-7) is known as the *lensmaker's equation.* It will be left for the reader to show that the same equation is obtained when one computes the distance from the second principal point to the second focal point.

Example. Find the focal length of the plano-convex thin lens in Fig. 4-7. The index of the lens is 1.50.

If the convex surface faces the light,

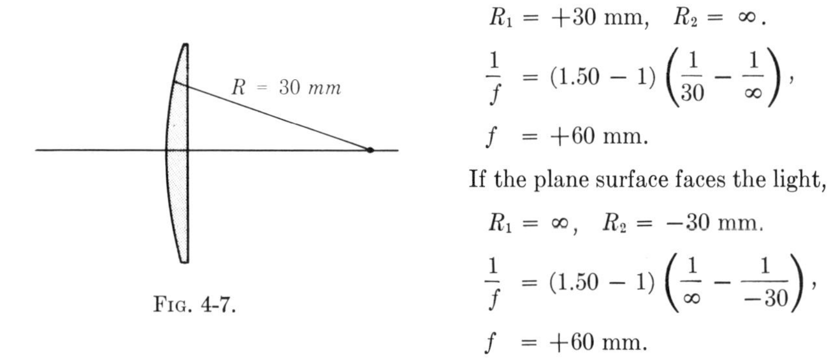

Fig. 4-7.

$$R_1 = +30 \text{ mm}, \quad R_2 = \infty.$$

$$\frac{1}{f} = (1.50 - 1)\left(\frac{1}{30} - \frac{1}{\infty}\right),$$

$$f = +60 \text{ mm}.$$

If the plane surface faces the light,

$$R_1 = \infty, \quad R_2 = -30 \text{ mm}.$$

$$\frac{1}{f} = (1.50 - 1)\left(\frac{1}{\infty} - \frac{1}{-30}\right),$$

$$f = +60 \text{ mm}.$$

It is therefore immaterial which surface is considered the first or, in other words, the first and second focal lengths are equal.

We now derive four important equations from which the image distance and the lateral magnification can be computed for an object at any arbitrary distance from a thin lens. In Fig. 4-6 there is shown an object

PQ of height y, together with its image $P'Q'$ of height y', formed by a thin lens. In the absence of aberrations, *all* rays from the point Q are imaged at Q'. (The entire deviation of all rays from Q is assumed to take place in a plane through the center of the lens.) Hence the point of intersection of *any two* rays from Q suffices to determine the position of the image Q'. When the position of Q' has been found in this way, the paths of all other rays from Q are known.

From the definitions of the second principal plane and the second focal point, ray 1 in Fig. 4-6, originally parallel to the axis, passes through the second focal point F' after refraction. From the definitions of the first focal point and the first principal plane, ray 2, incident on the lens after passing through the first focal point F, is parallel to the axis of the lens after refraction. A third ray that can readily be traced through a thin lens is numbered 3 in Fig. 4-6. This ray, directed toward the center of the lens, is undeviated, since the surfaces of the lens are parallel to one another at the lens axis and, as we have shown, there is no deviation of a ray passing through a plate whose surfaces are parallel. There is, of course, a transverse displacement, but this is negligible if the lens is thin.

Let us use the letter s, without a subscript, to represent the object distance *measured from the first principal point* H, and s' to represent the image distance *measured from the second principal point* H'. Since both H and H' coincide at the center of a thin lens, object and image distances can both be measured from the center of such a lens. In a thick lens, H and H' do not coincide, and this more general case is discussed in Sec. 4-9. Let x and x' represent object and image distances *measured from the respective focal points* F and F'. Object distances s and x are considered positive if the object lies at the left of its reference point, as in Fig. 4-6, and image distances s' and x' are considered positive if the image lies at the right of its reference point.

From the similar triangles PQF and FHB in Fig. 4-6,

$$\frac{PQ}{PF} = \frac{HB}{FH}.$$

But $PQ = y$, $PF = x$, $HB = -y'$, $FH = f$. (We set $HB = -y'$ since, by convention (6), the height y' of the inverted image $P'Q'$ is negative.) Hence

$$\frac{y}{x} = \frac{-y'}{f}. \tag{4-8}$$

In the same way, from the similar triangles $AH'F'$ and $F'P'Q'$,

$$\frac{y}{f} = \frac{-y'}{x'}. \tag{4-9}$$

Also, from the similar triangles QAB and FHB,

$$\frac{y - y'}{s} = \frac{-y'}{f}, \tag{4-10}$$

and from the triangles ABQ' and $AH'F'$,

$$\frac{y - y'}{s'} = \frac{y}{f}. \tag{4-11}$$

When Eqs. (4-10) and (4-11) are added, we get

$$\frac{y - y'}{s} + \frac{y - y'}{s'} = \frac{y}{f} - \frac{y'}{f} = \frac{y - y'}{f},$$

or

$$\boxed{\frac{1}{s} + \frac{1}{s'} = \frac{1}{f}.} \tag{4-12}$$

When Eq. (4-8) is divided by Eq. (4-9) we get

$$\frac{f}{x} = \frac{x'}{f},$$

or

$$\boxed{xx' = f^2.} \tag{4-13}$$

The lateral magnification m is the ratio of y' to y. Dividing Eq. (4-10) by Eq. (4-11) gives

$$\boxed{m = -\frac{s'}{s}.} \tag{4-14}$$

Also, from Eqs. (4-8) and (4-9),

$$\boxed{m = -\frac{f}{x} = -\frac{x'}{f}.} \tag{4-15}$$

Eq. (4-12) is known as the *Gaussian* form of the lens equation, after the mathematician Karl F. Gauss. (Gauss's law in electrostatics, as well as the unit of magnetic flux density in the electromagnetic system of units, the gauss, take their names from him also.) Eq. (4-13), first derived by

Sir Isaac Newton, is the *Newtonian* form of the lens equation. The Gaussian form is probably more familiar, but the Newtonian equation is algebraically simpler. Notice carefully that in the former equation object and image distances s and s' are measured from the *principal points* H and H' respectively (or from the center of a thin lens), while in the latter, object and image distances x and x' are measured from the *focal points* F and F'.

The lateral magnification m can be expressed either in terms of s and s', by Eq. (4-14), or in terms of x, x', and f, by Eq. (4-15).

Example. An object is located 30 cm to the left of a thin lens of focal length 20 cm, as in Fig. 4-6. Find the position and lateral magnification of its image, using both the Gaussian and Newtonian forms of the lens equation.

The object distance s, measured from the center of the lens, is +30 cm. From the Gaussian equation.

$$\frac{1}{30} + \frac{1}{s'} = \frac{1}{20},$$

$$s' = +60 \text{ cm}.$$

The image is real (s' is positive) and lies 60 cm to the right of the center of the lens.

The object distance x, measured from the first focal point, is +10 cm. From the Newtonian equation,

$$10x' = (20)^2,$$

$$x' = +40 \text{ cm}.$$

This is evidently in agreement with the answer above. The lateral magnification, by Eq. (4-14), is

$$m = -\frac{60}{30} = -2.$$

The image is inverted (m is negative) and twice the height of the object.

From Eq. (4-15),

$$m = -\frac{20}{10} = -\frac{40}{20} = -2.$$

In the seven parts of Fig. 4-8, a number of rays from the head of an arrow representing an object O have been traced through a thin lens of focal length f. Lens aberrations are neglected. The image of the head of the arrow has been located by using two (in some cases, all three) of the rays referred to in Fig. 4-6. In addition, the outermost rays incident on the lens have been drawn. The object distances are respectively, $+3f$, $+2f$, $+\frac{3}{2}f$, $+f$, $+\frac{3}{4}f$, 0, and $-2f$. In parts (1) to (5) inclusive, the object

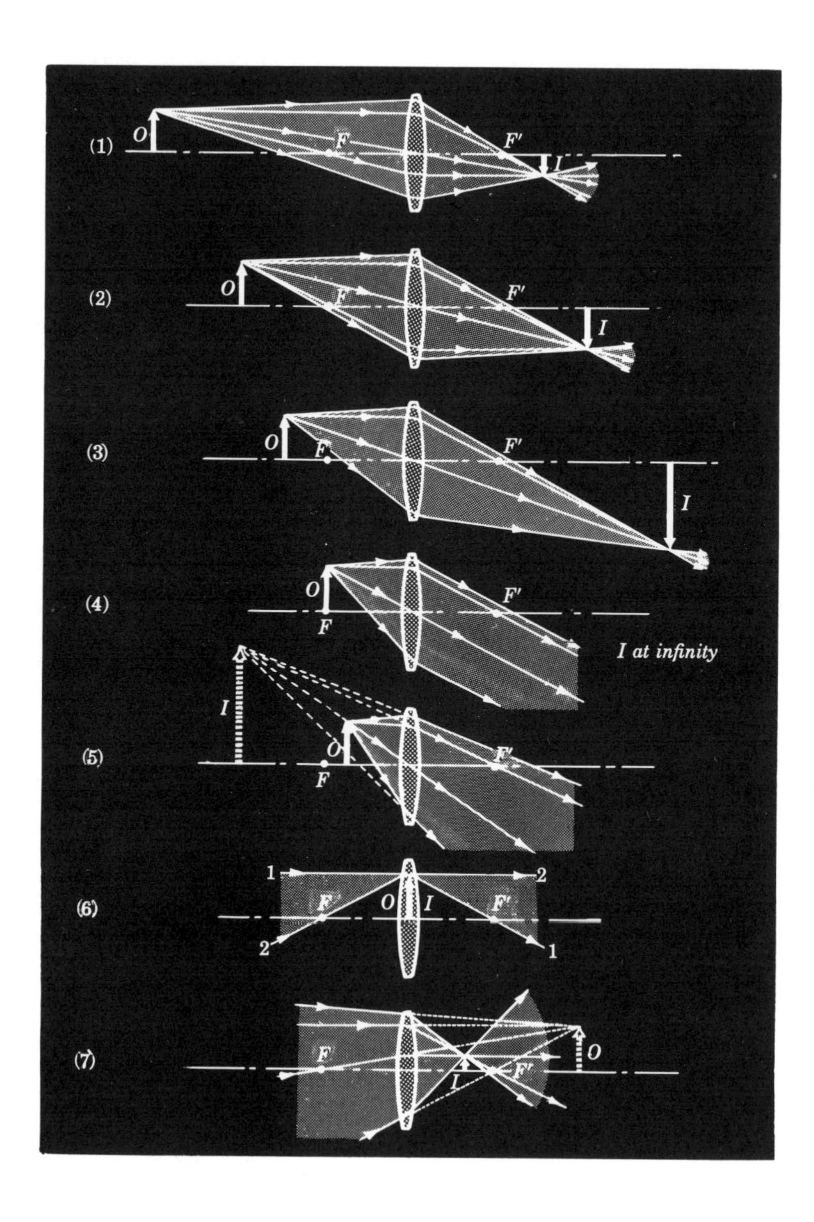

FIG. 4-8. Formation of an image by a thin lens.

Case	s	s'	x	x'	m	Object	Image	Erect or Inverted
1	$+3f$	$+\frac{3}{2}f$	$+2f$	$+\frac{1}{2}f$	$-\frac{1}{2}$	Real	Real	Inverted
2	$+2f$	$+2f$	$+f$	$+f$	-1	Real	Real	Inverted
3	$+\frac{3}{2}f$	$+3f$	$+\frac{1}{2}f$	$+2f$	-2	Real	Real	Inverted
4	$+f$	$\pm\infty$	0	$\pm\infty$	$\mp\infty$	Real	Real or Virtual	Erect or Inverted
5	$+\frac{2}{3}f$	$-2f$	$-\frac{1}{3}f$	$-3f$	$+3$	Real	Virtual	Erect
6	0	0	$-f$	$-f$	$+1$	Real or Virtual	Real or Virtual	Erect
7	$-2f$	$+\frac{2}{3}f$	$-3f$	$-\frac{1}{3}f$	$+\frac{1}{3}$	Virtual	Real	Erect

FIG. 4-9.

is real. It may be either a material object or the image formed by a preceding lens or lens system. In part (7) the object is virtual (a converging cone of rays is incident on the lens) and there must, of necessity, be a lens somewhere at the left of the one shown. In part (6) the object coincides with the lens. Here the distinction between real and virtual objects breaks down, and we may think of O as a real object just to the left of the lens, or as a virtual object just to the right of it. The diagram has been drawn as if O were a real image formed by a preceding lens, but it might equally well be a material object in contact with the lens. In the latter case, of course, there would be no rays at the left of the lens. Note that in this case object and image coincide.

Fig. 4-9 lists, for each part of Fig. 4-8, the values of s, s', x, x', and m, and also states whether object and image are real or virtual, and whether the image is erect or inverted.

Fig. 4-10 is a graph of the Gaussian lens equation,

$$\frac{1}{s} + \frac{1}{s'} = \frac{1}{f},$$

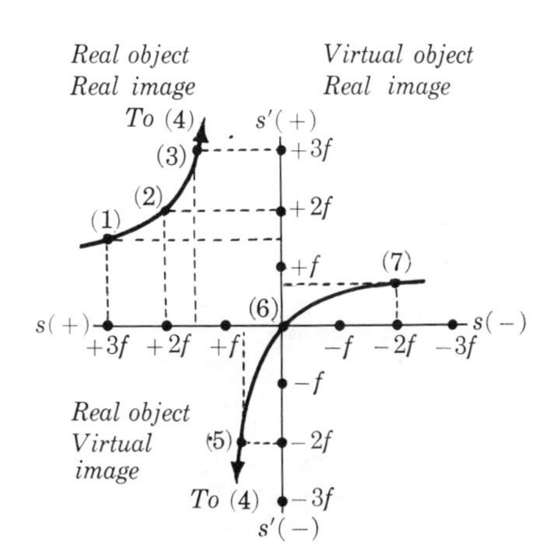

FIG. 4-10. Graph of the Gaussian lens equation.

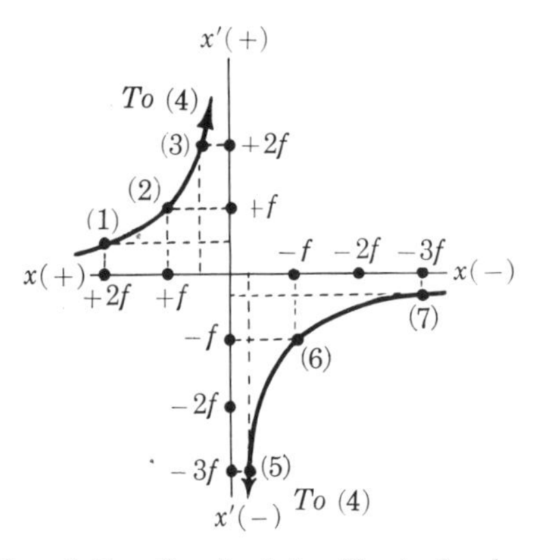

FIG. 4-11. Graph of the Newtonian lens equation.

in which object distances s are plotted horizontally and image distances s' vertically. To conform to our convention that object distances are positive when the object lies at the left of the lens, positive values of s are plotted at the left of the origin. The points numbered (1), (2), (3), etc., along the graph correspond to the respective parts of Fig. 4-8.

Fig. 4-11 is a similar graph of the Newtonian equation,

$$xx' = f^2.$$

The numbered points refer to the corresponding parts of Fig. 4-8.

The four preceding figures should be studied carefully.

4-6 Images as objects. It was shown in Sec. 3-6 and Fig. 3-17 how the image formed by any one *surface* in an optical system serves as the object for the next surface. In the majority of optical systems employing lenses, more than one lens is used and the image formed by any one *lens* serves as the object for the next lens. Fig. 4-12 illustrates the various possibilities. Lens 1, in Fig. 4-12, forms a real image at P of a real object at O. This real image serves as a real object for lens 2. The virtual image at Q formed by lens 2 is a real object for lens 3. If lens 4 were not present, lens 3 would form a real image at R. Although this image is never formed, it serves as a virtual object for lens 4, which forms a final real image at I.

4-7 Images in three dimensions. It has been pointed out in Sec. 2-6 that in most instances both an object and its image are three dimensional, and that it does not always suffice to represent an object by a

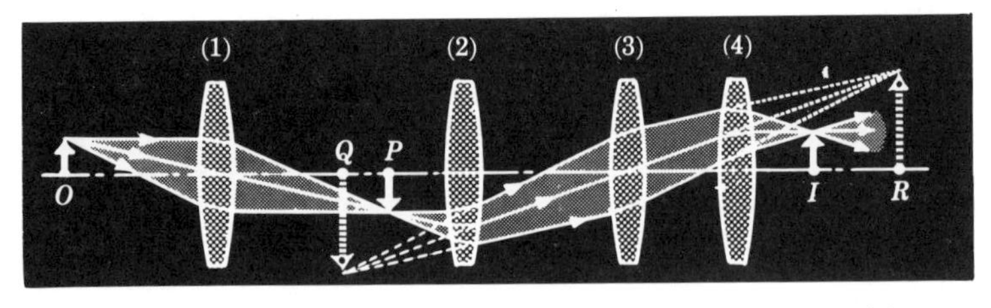

FIG. 4-12. The object for each lens, after the first, is the image formed by the preceding lens.

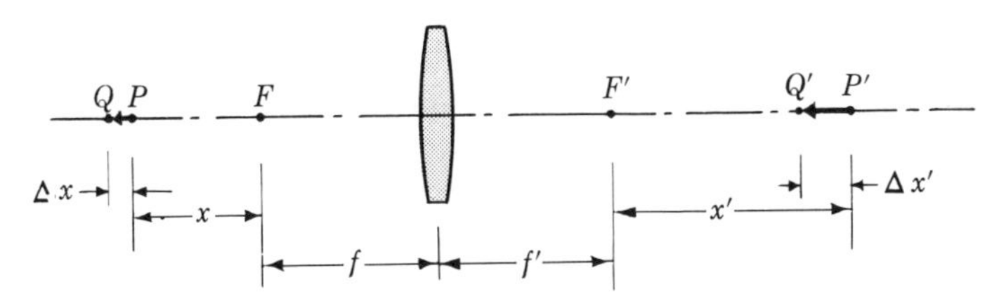

FIG. 4-13. The ratio $\Delta x'/\Delta x$ is the longitudinal magnification.

single arrow. We shall consider next some of the aspects of the formation of three-dimensional images by a lens.

The lateral or transverse magnification m, defined in Sec. 3-3, gives the ratio in which the transverse dimensions of an image are increased or reduced over those of the object. We now compute the corresponding ratio of the longitudinal or axial dimensions of an image to those of the object.

Fig. 4-13 shows a short arrow PQ, lying on the axis of a lens. The image of P is at P' and the image of Q is at Q'. Let x and x' be the object and image distances of P and P', measured from the focal points F and F' respectively. The length of the arrow PQ can be called Δx, and that of its image $P'Q'$ can be called $\Delta x'$. From Eq. (4-13),

$$x' = \frac{f^2}{x}.$$

Taking differentials of both sides, we get

$$dx' = -\frac{f^2}{x^2}\,dx,$$

or, approximately,

$$\Delta x' = -\frac{f^2}{x^2}\,\Delta x.$$

Since the transverse magnification, m, equals $-f/x$,

$$\Delta x' = -m^2 \Delta x.$$

The *longitudinal magnification* is defined as the ratio of the length of the image, $\Delta x'$, to that of the object, Δx. Hence,

$$\text{Longitudinal magnification} = \frac{\Delta x'}{\Delta x} = -m^2. \tag{4-16}$$

The longitudinal magnification is therefore equal to the (negative) square of the transverse magnification. It follows that except in the special case where $m = \pm 1$, the longitudinal and transverse magnifications are unequal. For example, if the transverse magnification equals $-2\times$, the longitudinal magnification equals $4\times$. If the object were a very small cube on the lens axis, it would not be imaged as a cube, but as a rectangular parallelepiped whose length is twice its width. In general (except when $m = \pm 1$) an object and its image are not geometrically similar.

It also follows from Eq. (4-16), since m^2 is necessarily positive, that Δx and $\Delta x'$ are always of opposite sign. That is, if an axial object in the form of an arrow points in the direction of increasing x, as in Fig. 4-13, its image points in the direction of decreasing x'. The reader can verify that, as a consequence, an axial object in the form of an arrow, and its image formed by a lens, always point in the same direction.

The three-dimensional image of a three-dimensional object, formed by a lens, is shown in Fig. 4-14. It must be emphasized again that the image exists in space and is not merely something "thrown on a screen." Arrows $o'b'$ and $o'c'$ are reversed in space, relative to ob and oc, but $o'a'$ points in the same direction as oa. Although we speak of the image as "inverted," only its transverse dimensions are reversed.

Fig. 4-14 should be compared with Fig. 2-15, showing the image formed by a plane mirror. Notice that the image formed by the lens, although it is inverted, is not perverted. That is, if the object is a left hand its image is a left hand also. This may be verified by pointing the left thumb along oa, the left forefinger along oc, and the left center finger along ob. A rotation of 180° about the thumb as an axis then brings the fingers into coincidence with $o'c'$ and $o'b'$. In other words, inversion of an image is equivalent to a rotation of 180° about the lens axis.

One other comparison with the image formed by a mirror should be made. Since the rays incident on a lens are not reversed in direction as they are by a mirror, an observer at the lens in Fig. 4-14, looking at the object, cannot see the image by merely turning around. Instead, he must go around to the other side of the lens to a position at least 10 inches beyond the image, and face in his original direction, in order to see the image. Then the arrow $o'a'$ points toward him as did the arrow oa, while $o'b'$ and $o'c'$ are both reversed. The situation differs from that in Fig. 2-15 in two respects; first, the images as they exist in space are different and, second, they must be viewed from opposite sides.

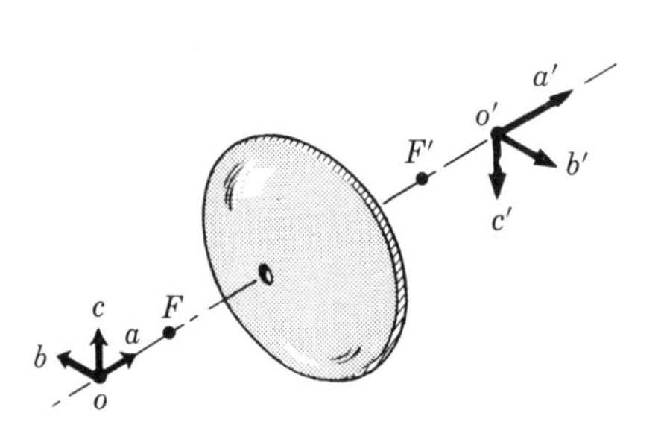

FIG. 4-14. A three-dimensional inverted image.

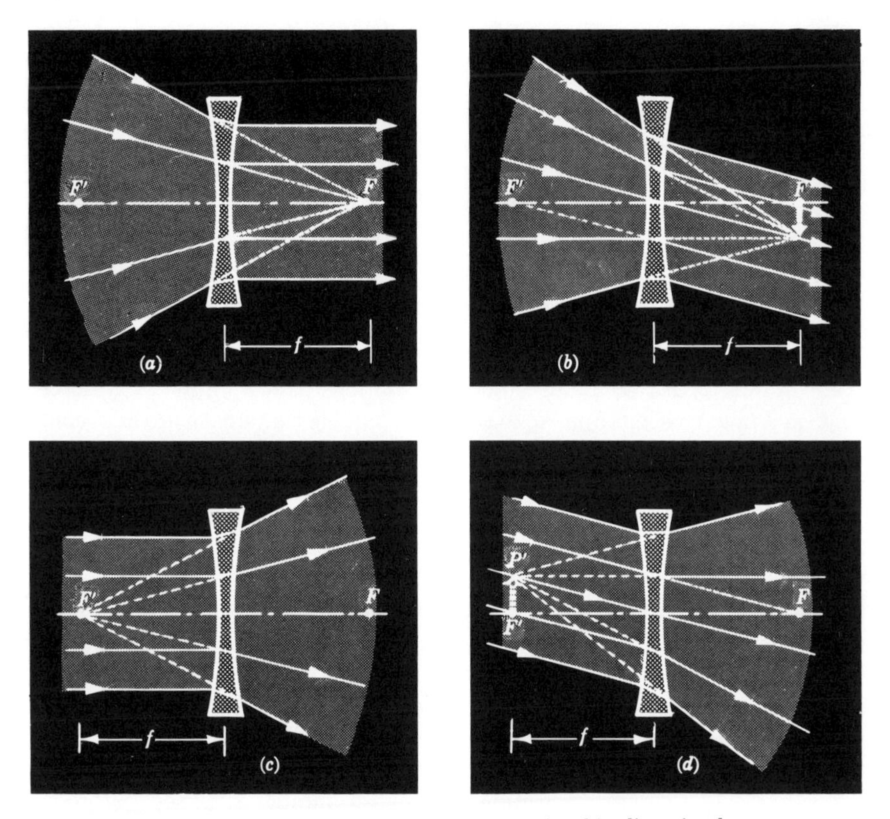

FIG. 4-15. Focal points and focal planes of a thin diverging lens.

4-8 Diverging lenses. A bundle of parallel rays incident on the lens in Fig. 4-2 converges to a real image after refraction by the lens. The lens is called a *converging lens.* Its focal length, as computed from Eq. (4-7), is a positive quantity and it is also called a *positive lens.*

A bundle of parallel rays incident on the lens in Fig. 4-15 becomes diverging after refraction and the lens is called a *diverging lens.* Its focal length, computed from Eq. (4-7), is a negative quantity and it is also called a *negative lens.*

The first focal point of any lens, whether negative or positive, may be defined as the axial object point which is imaged by the lens at infinity. Since the focal length of a diverging lens is negative, the object distance of this axial point is negative also or, in other words, it is a *virtual* object. Thus rays incident on the lens and *converging toward* the first focal point are parallel to the lens axis after refraction, as illustrated in Fig. 4-15 (a).

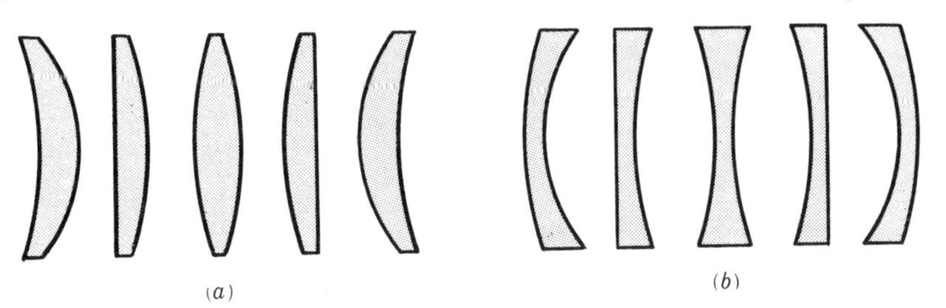

FIG. 4-16. (a) Meniscus, plano-convex, and double-convex converging lenses. (b) Meniscus, plano-concave, and double-concave diverging lenses.

The second focal point of any lens may be defined as the image point of an infinitely distant point object on the lens axis. Parallel rays incident on a negative lens appear to *diverge from* a virtual image at the second focal point as in Fig. 4-15 (c). Thus the focal points of a negative lens are reversed, relative to those of a positive lens.

The focal planes of a negative lens have the same significance as those of a positive lens and are illustrated in Fig. 4-15 (b) and (d). Rays converging toward a virtual object point in the first focal plane are parallel to one another after refraction; a bundle of parallel rays, not parallel to the lens axis, appears to diverge from a point in the second focal plane after refraction.

Any thin lens which is thicker at the center than at the rim, and whose index is greater than that of the medium in which it is immersed, is a positive or converging lens. If the lens is thicker at the rim than at the center, the lens is diverging. Sections through a number of lenses, both converging and diverging, are shown in Fig. 4-16.

In addition to lenses having spherical surfaces, use is frequently made of *cylindrical* lenses, particularly in spectacles, to correct a defect of vision known as astigmatism. One or both surfaces of a cylindrical lens are portions of cylinders. (See Fig. 4-17.) Since other defects of vision frequently accompany astigmatism, a spectacle lens may be cylindrical at one surface and spherical at the other surface, or one of its surfaces may be a combined sphere and cylinder. The use of spectacles to correct defects of vision will be discussed further in Sec. 6-3.

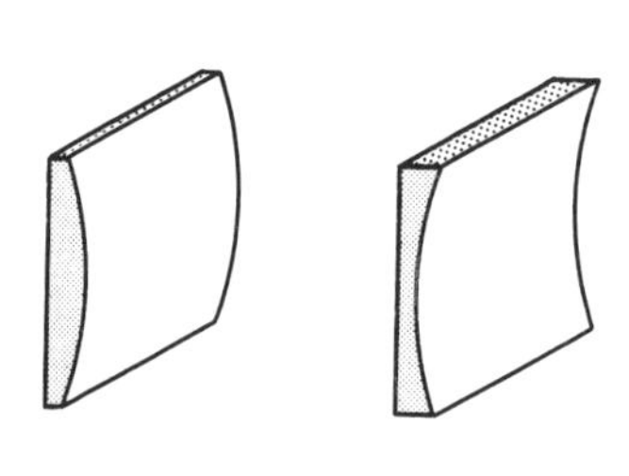

FIG. 4-17. Cylindrical lenses.

Lenses having *paraboloidal* surfaces are used in the illuminating system of a compound microscope. (See Sec. 6-6.)

4-9 Thick lens optics. The position and magnification of the image of a given object, formed by a thick lens, can be found by successive applications of the methods of Sec. 3-1 to each surface of the lens in turn. If but one such computation is to be made, this procedure is as good as any, but it becomes tedious when a large number of object and image distances are to be calculated. We shall show in this section that the Gaussian and Newtonian lens equations, which were derived for thin lenses only, can be applied to thick lenses as well.

To begin with, consider the *simple* thick lens illustrated in Fig. 4-18. The second focal point F' of this lens is located by finding the position of the image of an infinitely distant axial object point, as explained in Sec. 4-3. That is, we let s_1 in Fig. 4-18 equal infinity, find s_1', then s_2, and finally s_2'. This gives the position of the second focal point, measured from the second vertex.

The next step is to calculate the focal length of the lens. Fig. 4-18 shows a single ray originating at an infinitely distant axial object point. This ray is incident on the first surface of the lens at point A, at a height h above the axis, and it leaves the second surface at point D, at a height h'. Projections of the incident and emergent rays intersect at point E, which locates the second principal plane and the second principal point H'. From the similar triangles ABG and DCG, to within the precision of first order theory,

$$\frac{h}{s_1'} = \frac{h'}{-s_2}.$$

(The distance s_2 is negative, since the image formed by the first surface constitutes a virtual object for the second.)

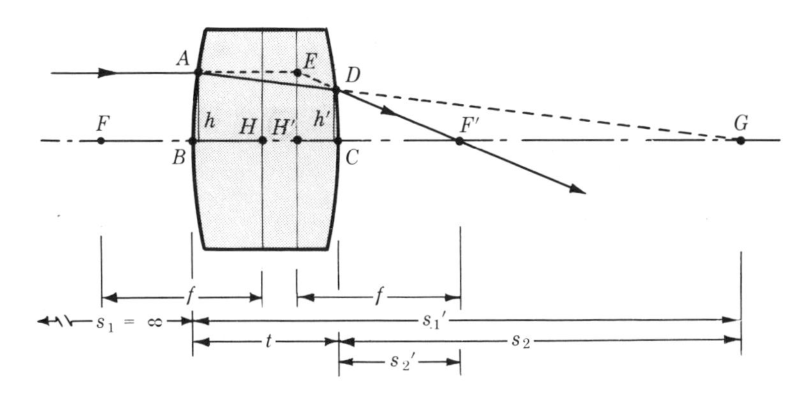

FIG. 4-18. A simple thick lens.

From the similar triangles $EH'F'$ and DCF',

$$\frac{h}{f} = \frac{h'}{s_2'}.$$

When the first of these equations is divided by the second, we get

$$\frac{f}{s_1'} = -\frac{s_2'}{s_2},$$

or

$$\boxed{f = s_1'\left(-\frac{s_2'}{s_2}\right).} \tag{4-17}$$

The distances s_1', s_2, and s_2' must all be computed in the process of locating the second focal point, so the calculation of the focal length requires little additional labor. If the lens is thin, Eq. (4-17) reduces to Eq. (4-7),

$$\frac{1}{f} = (n - 1)\left(\frac{1}{R_1} - \frac{1}{R_2}\right),$$

but note carefully that the latter equation is a special case and cannot be used to find the focal length of a thick lens.

Having found the position of the second focal point, and the focal length, we can now locate the second principal point by laying off a distance f to the left of the second focal point.

The positions of the first focal point and first principal point are found by a similar calculation. One can either find the position of an axial object point which is imaged at infinity, or turn the lens around and repeat the calculations used in locating the second focal point.

Not infrequently, the focal length of a lens is greater than one-half the distance between the focal points, so that the second principal point lies at the left of the first, rather than at the right as in Fig. 4-18. The order in Fig. 4-18 is considered normal; when the order is reversed the principal points are said to be *crossed.*

The focal points of a lens are easily located experimentally by finding the position of the image of a distant object, but this is not true of the principal points, for which special equipment is required. For this reason the manufacturer usually measures the focal length of a lens before it leaves the factory and stamps the value on the lens mount. The positions of the principal points are then readily determined from the positions of the focal points and the focal length.

Example. Find the focal length and the positions of the focal points and principal points of the simple thick lens in Fig. 4-18. The index of the lens is 1.50, its axial thickness is 25 mm, the radius of its first surface is 22 mm, and that of its second surface is 16 mm.

From Eq. (4-1), setting $s_1 = \infty$,

$$\frac{1}{\infty} + \frac{1.5}{s_1'} = \frac{1.5 - 1}{22},$$

$$s_1' = +66 \text{ mm}.$$

From Eq. (4-2),

$$s_2 = 25 - 66 = -41 \text{ mm}.$$

From Eq. (4-3),

$$\frac{1.5}{-41} + \frac{1}{s_2'} = \frac{1 - 1.5}{-16},$$

$$s_2' = +14.7 \text{ mm}.$$

That is, the second focal point F' lies 14.7 mm to the right of the second vertex.

The focal length of the lens, from Eq. (4-17), is

$$f = 66\left(-\frac{14.7}{-41}\right) = +23.6 \text{ mm}.$$

Hence the second principal point lies 23.6 mm to the left of the second focal point, or 8.9 mm to the left of the second vertex.

If the other side of the lens faces the light,

$$\frac{1}{\infty} + \frac{1.5}{s_1'} = \frac{1.5 - 1}{16},$$

$$s_1' = +48 \text{ mm}.$$

Hence

$$s_2 = 25 - 48 = -23 \text{ mm},$$

and

$$\frac{1.5}{-23} + \frac{1}{s_2'} = \frac{1 - 1.5}{-22},$$

$$s_2' = +11.4 \text{ mm}.$$

The first focal point is therefore 11.4 mm outside the vertex of the first surface. Since the two focal lengths are equal, the first principal point lies 23.6 mm to the right of the first focal point, or 12.2 mm inside the first vertex.

We now show that both the Gaussian and Newtonian forms of the lens equation are applicable to thick as well as to thin lenses. Fig. 4-19 shows an object PQ and its image $P'Q'$ formed by a thick lens. The focal points, principal points, and principal planes are indicated. The surfaces of the

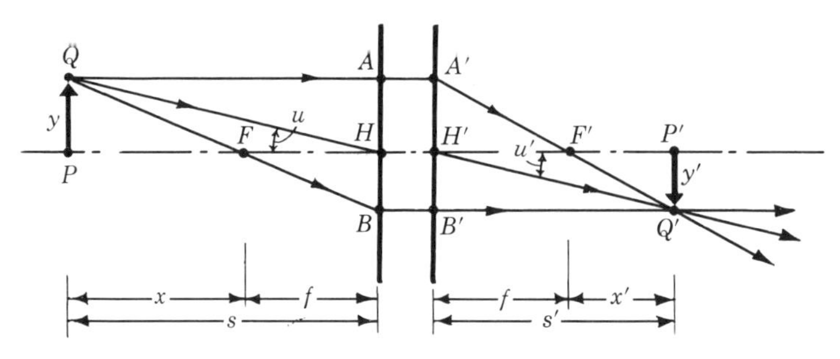

FIG. 4-19. Focal planes and principal planes of a thick lens.

lens are not shown, since they need not be considered once the points above have been located. A ray from Q, parallel to the axis, can be considered to proceed in a straight line to A', its point of intersection with the second principal plane, and from there through the second focal point F'. A ray from Q through the first focal point F is parallel to the axis from B, its point of intersection with the first principal plane.

Now compare Fig. 4-19 with Fig. 4-6. It will be seen that if the right and left sides of Fig. 4-19 are moved toward one another until the principal planes coincide, the two figures become identical. In other words, a thin lens is a special case of a thick lens in which the principal planes coincide.

The similar triangles PQF and FHB, in Fig. 4-6, correspond to the same triangles in Fig. 4-19, while the triangles $AH'F'$ and $F'P'Q'$, in Fig. 4-6, correspond to $A'H'F'$ and $F'P'Q'$ in Fig. 4-19. Hence Eqs. (4-8) to (4-11) apply also to a thick lens, and Eqs. (4-12) to (4-15) follow directly.

Fig. 4-19 also illustrates how the image formed by a thick lens is located graphically. In addition to the rays QA and QB, the ray QH can be readily traced through the lens. Since the lateral magnification y'/y is equal to $-s'/s$, it follows that

$$\frac{y}{s} = -\frac{y'}{s'}.$$

But

$$\frac{y}{s} = \tan\theta, \quad \frac{y'}{s'} = \tan\theta'.$$

Hence

$$\theta' = -\theta,$$

and the ray $H'Q'$ is parallel to the ray HQ. In other words, a ray from Q, proceeding toward the first principal point H, is undeviated by the lens and, after refraction, appears to come from the second principal point H'.

4-10 Compound lenses. For the purpose of minimizing aberrations, most lenses in optical instruments are *compound*, that is, they consist of several simple lenses having a common axis. The surfaces of adjacent lenses may be in contact, or there may be air spaces between them. Every compound lens, like a simple lens, has two focal points and two principal points. The distance between each focal point and its corresponding principal point is equal to the focal length. The Gaussian and Newtonian lens equations apply to compound as well as to simple lenses.

If the elements of a compound lens are too thick to be considered as thin lenses, the positions of the focal points must be found from a surface-by-surface computation as in Sec. 4-9. The focal length is then given by an extension of Eq. (4-17), namely,

$$f = s_1'\left(-\frac{s_2'}{s_2}\right)\left(-\frac{s_3'}{s_3}\right)\cdots\cdots, \tag{4-18}$$

where s_1', s_2, s_2', etc., refer to object and image distances measured from the respective surfaces. If the components are thin lenses, the focal lengths of which are known, the positions of the focal points can be found from a lens-by-lens computation. The focal length is still given by Eq. (4-18), the only difference being that s_1', s_2, s_2', etc., refer to object and image distances measured from the respective lenses.

For the special case illustrated in Fig. 4-20, where the compound lens consists of two thin lenses of focal lengths f_1 and f_2 respectively, separated by a distance t, the expression for the focal length f reduces to a rather

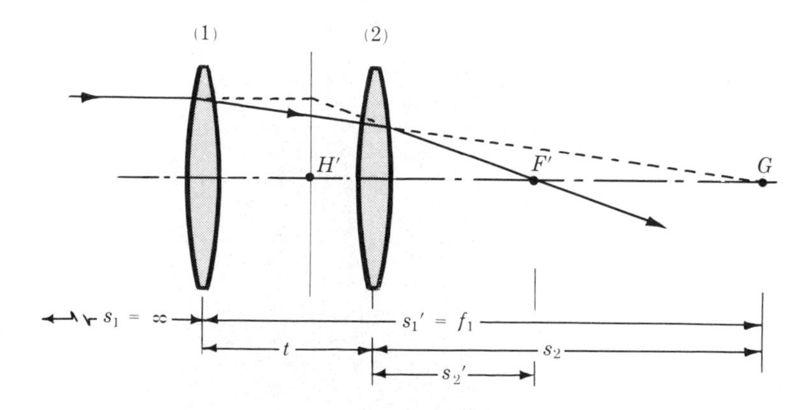

FIG. 4-20. A compound lens consisting of two thin lenses.

simple one. Fig. 4-20 shows a single incident ray parallel to the axis, passing through the second focal point F' of the compound lens, and undergoing an effective single deviation in the second principal plane. From a study of the figure it will be seen that

$$s_1 = \infty,$$

$$s_1' = f_1,$$

$$s_2 = -(s_1' - t),$$

and

$$\frac{1}{s_2} + \frac{1}{s_2'} = \frac{1}{f_2}.$$

It follows from these equations and Eq. (4-18) (the details are left as a problem), that

$$\frac{1}{f} = \frac{1}{f_1} + \frac{1}{f_2} - \frac{t}{f_1 f_2}. \tag{4-19}$$

In the special case where the lenses are in contact, the separation t is zero and Eq. (4-19) reduces to the useful relation

$$\boxed{\frac{1}{f} = \frac{1}{f_1} + \frac{1}{f_2}.} \tag{4-20}$$

Example. The compound lens in Fig. 4-21 consists of a thin positive lens of focal length +20 cm, separated by a distance of 10 cm from a thin negative lens of focal length −20 cm. Find the focal length of the compound lens, and the positions of its focal points and principal points.

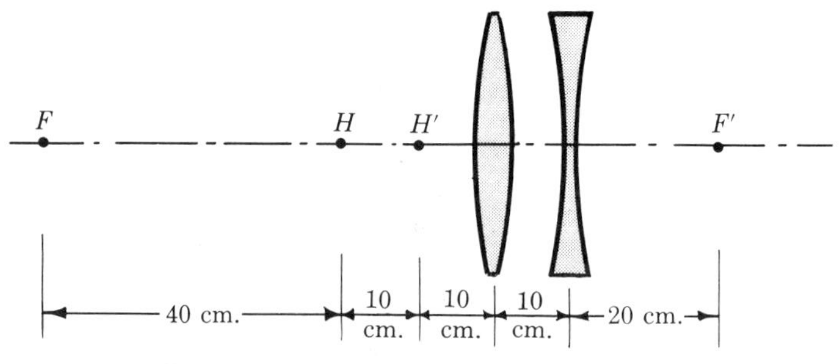

FIG. 4-21. Principle of the telephoto lens.

Following the method described above, we find

$$s_1' = +20 \text{ cm},$$
$$s_2 = -10 \text{ cm},$$
$$s_2' = +20 \text{ cm}.$$

The second focal point is therefore 20 cm to the right of the negative lens. If the other side of the lens faces the light,

$$s_1' = -20 \text{ cm},$$
$$s_2 = +30 \text{ cm},$$
$$s_2' = +60 \text{ cm},$$

and the first focal point is 60 cm to the left of the positive lens.

The focal length of the compound lens, from Eq. (4-19), is

$$\frac{1}{f} = \frac{1}{20} + \frac{1}{-20} - \frac{10}{20(-20)},$$
$$f = +40 \text{ cm}.$$

The first principal point lies 20 cm to the left of the positive lens, and the second principal point 10 cm to the left of this lens, as shown in Fig. 4-21.

This combination of a positive and a negative lens illustrates the principle of the telephoto camera lens. The lens has a long focal length, thus producing a large image ($m = -f/x$), but the second focal plane is not so far behind the lens mounting as to make the camera inconveniently long.

Problems—Chapter 4

(1) (a) Sketch the various possible thin lenses obtainable by combining two surfaces whose radii of curvature, in absolute magnitude, are 10 cm and 20 cm. (b) Which are positive lenses and which are negative lenses? (c) Find the focal length of each, if made of glass of index 1.50.

(2) An object in the form of an arrow ½ inch high is at right angles to the axis of a positive lens of focal length 1 inch and diameter 1½ inches. (a) Find graphically in a full size diagram the position and height of the image of the arrow for the following object distances s: 4 inches, 2 inches, 1 inch, ¾ inch. Include in each diagram those rays from the head and from the tail of the arrow that pass through the rim of the lens. (b) For each object distance above, compute the position and magnification of the image, using both the Gaussian and Newtonian forms of the lens equation.

(3) An object in the form of an arrow ½ inch high is 2 inches to the left of a positive thin lens of focal length 1 inch and diameter 1 inch. A second positive lens of focal length 1 inch is placed at a distance z to the right of the first. The axes of the lenses coincide. (a) Construct a full size diagram, and trace through the system the three rays from the head of the arrow representing the object that pass through the center and the rim of the first lens, for the following values of z: 4 inches, 3 inches, 2 inches, 1 inch. (The lens at a distance of 2 inches illustrates the function of the field lens of an ocular. See Sec. 6-5.) (b) Compute the position and magnification of the final image in each case, using both the Newtonian and Gaussian forms of the lens equation. (c) Find from your diagram the required diameter of the second lens in each case, if it is to transmit the entire cone of rays from the head of the arrow that passes through the first lens.

(4) Same as Prob. 3, except the second lens is a negative lens of focal length -1 inch.

(5) Find graphically the image formed by a thin converging lens of (a) a real object at the left of the first focal point; (b) a real object between the first focal point and the lens; (c) a virtual object between the lens and the second focal point; (d) a virtual object at the right of the second focal point.

(6) The radii of curvature of the surfaces of a thin lens are $+10$ cm and $+30$ cm. The index is 1.50. (a) Compute the position and size of the image of an object in the form of an arrow 1 cm high, perpendicular to the lens axis, 40 cm to the left of the lens. (b) A second similar lens is placed 160 cm to the right of the first. Find the position of the final image. (c) Same as (b) except the second lens is 40 cm to the right of the first. (d) Same as (c) except the second lens is diverging, of focal length -40 cm.

(7) (a) Derive an expression similar to Eq. (4-7) for the focal length of a thin lens of index n_1 in a medium of index n_2. (b) A thin lens of focal length 10 inches in air and refractive index 1.53 is immersed in water of refractive index 1.33. What is the focal length in water?

(8) Derive expressions (similar to Eq. (4-7)) for the first and second focal lengths of a thin lens of index n_2 having a medium of index n_1 on its left and a medium of index n_3 on its right.

(9) An equiconvex thin lens made of glass of index 1.50 has a focal length in air of 30 cm. The lens is sealed into an opening in one end of a tank filled with water (index = $\frac{4}{3}$). At the end of the tank opposite the lens is a plane mirror, distant 80 cm from the lens. (a) Find the position of the final image formed by the lens, water, mirror system, of a small object outside the tank on the lens axis and 90 cm to the left of the lens. (b) Is the image real or virtual? (c) Erect or inverted?

(10) Suppose the arrows *oa*, *ob*, and *oc* in Fig. 4-14 are each 1 mm long and are 30 cm to the left of a thin lens of focal length 20 cm. Find the length of the image of each arrow.

(11) A lens is used to project the image of a lantern slide on a screen located 160 feet from the projector, the image on the screen being 100 times larger (linear dimensions) than the slide. If the screen is now moved 100 inches closer to the lens, how much must the lantern slide be moved (relative to the lens) to keep the image on the screen in sharp focus?

(12) The lens of an enlarging camera has a focal length of 12 cm. When enlarging a negative to twice size (linear dimensions) the negative is 16 cm from the nearest point of the lens. How far from this point of the lens should the negative be when making an enlargement to three times size?

(13) The focal length of a certain camera lens is 120 mm. The distance from the nearest point of the lens to the film when photographing a distant object is 90 mm. How far and in what direction relative to the film must the lens be moved in order to make a life-size photograph of a small object?

(14) After focusing a camera on an object at a great distance, it is found that the lens must be moved 1 inch forward of the "infinity position" in order to focus on an object at a closer distance. The image of the object in this case is 1 inch high on the film. It is then desired to form an image (of the same object) 2 inches high on the film, which is done by refocusing the camera after moving the object still closer to the camera. In the latter case, what is the distance of the lens from its original infinity position?

(15) Find the focal lengths and the positions of the focal points and principal points of the lenses in Fig. 4-22 (a), (b), (c), and (d). The axial thickness of each lens is 10 mm and the index is 1.50.

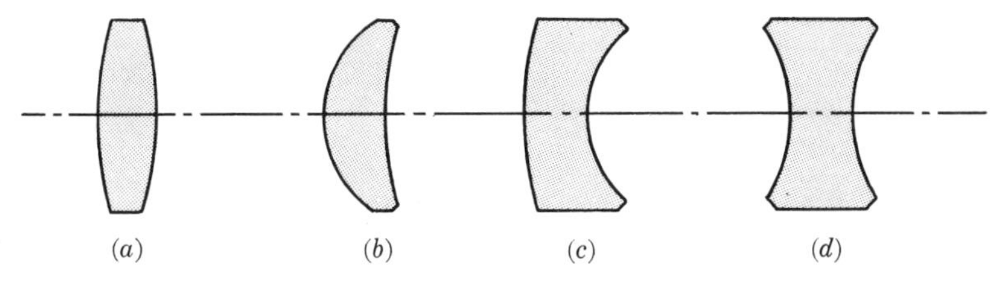

(a)	(b)	(c)	(d)
$R_1 = +100$ *mm*	$R_1 = +100$ *mm*	$R_1 = +200$ *mm*	$R_1 = -100$ *mm*
$R_2 = -100$ *mm*	$R_2 = +200$ *mm*	$R_2 = +100$ *mm*	$R_2 = +100$ *mm*

FIG. 4-22.

(16) The radii of curvature of the surfaces of a lens of index 1.50 are +10 cm and +8 cm. Make a sketch of the lens and find its focal length, if its thickness is (a) 4 cm, (b) 6 cm, (c) 8 cm.

(17) A glass sphere of radius R and index of refraction 1.50 is used as a thick lens. Locate its focal points, find its focal length, and locate its principal points.

(18) Locate the focal points, find the focal length, and locate the principal points of a rod of glass with convex ends whose radii of curvature are 30 cm. The index of refraction of the rod is 1.50 and its length is 45 cm.

(19) An ocular consists of two similar positive thin lenses having focal lengths of 2 inches, separated by a distance of 1 inch. Where are the focal points and principal points of the combination, and what is its focal length?

(20) A telephoto camera objective consists of a thin positive lens of 10 inches focal length followed at a distance of 5 inches by a thin negative lens of focal length 20 inches. Locate the focal points and principal points of the objective. What is its focal length?

(21) Two thin lenses whose focal lengths are $f_1 = -50$ mm and $f_2 = +100$ mm, respectively, are separated by 50 mm. (a) What is the focal length of the combination, considered as a single thick lens? (b) Where are the principal points of the combination?

(22) An object is located at the first principal point of a lens. (The object is necessarily virtual if this point is within the lens.) Find the position and magnification of the image.

(23) The focal length of a lens is 60 mm and the distance between its focal points is 125 mm. Compute the position of the image of (a) an object on the lens axis, 20 mm to the left of the first focal point; (b) an object on the axis, 20 mm to the right of the first focal point; (c) a virtual object 25 mm to the right of the second principal point.

(24) The focal length of a lens is 1.5 inches. The principal points are normal and are separated by 0.5 inch. (a) Show the principal points and focal points in a full size diagram. (b) Find graphically the image of an arrow 0.5 inch long, perpendicular to the lens axis, and 2.5 inches to the left of the first focal point. (c) Rays are converging toward a virtual object in the form of an arrow 0.5 inch long, perpendicular to the lens axis, and located at the second focal point. Find graphically the position of the image of the arrow. (d) Check the results of parts (a) and (b) by computation.

(25) Use both the Gaussian and Newtonian forms of the lens equation to compute the distance x at which an object O must be placed in Fig. 4-23 to be imaged at I. Distances are in centimeters.

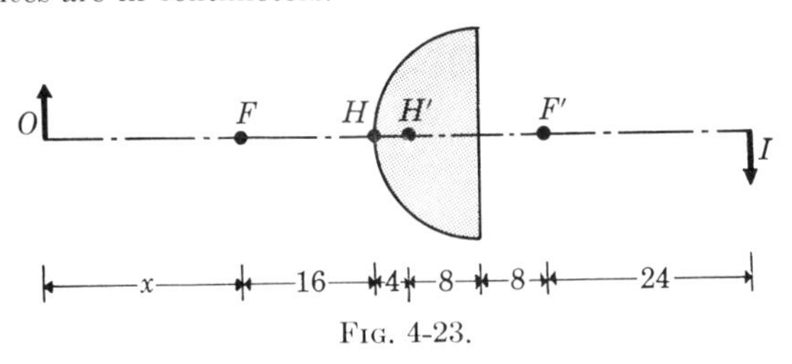

FIG. 4-23.

(26) The focal length of the lens in Fig. 4-24 is 10 cm. A distant object on the axis at the left of the lens is imaged at a point 7 cm to the right of vertex V_2. A distant object at the right of the lens is imaged 8 cm to the left of V_1. Find the position of the image of an object on the lens axis, 28 cm to the left of V_1.

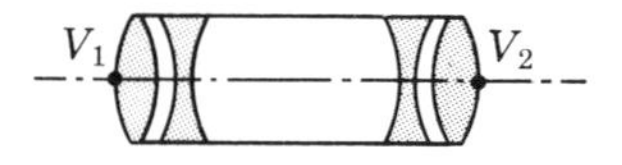

FIG. 4-24.

(27) An enlarging camera has a maximum extension of 4 ft between the negative and the paper. The focal length of the lens is 10 inches and the separation of its principal points, which are crossed, is 0.5 inch. (a) What is the maximum enlargement that can be made with this equipment? (b) What is the minimum reduction?

(28) Rays from a lens are converging toward a point image P as in Fig. 4-25. What thickness t of glass of index 1.50 must be interposed as in the figure in order that the image shall be formed at P'?

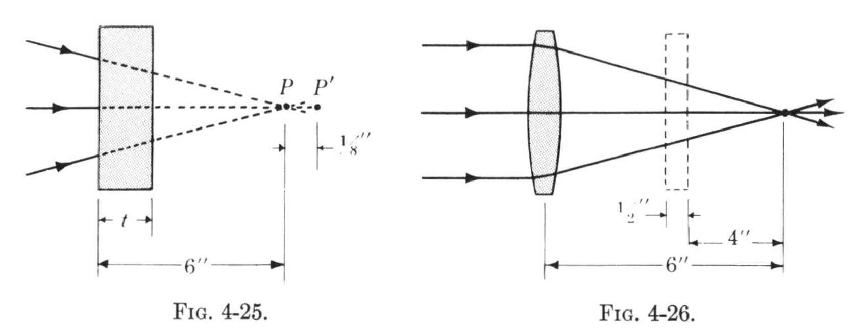

FIG. 4-25.

FIG. 4-26.

(29) An object 10 ft in front of a camera lens is sharply imaged on a photographic film 6 inches behind the lens. A glass plate 0.5 inch thick, of index 1.50, having plane parallel faces, is interposed between lens and film as shown in Fig. 4-26. (a) Find the new position of the image. (b) At what distance in front of the lens will an object be in sharp focus on the film with the plate in place, the distance from lens to film remaining 6 inches? Consider the lens as a simple thin lens.

CHAPTER 5

ABERRATIONS OF LENSES AND MIRRORS

5-1 Aberrations. The relatively simple relations we have shown to hold between object and image distances, focal lengths, radii of curvature, etc., were derived from the general expressions for the refraction of a ray at a spherical surface by making the approximation $\sin\theta = \theta$. That is, they apply only to rays originating at axial points and making small angles with the axis.

In general, however, the rays incident on a lens or a mirror originate not only at axial points, but at points which lie off the axis as well. Furthermore, because of the finite size of the lens or mirror, the cone of rays which forms the image of any point will not be the infinitesimal cone of first order theory. Nonparaxial rays proceeding from a given object point do not, in general, all intersect at the same point after refraction. Consequently, the image formed by these rays is not a sharp one. Furthermore, the focal length of a lens depends upon its index of refraction, which varies with wave length. Therefore, if the light proceeding from the object is not monochromatic, a lens will form a number of colored images, which lie in different positions and are of different sizes, even if formed by paraxial rays.

The departures of the actual image from the predictions of first order theory are called *aberrations*. Those caused by the variation of index with wave length are the *chromatic aberrations*. The others, which arise even if the light is monochromatic, are the *monochromatic aberrations*.

It may be worth emphasizing that aberrations are not caused by any faulty construction of a lens or mirror, such as the departure of its surfaces from a true spherical shape, but are simply consequences of the laws of refraction and reflection at spherical surfaces. Hence the term "aberrations" is to be preferred to "defects," which is sometimes used to describe these phenomena.

5-2 Spherical aberration of a lens. The effect known as spherical aberration is illustrated in Fig. 5-1. Paraxial rays from an axial point P are imaged at P'. Rays incident on the lens near its rim are imaged at P'', closer to the lens. Rays incident at intermediate zones of the lens are imaged between P' and P''. There is evidently no plane in which a sharp image of P is formed. If a screen S is placed perpendicular to the axis at

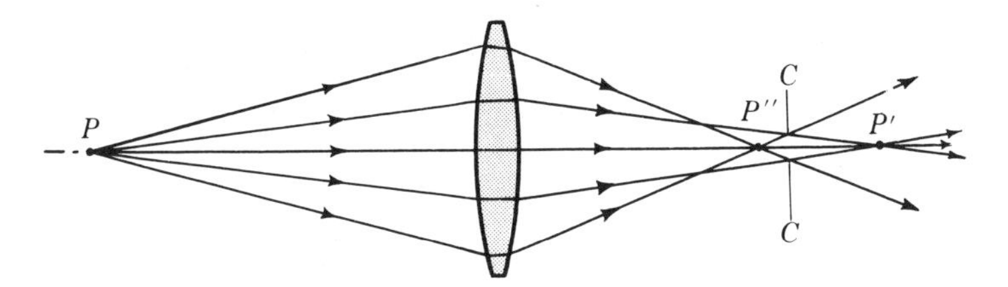

FIG. 5-1. Spherical aberration.

P', the position of the image formed by the paraxial rays, the image on the screen consists of a circular disk whose outline is the intersection with the screen of the outer cone of rays refracted by the lens. The refracted beam is everywhere of circular cross section and it will be seen from the figure that there is one plane, CC, at which the cross section of the beam is smallest. This smallest cross section is known as the *circle of least confusion* and the best image is secured if the screen is placed at this point.

The spherical aberration of a given lens may be reduced by "stopping down" the lens, that is, inserting a diaphragm in the beam so as to expose only the central portion of the lens. Of course, this entails a decrease in the amount of light transmitted.

A lens may be designed for minimum spherical aberration by proper choice of radii of curvature of the lens surfaces. As an example, consider a plano-convex thin lens forming an image of an object at infinity. The focal length of the lens as computed from Eq. (4-7) is the same whether the plane surface (as in Fig. 5-2 (a)) or the convex surface (as in (b)) faces the incident light. The spherical aberration, however, is much smaller in (b) than in (a). It is clear from the figure that in (a) all of the deviation is produced at the convex surface, while in (b) some deviation takes place

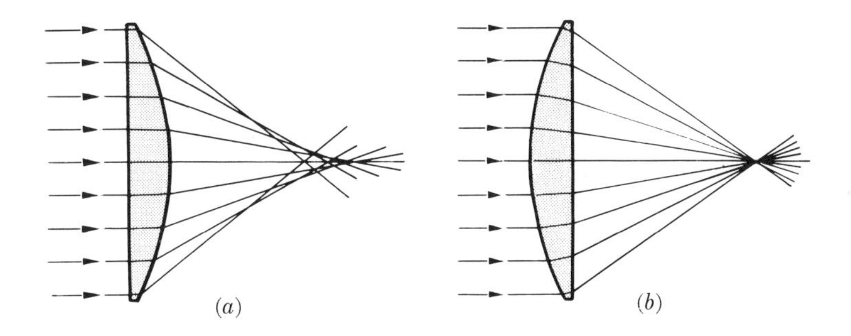

FIG. 5-2. Spherical aberration is minimized if the deviation is equally divided between the surfaces.

at each surface. As a general rule, spherical aberration will be minimized if a lens is designed or used in such a way that the deviation is equally divided between the two surfaces. Spherical aberration can never be entirely eliminated from a lens having spherical surfaces, if both object and image are real. However, if the object or image is virtual, it is possible to design a (thick) lens of zero spherical aberration, for one particular pair of object and image points.

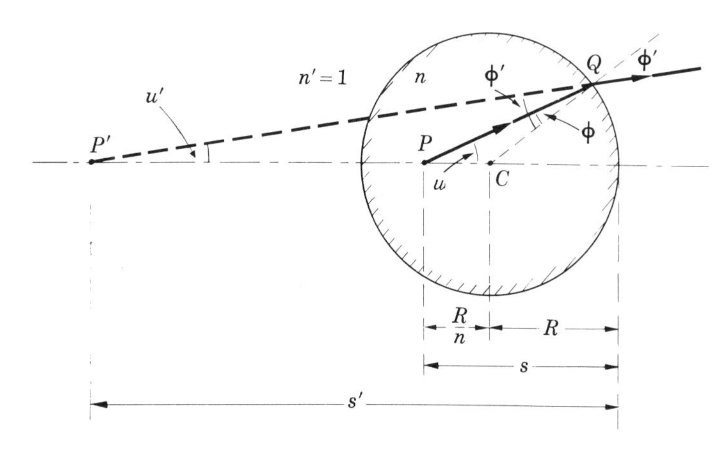

FIG. 5-3. All rays from P are imaged at P'.

Consider a ray which originates at P in Fig. 5-3, within a transparent sphere of radius R and index n. If the distance $PC = R/n$, all rays originating at P will, after refraction, appear to diverge from a point P', independent of the slope angle u of the ray. We may apply the general formulas for ray tracing, Eqs. (3-1) to (3-4). The lettering in Fig. 5-3 corresponds to that in Fig. 3-1. The object distance s in Fig. 5-3 is

$$s = -\left(R + \frac{R}{n}\right),$$

where the negative sign is introduced, since R is negative. Then

$$\frac{R + s}{R} = -\frac{1}{n},$$

and Eq. (3-1) becomes

$$\sin \phi = -\frac{1}{n} \sin u. \tag{5-1}$$

In Fig. 5-3, $n' = 1$, so from Eq. (3-2),

$$\sin \phi' = n \sin \phi. \tag{5-2}$$

From Eqs. (5-1) and (5-2),

$$\sin \phi' = -\sin u,$$

$$\phi' = -u. \tag{5-3}$$

From Eqs. (3-3) and (5-3),

$$u' = -\phi.$$

Therefore

$$\sin u' = -\sin \phi,$$

and from Eq. (5-1),

$$\sin u' = -\frac{1}{n} \sin \phi'. \tag{5-4}$$

Finally, from Eqs. (3-4) and (5-4),

$$s' = R - R \frac{\sin \phi'}{-\frac{1}{n} \sin \phi'} = R + nR.$$

Since R is negative, the image lies a distance $R + nR$ to the left of the vertex V, and is virtual. Furthermore, since the final expression for s' does not contain the angle u, all rays from P which are refracted by the surface appear to originate at P'.

A lens of this sort is frequently used as the front lens of a high-power microscope objective. Of course a real object cannot be embedded within a solid spherical lens, but it can be embedded optically by grinding away a portion of the lens and placing a drop of oil having the same index as the lens between the lens surface and the object to be examined. An objective used in this way is called an "oil immersion" objective. For further discussion of oil immersion objectives see Sec. 10-3.

5-3 Spherical aberration of a mirror. If a spherical mirror includes a relatively large fraction of the entire spherical surface of which the mirror is a part, the mirror is said to be of large aperture. If it includes only a relatively small fraction of the surface, the aperture is small. Thus the aperture of a mirror depends not only on its diameter, but on its radius of curvature as well. The *relative aperture* of a mirror is defined as the ratio of its diameter to its focal length.

In Fig. 5-4, a bundle of parallel rays is incident on a mirror of large relative aperture. The figure has been carefully drawn to show the paths

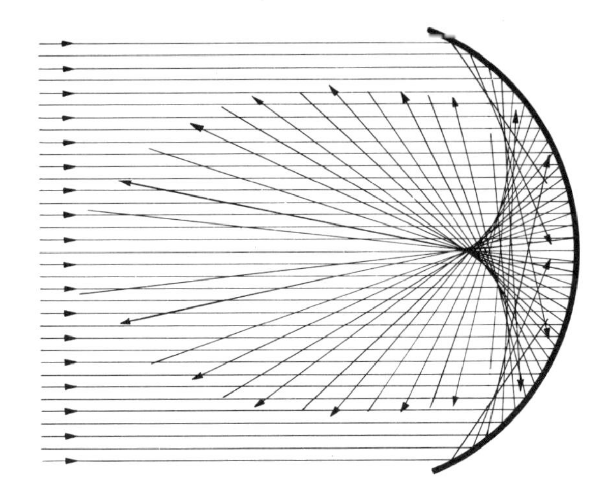

FIG. 5-4. Spherical aberration of rays originally parallel to the axis of a mirror of large aperture.

of the reflected rays and it will be seen that they do not all intersect at a common point. Rays incident near the vertex of the mirror intersect at F, the focal point for paraxial rays, but other rays cross the axis at points nearer the vertex. The failure of all the rays to intersect at a common point is another example of spherical aberration.

The envelope of the reflected rays is called a *caustic* and may frequently be seen when sunlight is reflected from the inner surface of a tumbler.

Conversely, rays from a point source at the focal point of a spherical mirror of large aperture are not all reflected parallel to the axis of the mirror, but as shown in Fig. 5-5.

Spherical aberration cannot, of course, be eliminated from a *spherical* mirror. However, it is always possible to find a surface of revolution of nonspherical or *aspherical* form, such that all rays diverging from any given axial point are imaged at a second axial point. It does not follow that rays from some other axial point will be sharply imaged also. In other words, by the proper choice of an aspherical surface, spherical aberration can be eliminated from a mirror *for any one pair of conjugate points.*

It is a property of a *paraboloid of revolution* (that is, the surface formed by the revolution of a parabola about its axis) that rays from an object at infinity are all imaged at the same point on its axis. Conversely, rays from a point source at this point are all reflected parallel to the mirror axis. The conjugate points for which spherical aberration is eliminated are then a point at infinity and the focal point of the mirror.

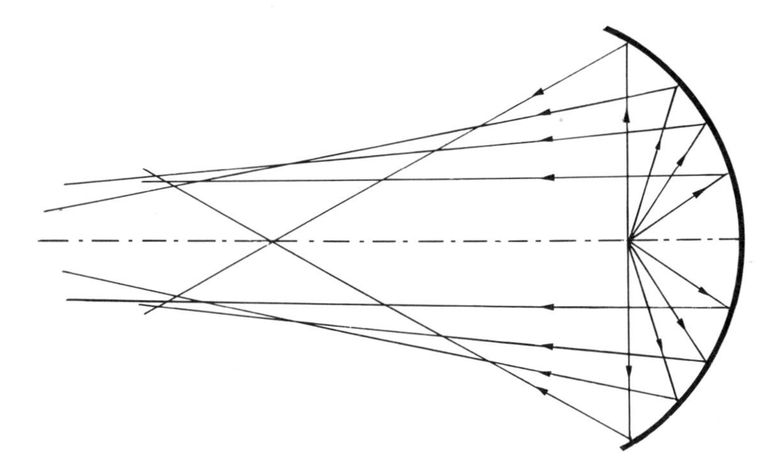

FIG. 5-5. Spherical aberration of rays from the focal point of a mirror of large aperture.

The reflecting mirrors used in astronomical telescopes, where all of the objects to be imaged are at extremely large distances from the mirror, are paraboloidal surfaces. The mirror is first ground to a spherical surface of approximately the correct curvature and then by a slow and tedious process of hand grinding and polishing it is gradually brought to the correct shape. This process is known as "figuring" the surface. Telescope mirrors are usually constructed of glass, ground and polished to the proper concave form and then silvered or aluminized to provide a reflecting surface. If the reflecting surface is injured by tarnishing, it can be removed and a fresh one applied without damage to the shape of the surface, which would not be possible were the mirror of metal.

Paraboloidal mirrors are also used in searchlights, where a beam as nearly parallel as possible is desired. In this case the reflectors are usually of metal, shaped by machine to approximately the correct curvature. High optical precision is not essential. Even with a paraboloidal reflector, however, a strictly parallel beam of light can never be obtained in practice, because only those rays diverging from the focal point are reflected parallel to the axis. Every actual light source must be of finite dimensions and hence some rays striking the reflecting surface come from points off the axis. These rays are reflected at an angle with the axis and hence the beam will always diverge to some extent. The smaller the light source, the less the spreading of the beam, but it can never be reduced to zero.

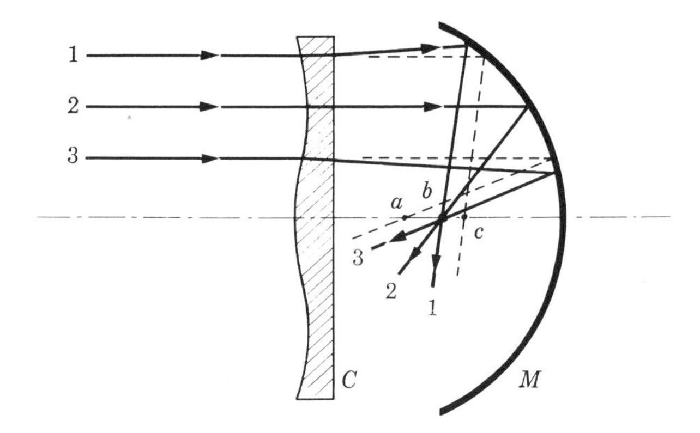

FIG. 5-6. The Schmidt corrector for minimizing spherical aberration of a mirror.

A spherical mirror can be corrected for spherical aberration by inserting a lens in the path of the light rays, either before or after reflection from the mirror. The function of the lens is not primarily to alter the focal length or magnification of the system, but to offset by its own spherical aberration the spherical aberration of the mirror. Fig. 5-6 illustrates the system devised by Schmidt in 1932. The spherical mirror M might be that of a reflecting telescope. In the absence of a correcting lens, rays 1, 2, and 3, proceeding from a distant object, would be reflected so as to cross the axis at points a, b, and c, as shown by the dotted lines. (See also Fig. 5-4.) The correcting lens (or plate, as it is often called) is shown at C. It is plane on one surface. The other surface is convex in the central region and concave in the outer portion. The outer portion thus functions as a diverging lens and the central portion as a converging lens. The corrected paths of rays 1, 2, and 3 are shown by full lines.

Because the curved surfaces of the Schmidt corrector are aspherical, they cannot be produced by machine grinding and polishing. However, a greater departure from the theoretically correct figure is permissible in the corrector than could be tolerated in the mirror if the latter were to be corrected by figuring. Hence the manufacture of correcting plates, although usually a hand operation, is less expensive and time-consuming than would be the figuring of the mirror itself. Schmidt correctors can be successfully constructed of a plastic material, formed under pressure between carefully ground and polished steel dies. These correctors are used in conjunction with a spherical mirror for projecting onto a viewing screen an enlarged image of the screen of a television tube.

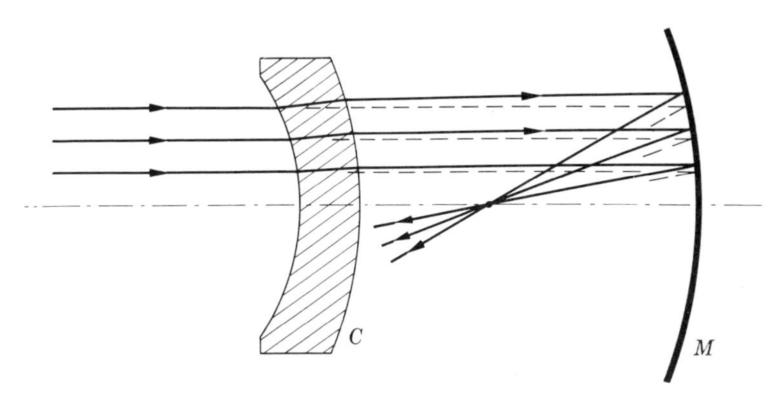

FIG. 5-7. The Maksutov corrector.

The complicated aspherical surfaces of the Schmidt corrector can be avoided and excellent correction can still be retained in the system invented by D. D. Maksutov, of the State Optical Institute of the U.S.S.R., in 1941. As shown in Fig. 5-7, a thick meniscus lens having *spherical* surfaces of approximately equal radii of curvature is placed in front of the mirror. All rays originally parallel to the axis are deviated outward by the lens, but the deviation of the outer rays is greater than that of the paraxial rays and, as indicated by the full lines, spherical aberration is minimized. Since the spherical surfaces of the meniscus lens can be ground and polished by machine methods, these lenses are simple and cheap to manufacture.

Any type of correcting lens necessarily introduces some chromatic aberration, from which the mirror alone, of course, is free. This effect, however, is not serious in either of the above systems.

5-4 Coma. The aberration known as coma affects rays from points not on the axis of the lens. It is similar to spherical aberration (which relates to points on the lens axis) in that both arise from the failure of the lens to image central rays and rays through outer zones of the lens at the same point. Coma differs from spherical aberration, however, in that a point object is imaged not as a circle but as a comet-shaped figure (whence the term "coma").

Fig. 5-8 is a drawing showing the effect of coma. It is assumed in this discussion that coma is the only aberration present. In general, the presence of other aberrations will modify the appearance of the image. The axis of the lens is OO', and P is an object point below the axis. A

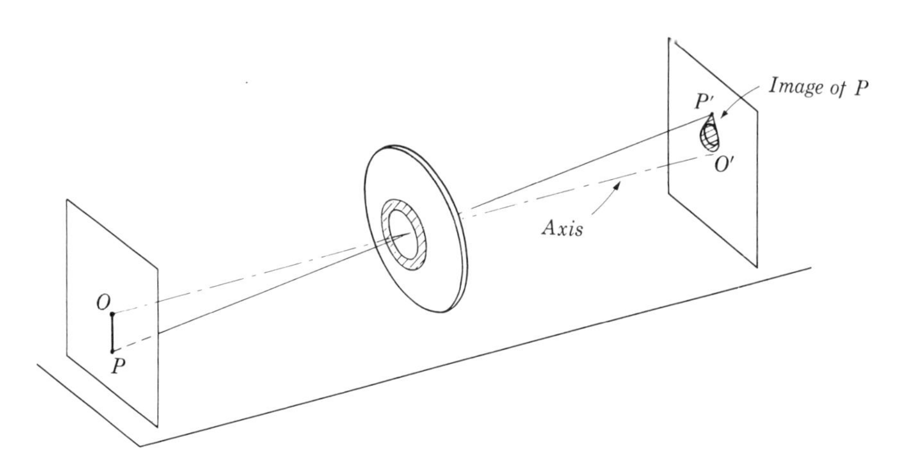

FIG. 5-8. Coma.

small pencil of rays from P through the center of the lens is imaged as a point at P'. The hollow cone of rays passing through the shaded zone is imaged as the circle below P'. Rays through inner zones are imaged as smaller circles above this one and those through outer zones as larger circles below it. The totality of all of these circles produces an image having the shape shown.

Coma, like spherical aberration, may be corrected by a proper choice of radii of curvature of the lens surfaces and, unlike spherical aberration, may be entirely eliminated from a single thin lens for any given pair of object and image points. (The lens will still exhibit coma for other object and image distances.) Unfortunately, the necessary curvature of the lens surfaces for zero coma is not the same as for minimum spherical aberration, so that a lens of minimum spherical aberration cannot be free from coma. Coma may also be eliminated by a stop or diaphragm of the proper size, located at the proper point on the lens axis.

The spherical lens illustrated in Fig. 5-3 can be shown to be free from coma for the same pair of object and image distances for which it is free from spherical aberration. Conjugate points free from both spherical aberration and coma are termed *aplanatic* and a lens possessing such a pair of points is called an *aplanatic lens.*

5-5 Astigmatism and curvature of field. These two aberrations will be considered together, as they are joint aspects of the same phenomenon. Astigmatism (not to be confused with the type of defective vision of the

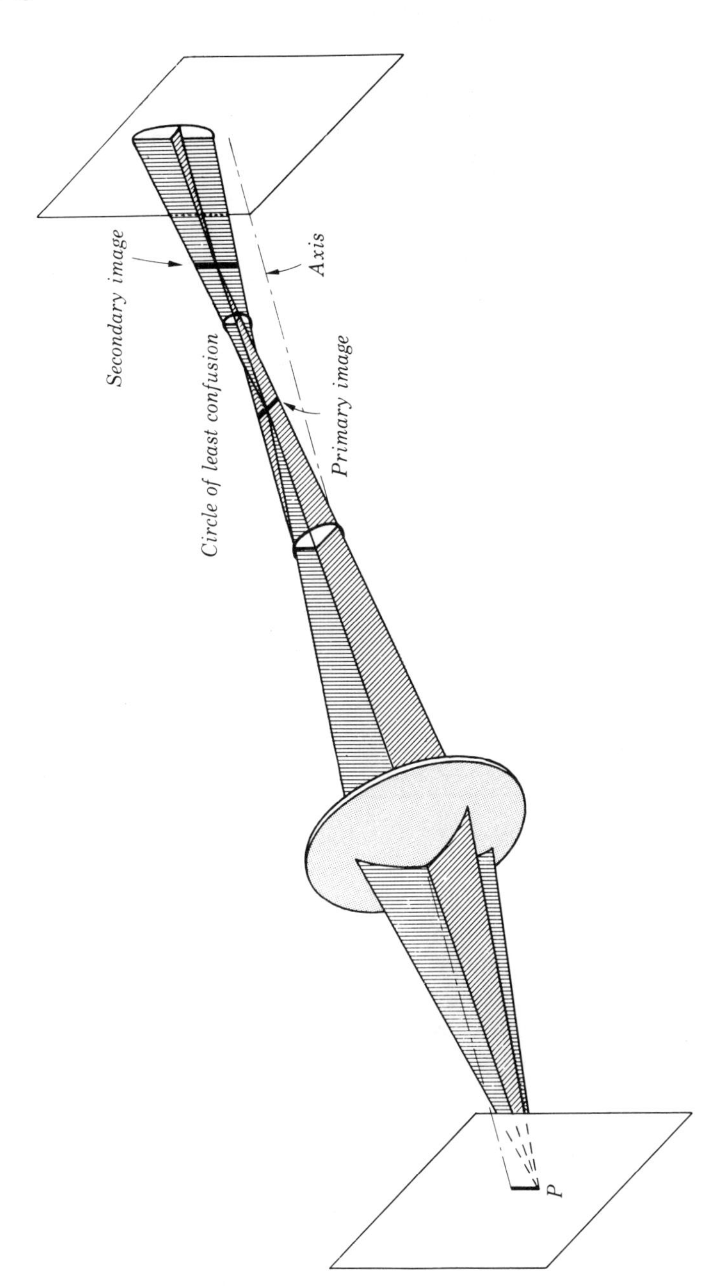

Fig. 5-9. Astigmatism.

same name), like coma, affects the image formed by a lens of points not on the lens axis. The two differ, however, in that coma results in spreading out the image of a point over a plane perpendicular to the lens axis, while astigmatism spreads the image in a direction along the axis. The effect is shown in Fig. 5-9. (It is assumed that astigmatism is the only aberration present.) Two sections of the cone of rays diverging from P and refracted by the lens are shown shaded. After refraction, all rays from P pass through a horizontal line, the *primary image*, and later through a vertical line, the *secondary image*. (See Sec. 2-7 for a discussion of another example of astigmation.) The cross section of the refracted beam is elliptical, the ellipse degenerating to a straight line at the primary and secondary images, and to a circle, the circle of least confusion, at some point between them. The cross section of the beam is smallest at the circle of least confusion and the best focus is at this point.

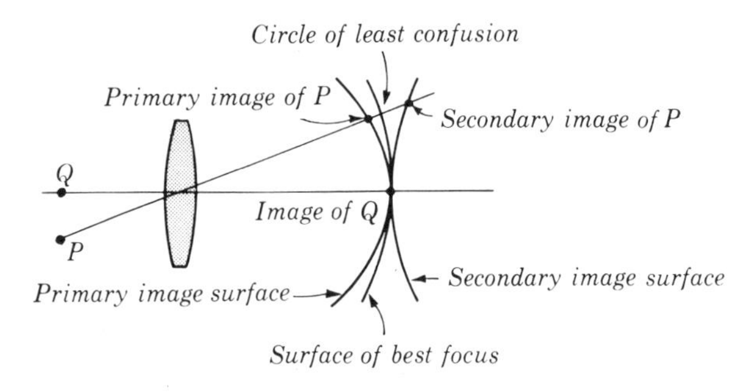

FIG. 5-10. Curvature of field.

If one considers the images of all points in the object plane, the locus of the primary images of these points is a surface of revolution about the lens axis, designated as the primary image surface. (Fig. 5-10.) Similarly, the secondary image surface is the locus of the secondary images. The surface of best focus is the locus of the circles of least confusion. All of these surfaces are tangent to one another at the axis of the lens. In general, the surface of best focus is not a plane, but a curved surface as shown, and this aberration is known as *curvature of field*. The failure of the primary and secondary images to coincide is termed *astigmatism*. The shape of the image surfaces depends upon the shape of the lens and the position of stops on the lens axis.

Complete elimination of astigmatism and curvature of field would require that the primary and secondary image surfaces should be planes, in which case the surface of best focus would also be a plane coincident with them. It is not possible to secure this result with a single lens. It is possible, however, to eliminate either curvature of field or astigmatism by proper location of stops on the lens axis. To eliminate curvature of field, the primary and secondary image surfaces are made to have equal and opposite curvatures, as in Fig. 5-11 (a). The surface of best focus is

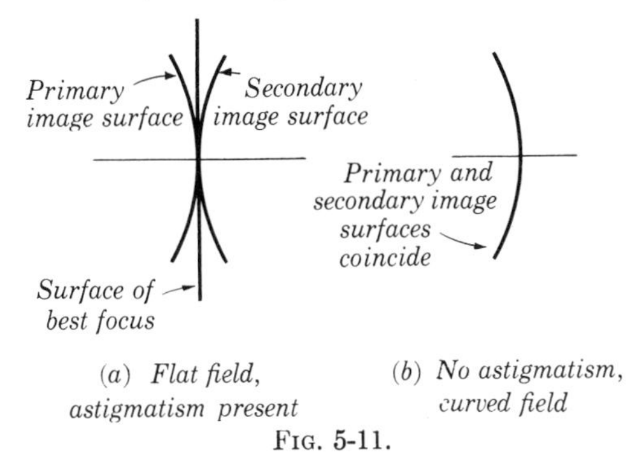

FIG. 5-11.

then a plane midway between them, perpendicular to the lens axis. Astigmatism is then necessarily present, the quality of the image becoming poorer as one proceeds outward from the axis. To eliminate astigmatism, both surfaces are made to have the same curvature.

For object points at relatively small angular distances from the axis, coma is more objectionable than astigmatism, while the reverse is true for large angles. Hence a telescope objective, whose field is small, would probably be corrected for coma rather than for astigmatism. In a wide-field camera lens, on the other hand, some correction would necessarily be made for astigmatism.

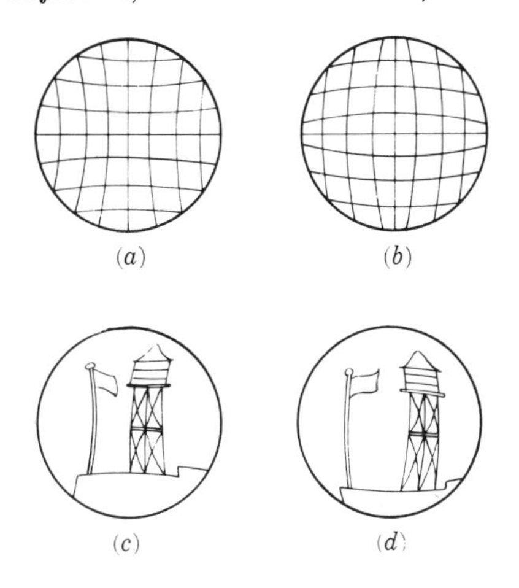

FIG. 5-12. (a), (c), pincushion distortion. (b), (d), barrel distortion.

5-6 Distortion.

Spherical aberration, coma, and astigmatism refer to the failure of a lens to form a point image of a point object. *Distortion* is an aberration arising not from a lack of sharpness of the image, but from a variation of magnification with axial distance. If the magnification increases with increasing axial distance, the outer parts of the field are disproportionately magnified. A square network then appears like Fig. 5-12 (a). This effect is referred to as "pincushion" distortion. If the magnification decreases with increasing axial distance, the opposite effect, known as "barrel" distortion, is obtained (Fig. 5-12 (b)). A moderate amount of distortion is not objectionable in an instrument intended solely for visual use, but it must evidently be eliminated from a camera lens used for aerial mapping or for copying drawings, where straight lines must be imaged as straigh lines.

A single thin lens is free from distortion for all object distances if there are no stops which limit the cone of rays striking the lens. If there are such stops on the lens axis, then in general the image will be distorted. Fig. 5-13 illustrates the way in which the position of a stop affects distortion. It will be seen that, depending upon whether the stop is in front of or behind the lens, a different cone of rays is utilized in the formation of

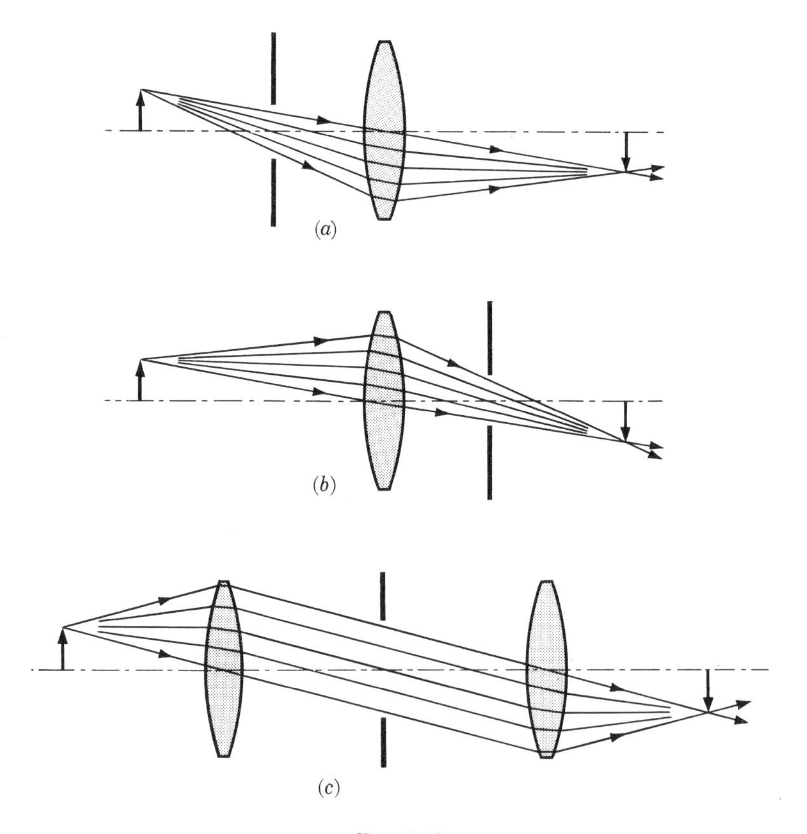

FIG. 5-13.

the image of the head of the arrow. If the stop is placed as in (a), the distortion is of the barrel type, while it is of the pincushion type if the stop is placed as in (b).

By constructing a lens of two symmetrical elements having a stop midway between them, as in Fig. 5-13 (c), the cone of rays passing through the first lens corresponds to Fig. 5-13 (b), while that passing through the second lens corresponds to Fig. 5-13 (a). Hence the distortion introduced by the second lens compensates for the distortion produced by the first. Many camera and projection lenses are constructed in this way.

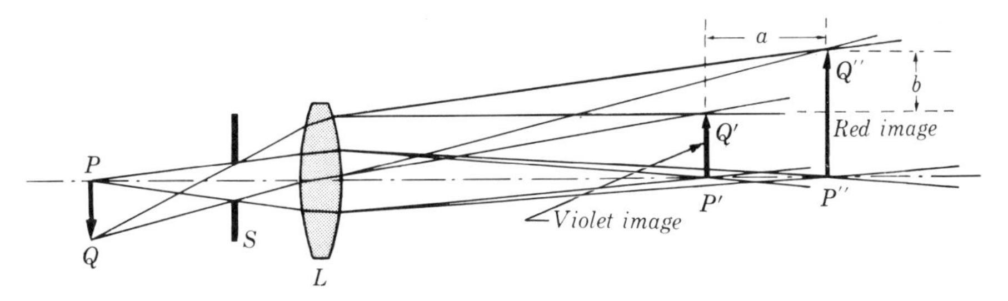

Fig. 5-14. Chromatic aberration.

5-7 The chromatic aberrations. The focal length of a lens is a function of the index of refraction of the material of which the lens is composed. Since the index of refraction of all optical substances varies with wave length, the focal length of a lens is different for different colors. As a consequence, a single lens forms not merely one image of an object, but a series of images at varying distances from the lens, one for each color present in the incident light. Furthermore, since the magnification depends upon the focal length, these images are of different sizes. The variation of image distance with index of refraction is called *axial* or *longitudinal chromatic aberration* and the variation of image size is called *lateral chromatic aberration.* These aberrations are illustrated in Fig. 5-14, which is much exaggerated for the sake of clearness. (The monochromatic aberrations are assumed to be absent.)

PQ is an object in the form of an arrow. It is imaged by a lens L, in front of which is a stop S. A cone of rays is shown diverging from each end of the arrow. Since the index is larger for the shorter wave lengths, the focal length is smaller for these wave lengths. The violet rays are therefore imaged nearest the lens and form the smallest image. The red rays are imaged farthest from the lens and form the largest image. The violet image is shown at $P'Q'$ and the red image at $P''Q''$. Images formed by other colors (not shown) lie at intermediate points and are of intermediate sizes. The axial chromatic aberration is measured by the distance a and the lateral chromatic aberration by the distance b. It will be seen that there is no one plane in which all images are simultaneously in sharp focus.

An "achromatic doublet" lens, consisting of two thin lenses of different kinds of glass in contact, may be designed to have the same focal length for any two colors. It is then said to be *achromatized for focal length* for these two colors. In effect, the chromatic aberration of one of the lenses compensates for the chromatic aberration of the other. Suppose the glasses to be used are the flint and crown whose dispersive powers were computed in Sec. 2-12. Let the wave lengths for which the focal length is to be achromatized be λ_F and λ_C and let the desired focal length for these wave lengths be 100 mm. Let primed quantities refer to the flint lens, unprimed to the crown lens, and let f_0 represent the focal length of the doublet. Then from the expression for the focal length of two thin lenses in contact, Eq. (4-20),

$$\frac{1}{f_0} = \frac{1}{f} + \frac{1}{f'}$$

$$= (n - 1)\left(\frac{1}{R_1} - \frac{1}{R_2}\right) + (n' - 1)\left(\frac{1}{R_1'} - \frac{1}{R_2'}\right).$$

For convenience, let

$$\left(\frac{1}{R_1} - \frac{1}{R_2}\right) = K, \qquad \left(\frac{1}{R_1'} - \frac{1}{R_2'}\right) = K'.$$

Then

$$\frac{1}{f_0} = K\,(n - 1) + K'\,(n' - 1).$$

But if f_0 (or $1/f_0$) is to have the same value for the wave lengths λ_F and λ_C,

$$K\,(n_F - 1) + K'\,(n'_F - 1) = K\,(n_C - 1) + K'\,(n'_C - 1).$$

After expanding and cancelling, this equation reduces to

$$\frac{K}{K'} = -\frac{n'_F - n'_C}{n_F - n_C}.$$

The focal lengths of the elements for the intermediate wave length λ_D are

$$\frac{1}{f_D} = (n_D - 1)\,K, \quad \frac{1}{f'_D} = (n'_D - 1)\,K',$$

or

$$\frac{K}{K'} = \frac{n'_D - 1}{n_D \ - 1}\frac{f'_D}{f_D}.$$

When the two expressions for K/K' are equated, we obtain

$$\frac{f'_D}{f_D} = -\frac{n'_F - n'_C}{n'_D - 1} \div \frac{n_F - n_C}{n_D \ - 1} = -\frac{\omega'}{\omega},$$

where ω and ω' are the dispersive powers of the crown and flint glasses. That is, the ratio of the focal lengths of the elements (for the wave length λ_D) should equal the negative of the ratio of the dispersive powers. Note that since ω and ω' are both positive, the ratio of the focal lengths is necessarily negative. That is, the doublet must consist of a positive and a negative lens. Let the crown be the positive and the flint be the negative lens, and assume the crown lens faces the light. If the faces of the lenses in contact are to be cemented together, the radius of curvature of the second surface of the crown lens, R_2, must equal the radius of curvature of the first surface of the flint lens, R_1'. We now have the following three conditions imposed on the four radii of curvature.

$$R_2 = R_1',$$

$$\frac{f'_D}{f_D} = -\frac{\omega'}{\omega} = \frac{(n_D \ - 1)\ (1/R_1 \ - 1/R_2\ )}{(n'_D - 1)\ (1/R_1' - 1/R_2')},$$

$$1/f_0 = 1/100 = (n_F - 1)\ (1/R_1 - 1/R_2)$$
$$+ (n'_F - 1)\ (1/R_1' - 1/R_2').$$

A fourth condition may still be imposed arbitrarily and may be utilized for other corrections, such as minimizing spherical aberration. For simplicity, however, let us assume that the crown lens is to be plano-convex, with the plane side facing the light. Then

$$R_1 = \infty.$$

Introducing the proper values of n_D, n'_D, ω, and ω', and solving the four equations simultaneously, one obtains

$$R_1 = \infty, \quad R_2 = -21.1 \text{ mm} = R_1',$$
$$R_2' = -40.4 \text{ mm}.$$

The doublet is shown in Fig. 5-15.

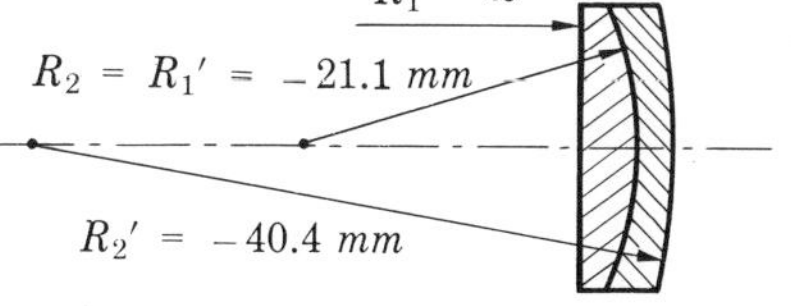

Fig. 5-15. An achromatic doublet.

It is also possible to achromatize for focal length without using two different kinds of glass, by constructing a lens of two elements separated from one another. The equivalent focal length f_0 of two thin lenses not in contact, from Eq. (4-19), is

$$\frac{1}{f_0} = \frac{1}{f} + \frac{1}{f'} - \frac{t}{ff'},$$

where f and f' represent the focal lengths of the lenses and t is their separation. Let

$$\left(\frac{1}{R_1} - \frac{1}{R_2}\right) = K, \qquad \left(\frac{1}{R_1'} - \frac{1}{R_2'}\right) = K'.$$

Then

$$\frac{1}{f_0} = K\,(n-1) + K'\,(n-1) - tKK'\,(n-1)^2.$$

We wish to make the focal length f_0 independent of wave length. Then $1/f_0$ is also independent of wave length. That is,

$$\frac{d}{d\lambda}\left(\frac{1}{f_0}\right) = K\frac{dn}{d\lambda} + K'\frac{dn}{d\lambda} - 2tKK'\,(n-1)\frac{dn}{d\lambda} = 0,$$

or

$$\frac{dn}{d\lambda}\left[K + K' - 2tKK'\,(n-1)\right] = 0.$$

But since $dn/d\lambda$, the slope of the index vs wave length curve (Fig. 2-27) is not zero, the bracketed term must be zero. Hence

$$t = \frac{K + K'}{2KK'\,(n-1)}.$$

Multiply numerator and denominator by $(n-1)$ and replace $K(n-1)$ by $1/f$ and $K'(n-1)$ by $1/f'$. This gives

$$t = \frac{\dfrac{1}{f} + \dfrac{1}{f'}}{\dfrac{2}{ff'}} = \frac{f + f'}{2}.$$

That is, the separation of the elements should be one-half the sum of their focal lengths. Many eyepieces are achromatized in this way. (See Sec. 6-5.)

It should not be inferred that a lens system corrected in this way is fully achromatic, in the sense that the focal length is the same for all wave lengths. This method of achromatizing makes the focal length f_0 a minimum for the particular wave length to which f and f' refer. As is the case with any function exhibiting a minimum, small changes in either direction from the minimum produce very small changes in the function.

5-8 Summary. It should be evident from the preceding discussion that it is not possible to eliminate all of the seven aberrations from a single thin lens, or even to minimize them all simultaneously. However, by constructing a compound lens of a number of individual lenses, it is possible to balance the aberrations of one part of the system against those of another. The greater the number of elements, the greater the degree of correction which may be secured. Hence all high quality lenses are compound lenses. Even then, no lens system ever contains a sufficient number of refracting surfaces to permit the complete elimination of all five monochromatic aberrations and the two chromatic aberrations. As a consequence, a lens designer is forced to select the aberrations most detrimental to the purpose for which the lens is destined and reduce these aberrations to negligible amounts. Thus in a telescope objective, which is required to cover only a small angular field, spherical aberration, coma, and axial chromatism are the most important. On the other hand, in a photographic objective, which must ordinarily cover a large field, only partial correction of these aberrations is possible because of the necessity of producing some degree of correction for such aberrations as astigmatism, curvature of field, and distortion. As a general rule, systems that need be corrected only for a small field or for a small aperture produce images of higher quality than systems with either a large field or a large aperture.

CHAPTER 6

OPTICAL INSTRUMENTS

6-1 The eye. Since the purpose of most optical instruments is to enable us to see better, the logical place to begin a discussion of such instruments is with the eye. The essential parts of the eye, considered as an optical system, are shown in Fig. 6-1.

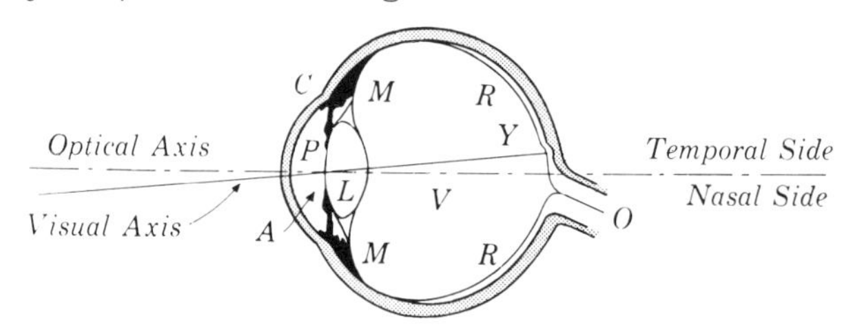

FIG. 6-1. The eye.

The eye is very nearly spherical in shape, and about an inch in diameter. The front portion is somewhat more sharply curved, and is covered by a tough, transparent membrane C, called the *cornea.* The region behind the cornea contains a liquid A called the *aqueous humor.* Next comes the *crystalline lens,* L, a capsule containing a fibrous jelly hard at the center and progressively softer at the outer portions. The crystalline lens is held in place by ligaments which attach it to the ciliary muscle M. Behind the lens, the eye is filled with a thin jelly V consisting largely of water, called the *vitreous humor.* The indices of refraction of both the aqueous humor and the vitreous humor are nearly equal to that of water, about 1.336. The crystalline lens, while not homogeneous, has an "average" index of 1.437. This is not very different from the indices of the aqueous and vitreous humors, so that most of the refraction of light entering the eye is produced at the cornea.

A large part of the inner surface of the eye is covered with a delicate film of nerve fibers, R, called the *retina.* A cross section of the retina is shown in Fig. 6-2. Nerve fibers branch out from the *optic nerve* O, and terminate in minute structures called rods and cones. The rods and cones, together with a bluish liquid called the visual purple which circulates among them, receive the optical image and transmit it along the optic nerve to the brain. There is a slight depression in the retina at Y called

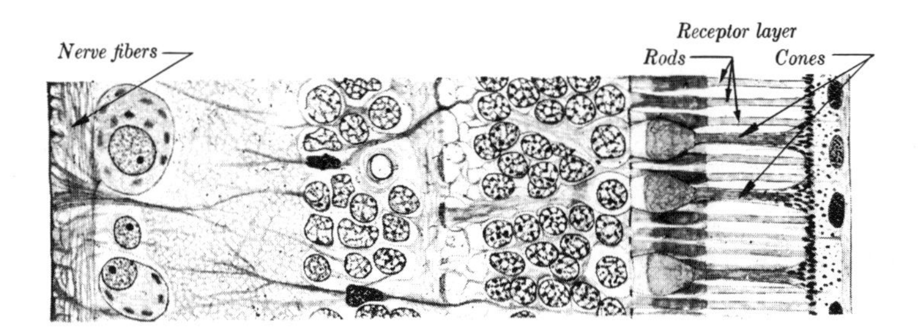

FIG. 6-2. Section of the human retina (500×). Light is incident from the left.

the yellow spot or macula. At its center is a minute region, about 0.25 mm in diameter, called the *fovea centralis*, which contains cones exclusively. Vision is much more acute at the fovea than at other portions of the retina, and the muscles controlling the eye always rotate the eyeball until the image of the object toward which attention is directed falls on the fovea. The outer portion of the retina merely serves to give a general picture of the field of view. The fovea is so small that motion of the eye is necessary to focus distinctly two points as close together as the dots in a colon. (:)

There are no rods or cones at the point where the optic nerve enters the eye and an image formed at this point cannot be seen. This region is called the *blind spot*. The existence of the blind spot can be demonstrated by closing the left eye and looking with the right eye at the cross in Fig. 6-3. When the diagram is about 10 inches from the eye, the square disappears. At a smaller distance, the square reappears while the circle disappears. At a still smaller distance, the circle again appears.

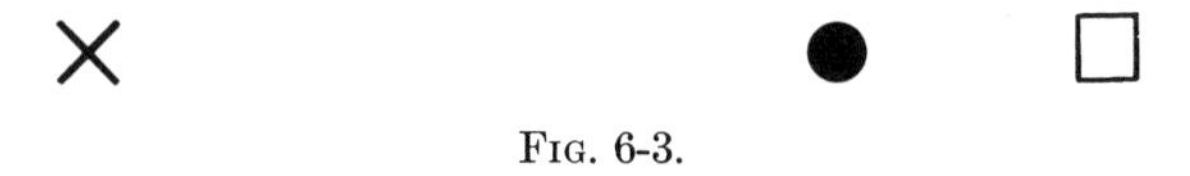

FIG. 6-3.

In front of the crystalline lens is the iris, at the center of which is an opening P called the *pupil*. The function of the pupil is to regulate the quantity of light entering the eye; the pupil automatically dilates if the brightness[1] of the field is low, and contracts if the brightness is increased. This process is known as *adaptation*. Fig. 6-4 illustrates the manner in which the size of a normal pupil varies with the field brightness. Observe that the range of pupillary diameter is only about fourfold (hence the

[1] The correct technical term for the quantity referred to here as "brightness," is *luminance*. See Chap. 13.

range in area is about sixteenfold) over a range of brightness from 10^{-2} to 10^3 candles/m^2 which is 100,000 fold. The relatively enormous variation in light entering the eye is far from compensated by the change in size of the pupil, the receptive mechanism of the retina being able to adapt itself to large differences in quantity of light.

In order to see an object distinctly, a sharp image of it must be formed on the retina. If all the elements of the eye were rigidly fixed in position, there would be but one object distance for which a sharp retinal image would be formed while, in fact, the normal eye can focus sharply on an object at any distance from infinity up to about 10 inches in front of the eye. This is made possible by the action of the crystalline lens and the ciliary muscle to which it is attached. When relaxed, the normal eye is focused on objects at infinity, i.e., the second focal point is at the retina. When it is desired to view an object nearer than infinity, the ciliary muscle tenses and the crystalline lens assumes a more nearly spherical shape. This process is called *accommodation.*

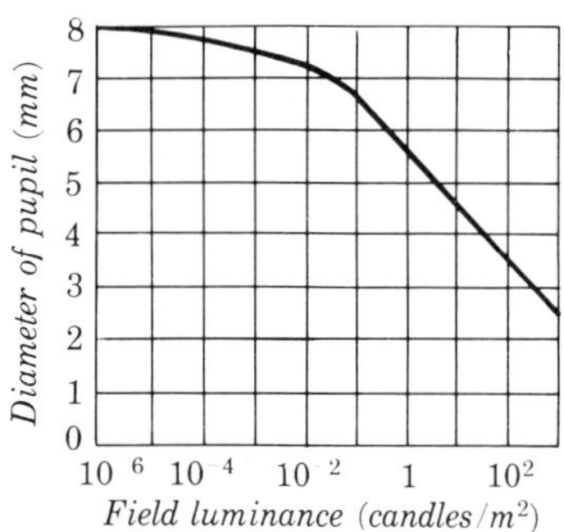

FIG. 6-4. Pupillary diameter as a function of field luminance ("brightness").

The extremes of the range over which distinct vision is possible are known as the *far point* and the *near point* of the eye. The far point of a normal eye is at infinity. The position of the near point evidently depends on the extent to which the curvature of the crystalline lens may be increased in accommodation, and the range of accommodation gradually diminishes with increasing age as the crystalline lens loses its flexibility. For this reason the near point gradually recedes as one grows older. This recession of the near point with age is called *presbyopia,* and should not be considered a defect of vision, since it proceeds at about the same rate in all normal eyes. A table of the approximate position of the near point at various ages is given below.

Age (years)	Near Point (cm)
10	7
20	10
30	14
40	22
50	40
60	200

6-2 Defects of vision. There are a number of common defects of vision which have to do simply with an incorrect relation between the various parts of the eye considered as an optical system. A normal eye forms on the retina an image of an object at infinity when the eye is relaxed, and is called *emmetropic.* If the far point of an eye is not at infinity, the eye is *ametropic.* The two simplest forms of ametropia are *myopia* (nearsightedness), and *hyperopia* (farsightedness). These are illustrated in Fig. 6-5 (b) and (c).

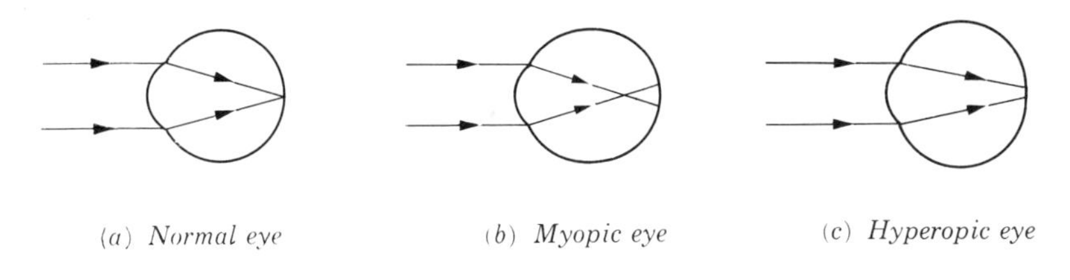

(*a*) *Normal eye* (*b*) *Myopic eye* (*c*) *Hyperopic eye*

Fig. 6-5.

In the myopic eye, the eyeball is too long in comparison with the radius of curvature of the cornea, and rays from an object at infinity are focused in front of the retina. The most distant object for which an image will be formed on the retina is then nearer than infinity, or the far point is nearer than infinity. On the other hand, the near point of a myopic eye, if the accommodation is normal, is even closer to the eye than is that of a person with normal vision.

In the hyperopic eye, the eyeball is too short, and the image of an infinitely distant object would be formed behind the retina. By accommodation, these parallel rays may be made to converge on the retina but, evidently, if the range of accommodation is normal, the near point will be more distant than that of an emmetropic eye. These defects may be stated in a somewhat different way. The myopic eye produces too much convergence in a parallel bundle of rays for an image to be formed on the retina; the hyperopic eye, not enough.

Astigmatism refers to a defect in which the surface of the cornea is not spherical, but is more sharply curved in one plane than another. (It should not be confused with the lens aberration of the same name, which applies to the behaviour, after passing through a lens having spherical surfaces, of rays making a large angle with the axis.) Astigmatism makes it impossible, for example, to focus clearly on the horizontal and vertical bars of a window at the same time.

6-3 Spectacles. All of the types of defective vision mentioned above may be corrected by the use of spectacles. We shall illustrate with the aid of numerical examples.

Presbyopia and hyperopia. The near point of either a presbyopic or a hyperopic eye is farther from the eye than normal. Then in order to see clearly an object at normal reading distance (this distance is usually assumed to be 25 cm or 10 inches), we must place in front of the eye a lens of such focal length that it forms an image of the object, at or beyond the near point. Thus the function of the lens is not to make the object appear larger, since the object and its image subtend equal angles at the lens, but in effect to move the object farther away from the eye to a point where a sharp retinal image can be formed.

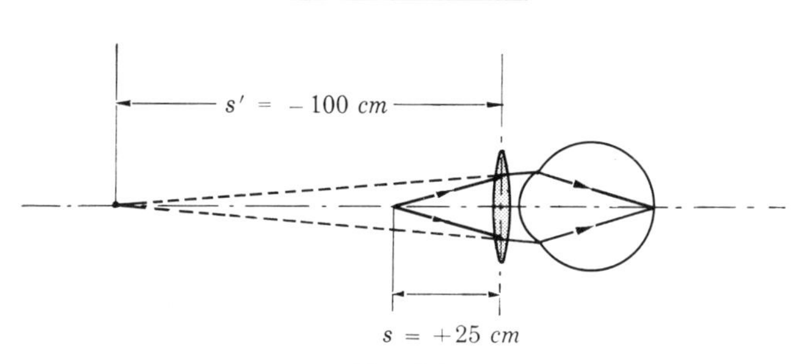

FIG. 6-6.

Example. The near point of a certain eye is 100 cm in front of the eye. What lens should be used to see clearly an object 25 cm in front of the eye? (See Fig. 6-6.)

We have

$$s = +25 \text{ cm}, \quad s' = -100 \text{ cm}.$$

$$\frac{1}{f} = \frac{1}{s} + \frac{1}{s'} = \frac{1}{+25} + \frac{1}{-100},$$

$$f = +33 \text{ cm}.$$

That is, a converging lens of focal length 33 cm is required.

Myopia. The far point of a myopic eye is nearer than infinity. To see clearly objects beyond the far point, a lens must be used which will form an image of such objects, not farther from the eye than the far point.

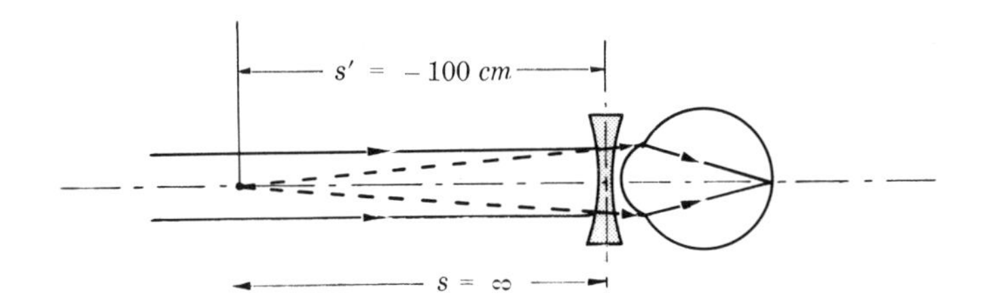

Fig. 6-7.

Example. The far point of a certain eye is 1 meter in front of the eye. What lens should be used to see clearly an object at infinity?

Assume the image to be formed at the far point. Then

$$s = \infty, \quad s' = -100 \text{ cm}.$$

$$\frac{1}{f} = \frac{1}{s} + \frac{1}{s'} = \frac{1}{\infty} + \frac{1}{-100},$$

$$f = -100 \text{ cm}.$$

A diverging lens of focal length 100 cm is required, as in Fig. 6-7.

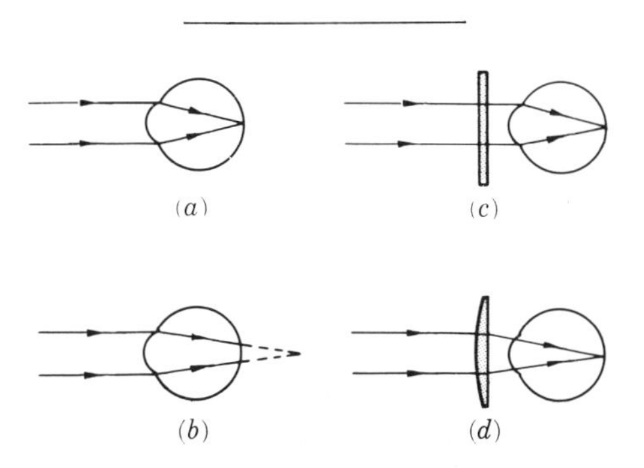

Fig. 6-8.

Astigmatism. The correction of astigmatism by means of a cylindrical lens is illustrated in Fig. 6-8, in which (a) and (b) represent the top and side views respectively of an astigmatic eye. The curvature of the cornea in a horizontal plane, as seen in (a), has the proper value such that rays from infinity are focused on the retina. In the vertical plane, however, as seen in (b), the curvature is not sufficient to form a sharp retinal image.

By placing before the eye a cylindrical lens with axis horizontal, as in (c) and (d), the rays in a horizontal plane are unaffected, while the additional convergence of the rays in a vertical plane, shown in (d), now causes these to be sharply imaged on the retina.

It is customary for the optometrist to describe the converging or diverging effect of spectacle lenses in terms, not of the focal length, but of its reciprocal. The reciprocal of the focal length of a lens is called its *dioptric power*, and if the focal length is in meters the power is in *diopters*. Thus the power of a positive lens whose focal length is 1 meter, is 1 diopter; if the focal length is 2 meters the power is 0.5 diopter and so on. If the focal length is negative, the power is negative also. For example, a lens of power -0.5 diopter is a diverging lens of focal length -2 meters.

It is useful to consider not only the power of a lens, but the powers of the individual lens surfaces. For instance, a double convex thin lens may be thought of as two plano-convex thin lenses with their plane surfaces in contact. The power D_1 of the first lens is

$$D_1 = \frac{1}{f_1} = (n-1)\left(\frac{1}{R_1} - \frac{1}{\infty}\right) = \frac{n-1}{R_1}.$$

The plane surface contributes nothing to the power, which may be considered as associated with the convex surface only.

The power of the second lens is

$$D_2 = \frac{1}{f_2} = (n-1)\left(\frac{1}{\infty} - \frac{1}{R_2}\right) = -\frac{n-1}{R_2}.$$

Treating the lens as a double convex thin lens, its power is

$$D = \frac{1}{f} = (n-1)\left(\frac{1}{R_1} - \frac{1}{R_2}\right).$$

Hence

$$D = D_1 + D_2,$$

and the power of the lens equals the sum of the powers of its surfaces. An optometrist who grinds his own lenses usually purchases from the manufacturer lens blanks, one surface of which has been ground and

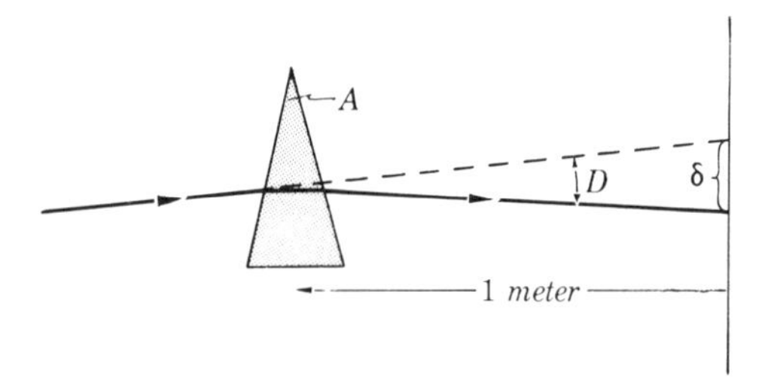

FIG. 6-9.

polished to a specified radius or power. By subtracting the power of this surface from the desired power of the completed lens, the power to which the other surface should be ground is given at once.

For the correction of certain types of defective vision a prism, usually of small angle, is combined with a spectacle lens. The deviation produced by the prism is expressed in *prism diopters.* The power of a prism in prism diopters is defined as the linear deviation, Δ, in centimeters which the prism produces at a distance of 1 meter, when a ray passes through the prism at minimum deviation. See Fig. 6-9.

6-4 The simple microscope or magnifier. The apparent size of an object is determined by the size of its retinal image which, in turn, if the eye is unaided, depends upon the angle subtended by the object at the eye. When one wishes to examine a small object in detail one brings it close to the eye, in order that the angle subtended and the retinal image may be as large as possible. Since the eye cannot focus sharply on objects closer than the near point, a given object subtends the maximum possible angle at an unaided eye when placed at this point. (We shall assume hereafter that the near point is 25 cm from the eye.) By placing a converging lens in front of the eye, the accommodation may, in effect, be increased. The object may then be brought closer to the eye than the near point, and will subtend a correspondingly larger angle. A converging lens used for this purpose is called a *magnifying glass,* a *simple microscope,* or a *magnifier.* The magnifier forms a virtual image of the object and the eye "looks at" this virtual image. Since a (normal) eye can focus sharply on an object anywhere between the near point and infinity, the image can be seen equally clearly if it is formed anywhere within this range.

The magnifier is illustrated in Fig. 6-10. In part (a), the object is at the near point and subtends an angle u at the eye. In part (b), a magnifier in front of the eye forms an image at infinity and the angle subtended is u'. The height of the retinal image in either case is proportional to the tangent of the angle subtended, and the *angular magnification* γ (not to be confused with the *lateral* magnification m) is defined as

$$\gamma = \frac{\tan u'}{\tan u}.$$

The angular magnification can be expressed as follows. Let y represent the height of the object and f the focal length of the magnifier, both expressed in centimeters. Then from Fig. 6-10 (a),

$$\tan u = \frac{y}{25},$$

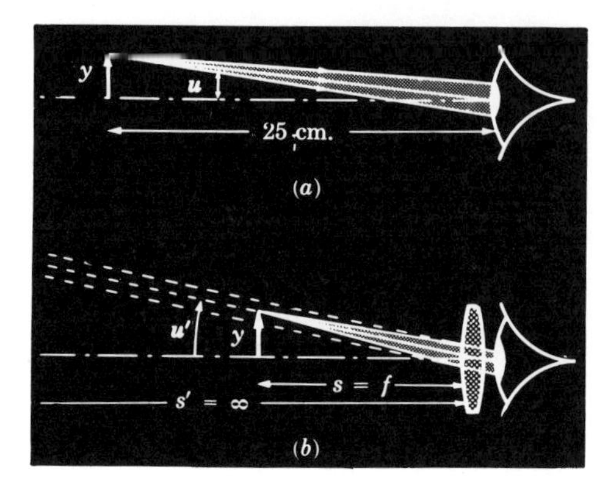

FIG. 6-10. The magnifier. Angular magnification $\gamma = \tan u'/\tan u$.

and from 6-10 (b),

$$\tan u' = \frac{y}{f}.$$

Hence

$$\gamma = \frac{y/f}{y/25},$$

$$\boxed{\gamma = \frac{25}{f}.} \quad (f \text{ in centimeters}). \tag{6-1}$$

That is, the angular magnification of a simple magnifier of focal length 10 cm is 2.5× (2.5 times). The height of the retinal image of an object viewed through the magnifier, as in Fig. 6-10 (b), is 2.5 times as great as when viewed with the unaided eye at the minimum distance of distinct vision.

While it might appear that the angular magnification could be made as large as desired by decreasing the focal length f, the aberrations of a simple lens set a limit to γ of about 2× or 3×. If these aberrations are minimized, the magnification may be carried as high as 20×.

6-5 Oculars. An *ocular* or *eyepiece* is a magnifier used for viewing an image formed by a lens or lenses preceding it in an optical system. Thus in both the compound microscope and the telescope, a real image of the object under observation is formed by a lens called the objective, and an ocular is used to view this image.

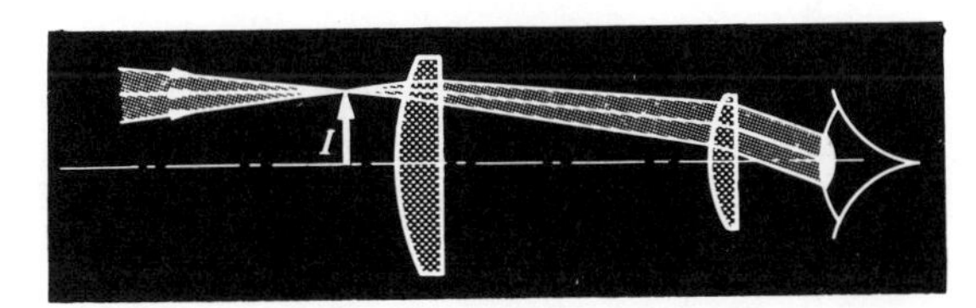

Fig. 6-11. The Ramsden ocular.

The Ramsden ocular is illustrated in Fig. 6-11. It is constructed of two plano-convex lenses of equal focal length, separated by a distance of about two-thirds of this length. The image to be examined is shown at I; the final image is at infinity. Since four refracting surfaces are available, the aberrations of a simple magnifier may be greatly reduced.

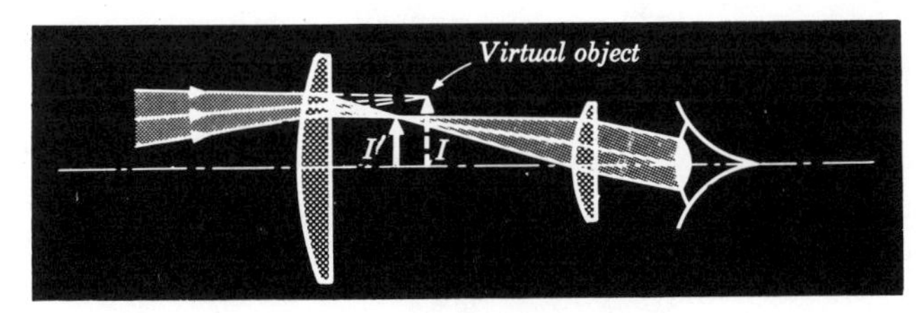

Fig. 6-12. The Huygens ocular.

The Huygens ocular, another common type, is shown in Fig. 6-12. The ratio of the focal lengths of the elements varies from about 3:1 to 1.5:1 and the ocular is achromatized as explained in Sec. 5-7 by making the separation equal to half the sum of the focal lengths. Rays from a preceding lens (not shown) are converging toward an image I which serves as a virtual object for the first lens. This lens forms a real image I' which is then imaged at infinity by the second lens.

Evidently the Ramsden ocular can be used as a magnifier to examine a real object (which should be placed at I), while the Huygens type can be used only to examine an image. The former is called a positive and the latter a negative ocular.

The front lens of an ocular is called the *field* lens, the other, the *eye* lens. The function of the field lens is to deviate inward the cone of rays converging toward each point of the image I. If the field lens is omitted, as has been done for simplicity in Fig. 6-13, an eye lens of much greater diameter is required in order to intercept those rays forming images of the outer portions of the field of view.

It is often desirable to mount cross hairs in an ocular to provide reference lines in the field of view or to make measurements by moving the cross hairs across an image. The cross hairs must evidently be in focus at the same time as the object under observation. They should therefore coincide with the image I in Fig. 6-11 and with the image I' in Fig. 6-12.

A highly corrected ocular is shown in the diagram of a prism binocular, Fig. 6-20.

6-6 The compound microscope. When an angular magnification higher than that attainable with a simple magnifier is desired, it is necessary to use a *compound microscope*, usually called merely a *microscope*. The essential elements of a microscope are illustrated in Fig. 6-13. The object to be examined is placed just beyond the first focal point F of the *objective* lens, which forms a real and enlarged image in the first focal plane of the ocular. The latter then forms a virtual image of this image at infinity. While both the objective and ocular of an actual microscope are highly corrected compound lenses, they are shown as simple thin lenses for simplicity.

The overall magnification M of a compound microscope, like the angular magnification of a simple microscope, is defined as the ratio of the tangent of the angle u' subtended at the eye by the final image, to the tangent of the angle u that would be subtended at the unaided eye by the object at a distance of 25 cm. Let y represent the height of the object and y' the height of its image formed by the objective. Then

$$\tan u = \frac{y}{25},$$

and

$$\tan u' = \frac{y'}{f_2},$$

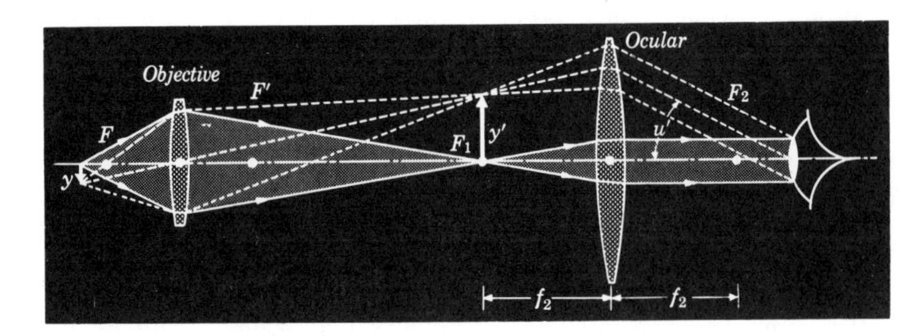

FIG. 6-13. The compound microscope.

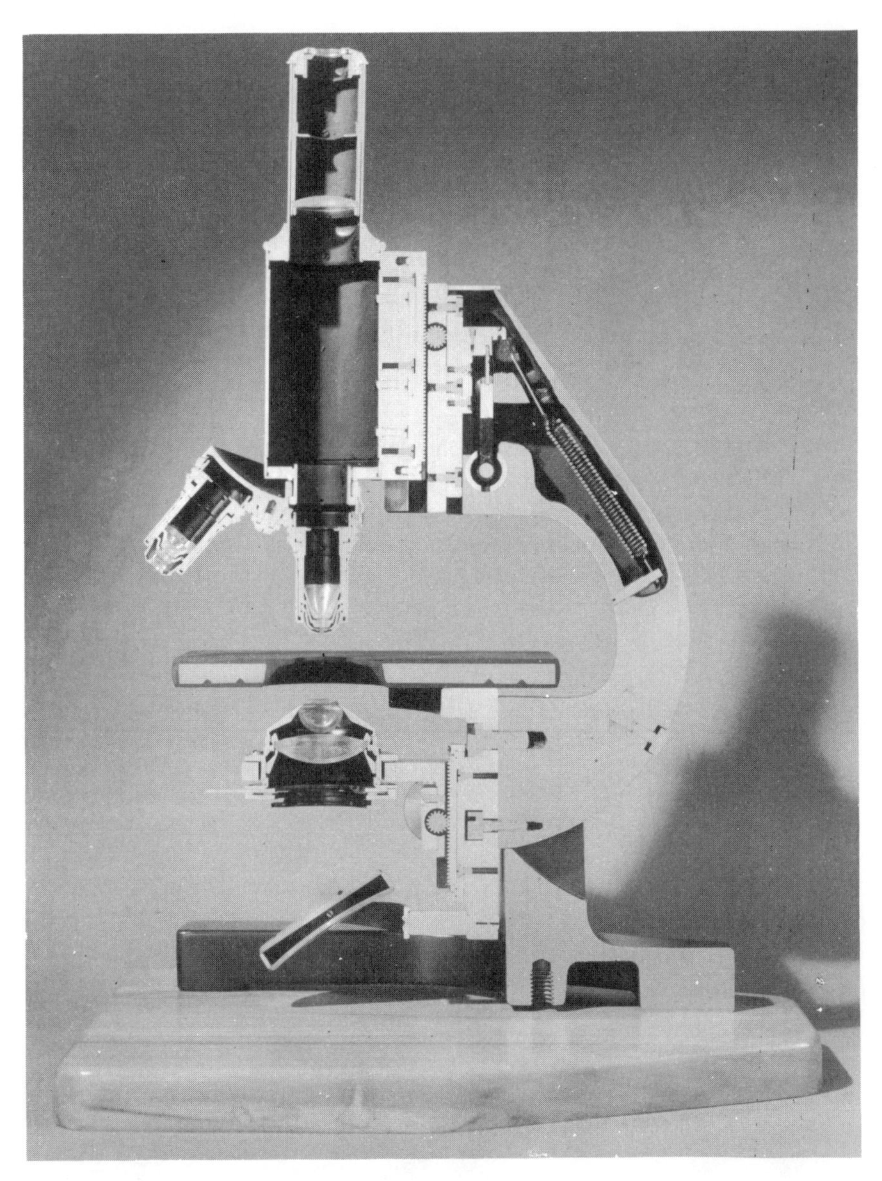

FIG. 6-14. Sectional view of a modern compound microscope.
(Courtesy of Bausch & Lomb Optical Co.)

where f_2 is the focal length of the ocular. Hence

$$M = \frac{\tan u'}{\tan u} = \frac{y'/f_2}{y/25}$$

$$= \frac{y'}{y} \times \frac{25}{f_2}.$$

But y'/y is the lateral magnification m produced by the objective, and $25/f_2$ is the angular magnification γ produced by the ocular. The overall magnification M is therefore the product of the lateral magnification of the objective, and the angular magnification of the ocular.

$$\boxed{M = m\gamma.} \tag{6-2}$$

Fig. 6-14 is a cut-away view of a modern microscope. The instrument is provided with a rotating nosepiece to which are permanently attached three objectives of different focal lengths. This construction makes possible a rapid and convenient exchange of one objective for another. Cross sections of three modern microscope objectives are shown in Fig. 6-15. The one on the right is of the oil immersion type and utilizes as its front lens the spherical lens shown in Fig. 5-3.

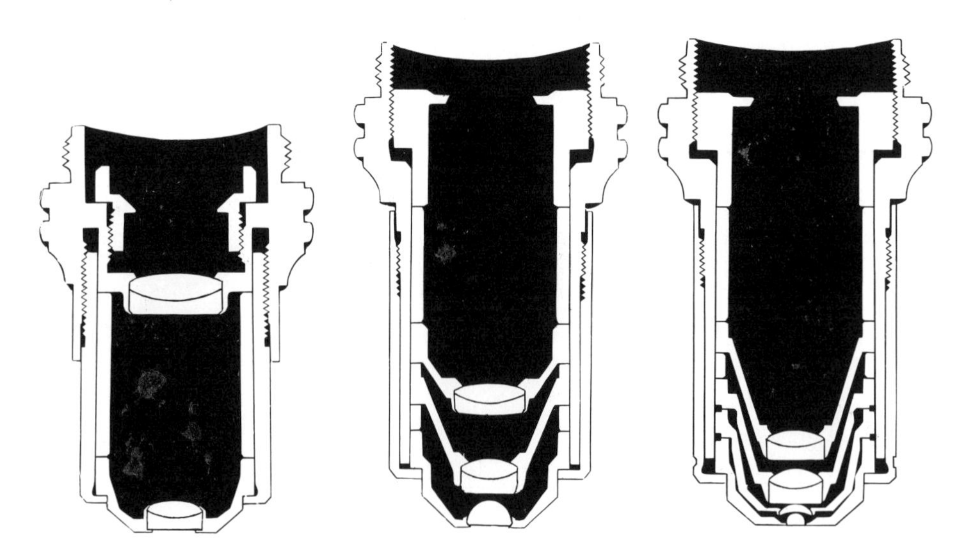

FIG. 6-15. Sectional views of three modern microscope objectives.

(Courtesy of Bausch & Lomb Optical Co.)

The mirror and condensing lens below the microscope stage are for the purpose of illuminating the object. The mirror is plane on one side and concave on the other. With low power objectives the lens may be swung to one side and the concave mirror alone provides sufficient illumination. (See Fig. 6-16 (a).) For objectives of power higher than 10×, the condenser and plane mirror are used as in Fig. 6-16 (b).

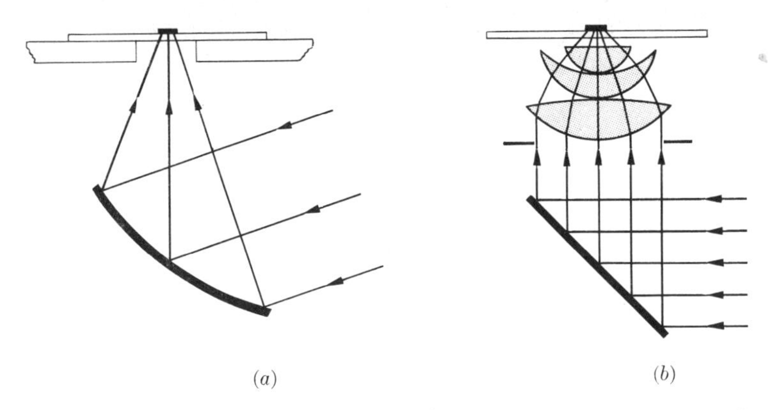

FIG. 6-16. Use of concave mirror (a) and condensing lens (b) for illuminating the object in a compound microscope.

6-7 Refracting telescopes. The optical system of a refracting telescope is essentially the same as that of a compound microscope. In both instruments, the image formed by an objective is viewed through an ocular. The difference is that the telescope is used to examine large objects at large distances and the microscope to examine small objects close at hand.

The *astronomical* telescope is illustrated in Fig. 6-17. The objective lens forms a real, reduced image I of the object O. I' is the virtual image of I formed by the ocular.

In practice, the objects examined by a telescope are at such large distances from the instrument that the image I is formed very nearly at the second focal point of the objective. Furthermore, if the image I' is at

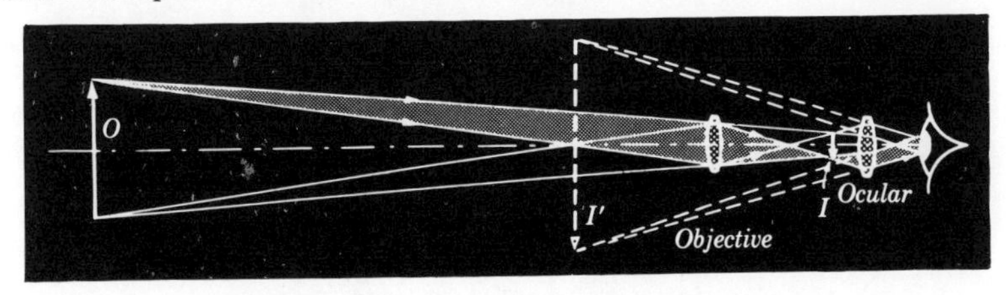

FIG. 6-17. The astronomical telescope.

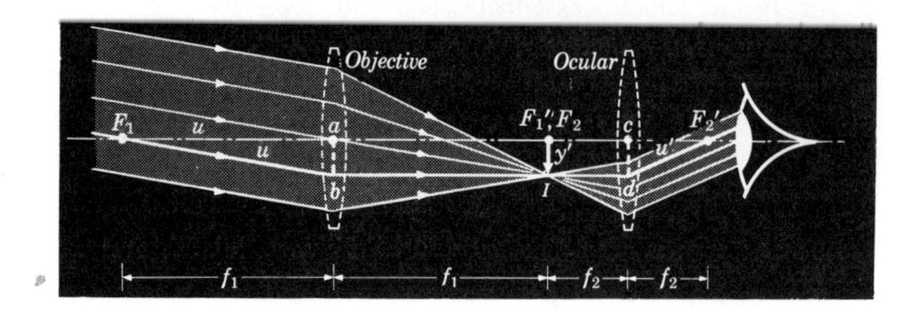

FIG. 6-18.

infinity, the image I is at the first focal point of the ocular. (This is not the case in Fig. 6-17, which has been drawn to show all the essential elements in a finite diagram.) The distance between objective and ocular, or the length of the telescope, is therefore the sum of the focal lengths of objective and ocular, $f_1 + f_2$.

The angular magnification γ of a telescope is defined as the ratio of the tangent of the angle subtended at the eye by the final image I', to the tangent of the angle subtended at the (unaided) eye by the object. This ratio may be expressed in terms of the focal lengths of objective and ocular as follows. The shaded bundle of rays in Fig. 6-18 corresponds to that in Fig. 6-17, except that the object and the final image are both at infinity. The ray passing through F_1, the first focal point of the objective, and through F_2', the second focal point of the ocular, has been emphasized. The object (not shown) subtends an angle u at the objective and would subtend essentially the same angle at the unaided eye. Also, since the observer's eye is placed just to the right of the focal point F_2', the angle subtended at the eye by the final image is very nearly equal to the angle u'. The distances ab and cd are evidently equal to one another and to the height y' of the image I. From the right triangles F_1ab and $F_2'cd$,

$$\tan u = \frac{-y'}{f_1},$$

$$\tan u' = \frac{y'}{f_2}.$$

Hence

$$\boxed{\gamma = -\frac{y'/f_2}{y'/f_1} = -\frac{f_1}{f_2}.} \tag{6-3}$$

The angular magnification of a telescope is therefore equal to the ratio of the focal length of the objective to that of the ocular. The negative sign denotes an inverted image. While an inverted image is not a disadvantage if the instrument is to be used for astronomical observations, it is desirable that a *terrestrial telescope* shall form an erect image. This may be accomplished by the insertion of an erecting lens or lens system between the objective and ocular as in Fig. 6-19. The erecting lens simply serves to

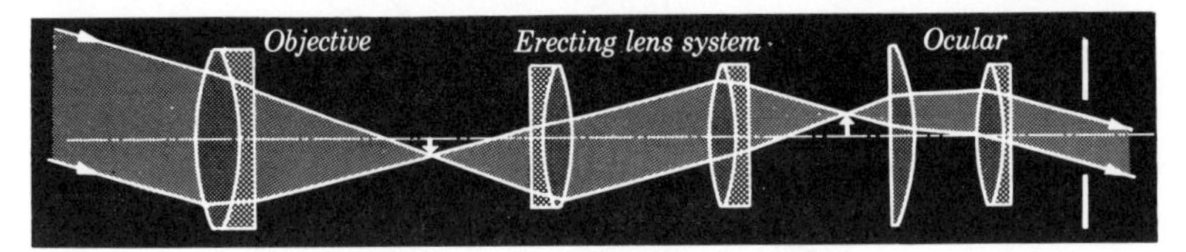

Fig. 6-19. A terrestrial telescope.

invert the image formed by the objective. This is the optical system of the "spyglass." It has the disadvantage of requiring an unduly long tube, since four times the focal length of the erecting lens must be added to the sum of focal lengths of objective and ocular.

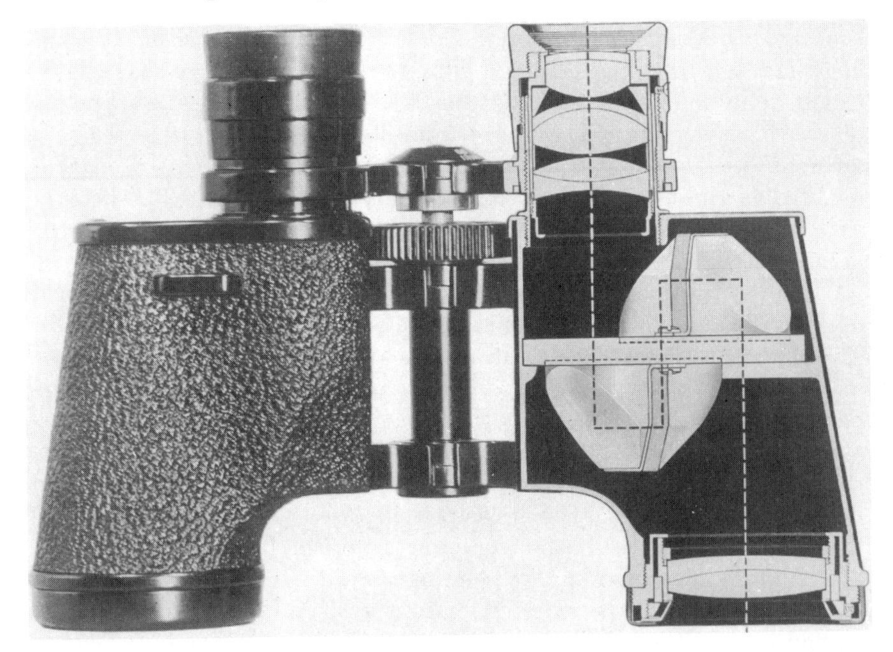

Fig. 6-20. The prism binocular.
(Courtesy of Bausch & Lomb Optical Co.)

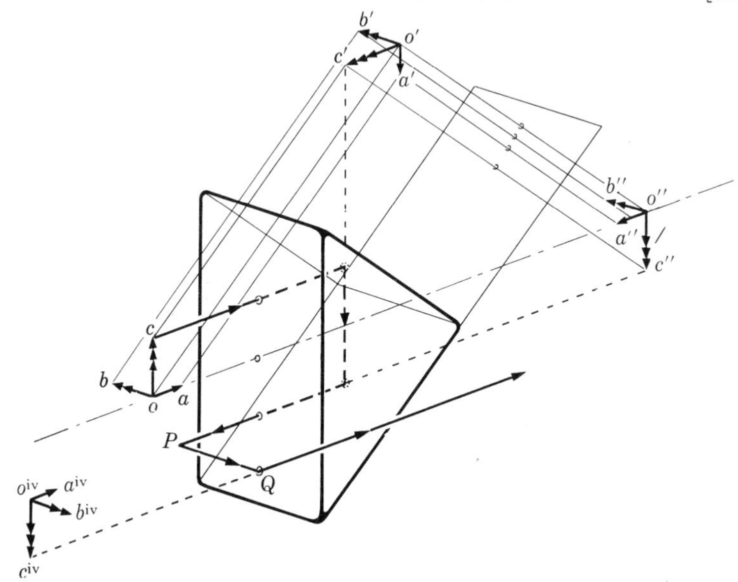

FIG. 6-21. Inversion of an image by the prism system of a prism binocular.

The long draw tube of the terrestrial telescope is avoided in the *prism binocular*, of which Fig. 6-20 is a cut-away view. A pair of 45°-45°-90° totally reflecting prisms, arranged as in Fig. 2-22, are inserted between objective and ocular. The image formed by the objective serves as a virtual object for the pair of prisms. The process by which the image is inverted as a result of four reflections is illustrated in Fig. 6-21, where for simplicity a real object *oabc* is shown, and where one of the prisms has been omitted from the diagram to avoid confusion.

The first (virtual) image formed by reflection at the upper inclined face of the prism is at $o'a'b'c'$. As explained in Sec. 2-6, this image is perverted. The plane of the lower inclined face of the prism is extended by light lines, and $o''a''b''c''$ is the virtual image of $o'a'b'c'$ formed by this face. Comparison of $o''a''b''c''$ with the object *oabc* shows that the second image is inverted but not perverted. The orientation of the image *in space* is not the same as that of an inverted image formed by a lens (see Fig. 4-14) but it must be remembered that in the present case the direction of travel of light has been reversed. An observer at the prism, facing the object *oabc*, sees *oa* pointing toward him, *ob* pointing toward his right, and *oc* pointing up. In order to see the reflected image $o''a''b''c''$ the observer must turn around so that the reflected light comes toward him. Then $o''a''$ points toward him, $o''b''$ toward his left, and $o''c''$ points down. Thus *ob* and *oc* have been reversed while *oa* has not.

The path of a single ray from the head of arrow *oc* has been traced through the system. This ray is initially horizontal, is reflected down at the first inclined prism face, then back horizontally. The second prism (not shown in the figure) is mounted with its hypotenuse against that of the first prism but with its triangular faces horizontal instead of vertical. The ray from the head of arrow *oc* enters the second prism through its hypotenuse, strikes one of the shorter faces at point P at an angle of 45°, is reflected across horizontally to point Q where it strikes the other face, and then emerges from the hypotenuse horizontally.

The image is again perverted twice by these two reflections. Image $o'''a'''b'''c'''$, formed by the reflection at P, is not shown. The final image $o^{iv}a^{iv}b^{iv}c^{iv}$ is shown in its correct orientation but, to save space, not in its actual position, which would be much farther to the left. It will be seen that as a consequence of the four reflections and the two reversals in direction, the final image is oriented in space exactly as is the inverted image formed by a lens in Fig. 4-14, and that also the light is traveling in its original direction. Hence if *oabc* represents the once inverted image formed by the objective, the image $o^{iv}a^{iv}b^{iv}c^{iv}$ is erect relative to the object being viewed. Finally, since the ocular produces an erect (virtual) image, the image seen by the eye is erect also.

The repeated reflections within the instrument give an optical path length considerably greater than the overall length of the instrument.

The *Galilean telescope* owes its name to Galileo, who constructed one of the first telescopes of this type in 1609. A diverging rather than a converging lens is used as an ocular. The optical system is illustrated in Fig. 6-22. Rays from a distant object (not shown) are made converging

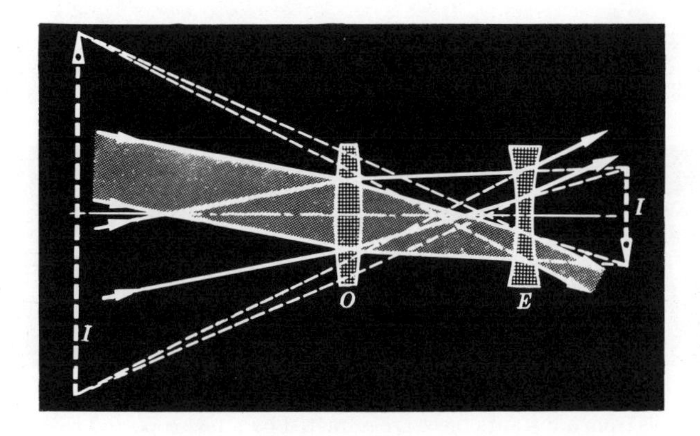

FIG. 6-22. The Galilean telescope.

by the objective O. The image I serves as a virtual object for the ocular E. The final image I' is virtual and erect, as indicated. The angular magnification of this telescope is also given by

$$\gamma = -\frac{f_1}{f_2},$$

but since f_2 is negative, γ is a positive quantity and the image is erect. The distance between objective and ocular is the difference between (the absolute values of) their focal lengths. Consequently this instrument may be made much more compact than the astronomical type. Its chief disadvantage is that it cannot cover as wide a field of view without the use of objectives of unduly large diameter. The "opera glass" is a Galilean telescope.

6-8 Normal magnification. Thus far nothing has been said regarding the diameters of the lenses in a telescope; the magnification involves only the ratio of focal lengths. To see how the diameter of the objective sets a limit to the useful magnification, let us consider the optical system of a refracting telescope from a somewhat different viewpoint.

The ocular of a telescope, as well as imaging the *image* formed by the objective, also forms a real, reduced image of the objective lens itself in

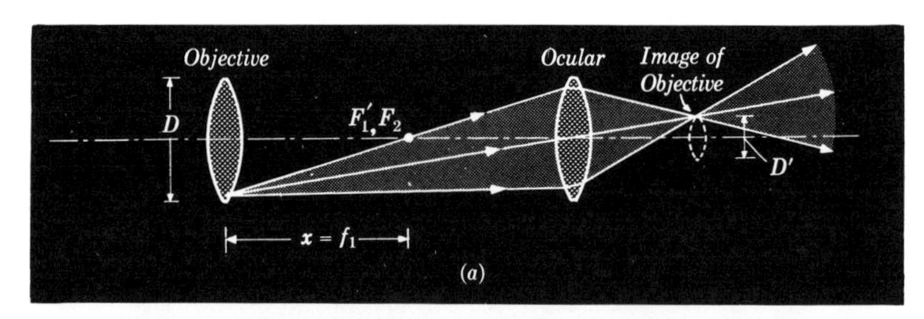

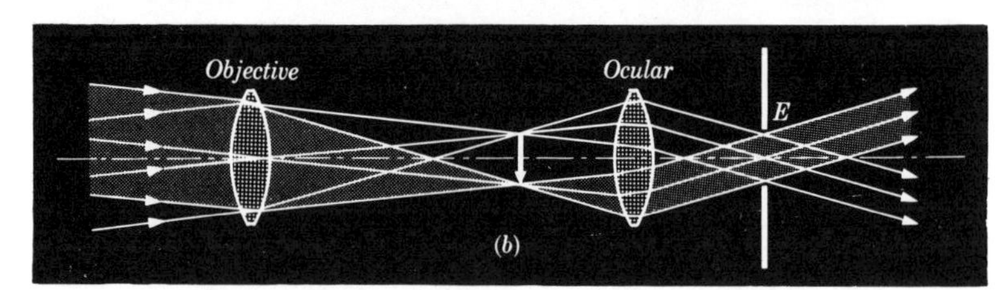

FIG. 6-23. (a) The ocular of a telescope forms a real, reduced image of the objective lens. (b) Rays from a distant object are refracted by a telescope and pass through the exit pupil E. The exit pupil lies at the same point, and has the same diameter, as the image of the objective lens.

the space beyond the ocular. All of the light that enters the objective, and is refracted by the ocular, must pass through this image of the objective, which is called the *exit pupil* of the telescope. The diameter of the transmitted beam is a minimum in the plane of the exit pupil. (See Fig. 6-23 (b).) If all of the transmitted light is to enter the pupil of the observer's eye, the diameter of the exit pupil of the telescope should be no larger than the pupillary diameter of the eye. In practice, the eye is usually placed at the exit pupil which is also called the *eye point* of the telescope.

Let us assume that the object being viewed and the virtual image formed by the ocular are both at infinity. If the objective lens is considered as the object for the ocular, the object distance x in Fig. 6-23 (a), measured from the first focal point F_2 of the ocular, is

$$x = f_1,$$

where f_1 is the focal length of the objective lens. Let D be the diameter of the objective and D' the diameter of its image. By definition, the lateral magnification of the image of the objective is

$$m = \frac{D'}{D},$$

and from Eq. (4-15),

$$m = -\frac{f_2}{x},$$

where f_2 is the focal length of the ocular. Then (disregarding algebraic signs),

$$\frac{D}{D'} = \frac{x}{f_2} = \frac{f_1}{f_2}.$$

But $\frac{f_1}{f_2}$ equals the angular magnification of the telescope, γ. Hence

$$\gamma = \frac{D}{D'}, \tag{6-4}$$

and the angular magnification equals the ratio of the diameter of the objective to the diameter of its image formed by the ocular, or the exit pupil.

Incidentally, Eq. (6-4) indicates a convenient method for measuring the angular magnification of a telescope. The instrument may be directed toward a bright sky and a screen moved along the axis until the minimum diameter of the transmitted beam is found. The magnification is then the ratio of the diameter of the objective to the minimum diameter of the transmitted beam.

The *normal magnification* of a telescope is defined as that at which the diameter of its exit pupil is just equal to the pupillary diameter of the eye, usually assumed to be 2 mm.

Example. The objective lens of a telescope is 20 mm in diameter and its focal length is 250 mm.

(a) What is the normal magnification of the telescope?

(b) What focal length ocular should be used?

(c) Find the position of the exit pupil.

(d) What would be the diameter of the exit pupil if an ocular were used which gave a magnification 50% in excess of normal?

(e) What would be the diameter of the exit pupil if the magnification were 50% of normal?

Assume all lenses to be thin.

(a) From Eq. (6-4), $\gamma = \dfrac{D}{D'} = \dfrac{20}{2} = 10\times.$

(b) $\gamma = \dfrac{f_1}{f_2}, \quad f_2 = \dfrac{f_1}{\gamma} = \dfrac{250}{10} = 25 \text{ mm}.$

(c) $x' = \dfrac{f_2^2}{x} = \dfrac{f_2^2}{f_1} = \dfrac{f_2}{\gamma} = \dfrac{25}{10} = 2.5 \text{ mm}.$

That is, the exit pupil is 2.5 mm to the right of the second focal point of the ocular, or 27.5 mm to the right of the ocular itself.

(d) If $\gamma = 15\times$, $D' = \dfrac{D}{\gamma} = \dfrac{20}{15} = 1.33 \text{ mm}.$

(e) If $\gamma = 5\times$, $D' = \dfrac{20}{5} = 4 \text{ mm}.$

The results of the preceding example will be used to illustrate an important point. Let us compare the quantities of light reaching the retinal image from an object on the telescope axis, when the eye is unaided and when the telescope is used. The diameter of the objective (20 mm) is 10 times that of the pupil of the eye and its area is 100 times as great. It therefore admits 100 times as much light. Suppose, first, that the normal magnification of $10\times$ is used. The diameter of the exit pupil then equals the diameter of the pupil of the eye, and all of the light admitted by the objective can enter the eye. The linear dimensions of the retinal image are increased by a factor of 10 and its area by a factor of 100. Hence 100 times the light is distributed over 100 times the area and the apparent brightness of the object viewed is the same with the telescope as with the unaided eye.

When a magnification of 15× is used, the area of the retinal image is increased by a factor of 225. The objective admits 100 times as much light as does the unaided eye and since the diameter of the exit pupil is only 1.33 mm, all of this light can enter the eye. However, since the light is distributed over an area 225 times as great, the apparent brightness is less than with the unaided eye.

When the magnification is 5×, the area of the retinal image is increased by a factor of 25. The diameter of the exit pupil is 4 mm (twice that of the pupil of the eye) and its area is 4 times that of the pupil. Hence only one-quarter of the light passing through the exit pupil can enter the eye. Therefore, although the objective admits 100 times as much light as the unaided eye, only one-quarter of this light reaches the retinal image, and 25 times the light is distributed over 25 times the area. The apparent brightness is the same as with normal magnification or as with the unaided eye.

This special case illustrates the general principle that *no optical instrument can increase the apparent brightness of an object.*[1]

6-9 The reflecting telescope. Most astronomical observations today are made not visually but photographically. By exposing a photographic plate for a long period of time, objects too faint ever to be seen can be recorded in a photograph. Hence the modern astronomical telescope is more accurately described as a "camera" than as a telescope. Although the term "astronomical telescope" is applied to the type of instrument described in Sec. 6-7, most telescopes in use today for astronomical purposes, particularly those of large size, make use of a concave mirror, rather than a lens, as the objective. There are several reasons for this, the chief one being the difficulty of manufacturing optical glass of high quality in large pieces. Since light does not travel *through* the glass of a reflector, the quality is not important. Furthermore, a refracting objective must be constructed of at least two lenses to correct for chromatic aberration, which requires the grinding and polishing of four surfaces, while for a mirror only a single surface need be finished. The mirror, of course, is inherently free from chromatic aberration. Offsetting these advantages are the facts that the mirror of the reflecting telescope is more susceptible to temperature changes, and the telescope does not cover as wide a field of view.

---

[1] This is true for an *extended* object, but if the object is so small or so distant (a star, for example) that the size of its image is determined by diffraction effects, the apparent brightness can be increased by a telescope. See Chap. 10 for a further discussion of diffraction.

FIG. 6-24. Testing the concave mirror of the Mt. Palomar telescope.

The largest *refracting* telescope in use today is that at the Yerkes Observatory, the objective of which is 40 inches in diameter. Fig. 6-24 is a photograph of the concave mirror of the largest reflecting telescope, located at Mt. Palomar in southern California. The diameter of this mirror is about 16 feet. The reflecting surface of the mirror is toward the reader. It has not yet been metallized, and through it can be seen the supporting ribs cast on its rear surface for mechanical rigidity.

The optical system of the reflecting telescope is essentially the same as that of the astronomical refracting telescope, the real image formed by the objective being examined by an ocular. However, since this image would be formed in the space in front of the mirror, where it would obviously be inconvenient to mount an ocular with which to observe it, various expedi-

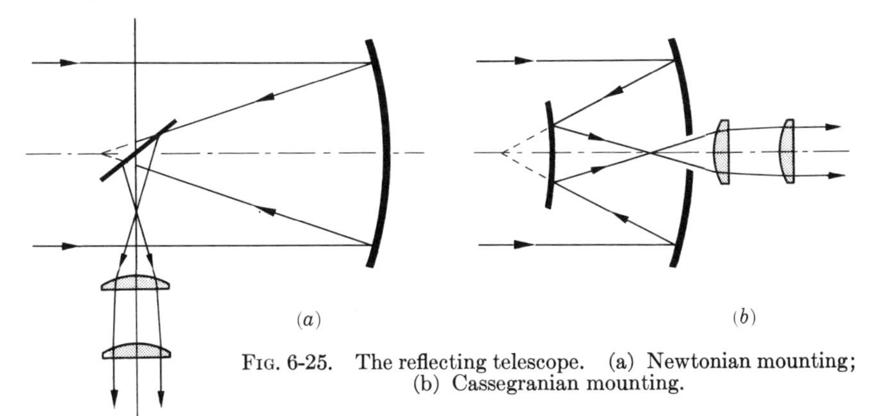

Fig. 6-25. The reflecting telescope. (a) Newtonian mounting; (b) Cassegranian mounting.

ents such as those illustrated in Fig. 6-25 are adopted. In the Newtonian telescope, Fig. 6-25 (a), a small plane mirror or totally reflecting prism is mounted on the axis of the telescope. The image may then be viewed in a direction at right angles to the axis. In the Cassegranian mounting, Fig. 6-25 (b), a small convex hyperboloidal mirror intercepts the rays proceeding toward the image, and reflects them back through an opening in the center of the objective.

6-10 The projection lantern. The optical system of the projection lantern, or the motion picture projector, is illustrated in Fig. 6-26. The arrow at the left represents the light source; for example, the filament of a projection lamp. For simplicity, the slide to be projected is represented as opaque except for a single transparent aperture.

Of course, light is emitted from all points of the source and in all directions. The diagram traces the course of three pencils of rays originating

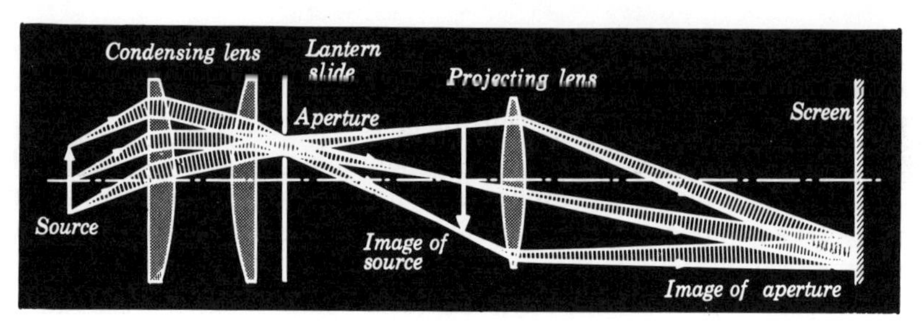

FIG. 6-26. The projection lantern.

at the ends and at the mid-point of the source. The function of the condensing lens is to deviate the light from the source inward, so that it can pass through the projecting lens. If the condensing lens were omitted, light passing through the outer portions of the slide would not strike the projecting lens and only a small portion of the slide near its center would be imaged on the screen.

A study of the figure will show that (a) for the three selected points of the source, only those rays within the shaded pencils can pass through the aperture, all others striking the condensing lens being intercepted by the opaque portions of the slide, and (b) similar pencils of rays could be drawn from every other point of the source.

Each of these pencils converges, after passing through the aperture, to form an image of its point of origin just to the left of the projecting lens. In practice, this image would be formed *at* the projecting lens, but for clarity in the diagram the image and the lens have been displaced slightly. The focal length of the condensing lens should be such that the image of the source just fills the projecting lens. If the image of the source is larger than the projecting lens, some of the light passing through the slide is wasted. If it is smaller, the area of the projecting lens is not being fully utilized. Thus in the diagram, the outer portions of the projecting lens serve no useful purpose. We could have saved money by buying a projecting lens of smaller diameter or, with the one illustrated available, we could increase the brightness of the image on the screen by using a condenser of shorter focal length to produce a larger image of the source.

Three rays tangent to the upper edge of the aperture have been emphasized in the figure. These rays originate at *different* points of the source. Hence, although they intersect at the edge of the aperture, this point of intersection does not constitute an image of any point of the source. But these three rays diverge from a common point of the lantern slide, and therefore this point of the slide is imaged as shown on the screen. Similarly, rays tangent to any point of the edge of the aperture are imaged at a conjugate point on the screen. Thus if the aperture is circular, a circular spot of light appears on the screen.

Notice that light from *all* points of the source illuminates every point of the image of the aperture, and would do the same were the aperture at any other point of the slide.

The preceding discussion has explained the conditions that determine the *focal length* of the condensing lens and the *diameter* of the projecting lens (the image of the source formed by the condensing lens should just fill the projecting lens). The *diameter* of the condensing lens must evidently be at least as great as the diagonal of the largest slide to be projected, while the *focal length* of the projecting lens is determined by the magnification desired between the slide and its image, and the distance of the lantern from the screen.

6-11 The camera. The essential elements of a camera are a lens, a light tight box, and a sensitized plate or film for receiving the image. In contrast with a telescope objective, for which the field of view is limited to a range of about 6°, a photographic objective is called upon to cover a field of 50° or even more. Furthermore, the relative aperture of the lens must be large in order that it may collect sufficient light to permit short exposures. The combination of wide field and large aperture makes the problem of correcting a photographic lens a difficult one. Nevertheless, even the simplest lenses, such as those used in Brownie cameras, are corrected for chromatic aberration and curvature of field. The "rapid rectilinear" lens, which consists of two achromats with a stop between them, is sensibly free from coma and distortion, as well as both lateral and axial chromatism. Spherical aberration is corrected and the lens has a flattened field. The "anastigmatic" lenses are corrected for spherical and chromatic aberration, coma, astigmatism, curvature of field, and distortion. Most modern high speed, short focal length lenses are modifications of the Zeiss "Tessar" lens, illustrated in Fig. 6-27.

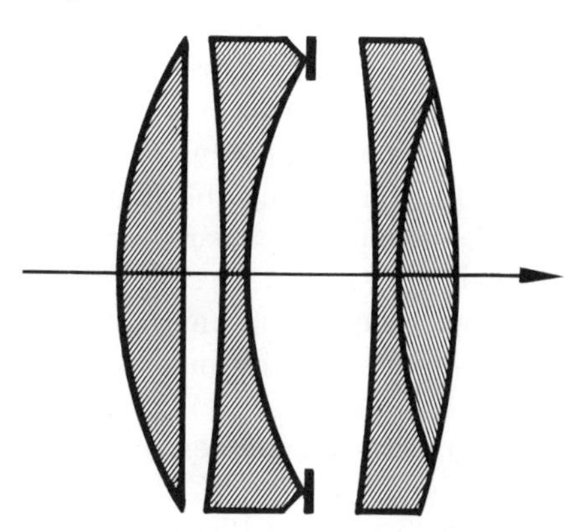

FIG. 6-27. Zeiss "Tessar" lens.

The light-gathering power of a photographic objective is usually stated in terms of its $f/$-number, which is determined by the focal length of the lens and by its diameter or the diameter of the aperture which effectively determines the lens area. Thus the notation $f/4.5$ means that the focal length of the lens is 4.5 times its effective diameter. The smaller the $f/$-number, the larger the lens diameter for a given focal length and the greater the light-gathering power or "speed" of the lens. Extremely fast lenses may have $f/$-numbers as small as $f/1.9$ or $f/1.5$. That of the Brownie lens is approximately $f/11$. The required time of exposure increases with the square of the $f/$-number. For example, the ratio of the exposure time at $f/6.3$ to the exposure time at $f/4.5$ is $(6.3)^2/(4.5)^2 \approx 2$.

For a given position of the photographic film, only those objects lying in the plane conjugate to that of the film are sharply focussed upon it, objects at a greater or less distance appearing somewhat blurred. However, because of lens aberrations, a point of a given object will be imaged as a small circle, called the *circle of confusion*, even with the best focussing. The circles of confusion of points at other distances will be larger. If extremely sharp definition of the image is not essential, there is evidently a certain range of object distances, called the *depth of field*, such that all objects within this range are simultaneously "in focus" on the plate. That is, the circles of confusion of points within this range are not so large that the image is unsatisfactory. The so-called *fixed-focus* camera is one with a large depth of field, so that all objects beyond a certain distance are simultaneously in satisfactory focus.

6-12 Stops. A *stop*, in an optical instrument, is a diaphragm, lens mounting, or some similar obstacle which limits the rays that can be transmitted through the instrument. There are two properties of stops that are of importance: first, their limitation of the quantity of light transmitted by the instrument and second, the limits they impose on the field of view. The stop which controls the quantity of light transmitted is called the *aperture stop*. The stop which controls the field of view is the *field stop*. The adjustable diaphragm of a camera lens is an aperture stop. The field stop, in a camera, is the photographic film itself. In a telescope, the field stop is usually the rim of one of the lenses or a diaphragm in the telescope tube. Stops may also be used to eliminate unwanted rays which would produce aberrations.

6-13 The rangefinder. A surveyor at point A in Fig. 6-28 can determine the distance AC to point C, on the opposite bank of a river, by laying off a measured base line AB at right angles to AC, then setting up his transit at point B and measuring the angle θ. The distance AC is computed from the relation $AC = AB \tan \theta$. It is this problem in trigonometry which is solved by the rangefinder.

The essential elements of a coincidence type rangefinder are shown in Fig. 6-29. Two penta prisms (see Fig. 2-23) are mounted at the ends of a horizontal tube. The distance between the prisms, AB, is the base line of the instrument; it may vary from a few feet to 100 feet. Each prism deflects rays from the object through exactly 90°. Paths of two rays proceeding from a distant object are shown by full lines. C and C' are two identical telescope objectives. The upper half of the image formed by one objective and the lower half of the image formed by the other are reflected by the two right-angle prisms P and P' into the ocular E. An observer thus sees a divided image, the upper half being displaced laterally relative to the lower.

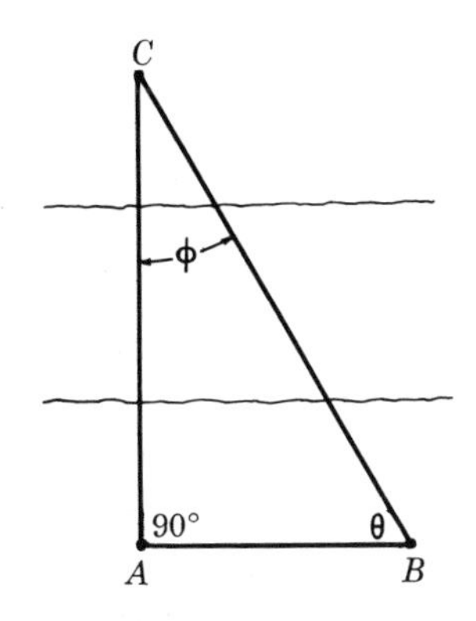

FIG. 6-28.

The dotted line indicates the path that would be followed by a ray striking the right prism, if the object were infinitely distant. The angle ϕ is evidently equal to the angle ϕ in Fig. 6-28, or to $90° - \theta$. The lateral displacement of one half of the image relative to the other depends on the angle ϕ or θ. A calibrated scale might be provided in the ocular from which the angle could be determined by observation of this displacement, and the object distance could be computed from this angle and the base line AB. Instead, however, the two halves of the image are brought into coincidence by displacing the prism D to the left or right, and the distance to the object is read directly from the calibrated scale S.

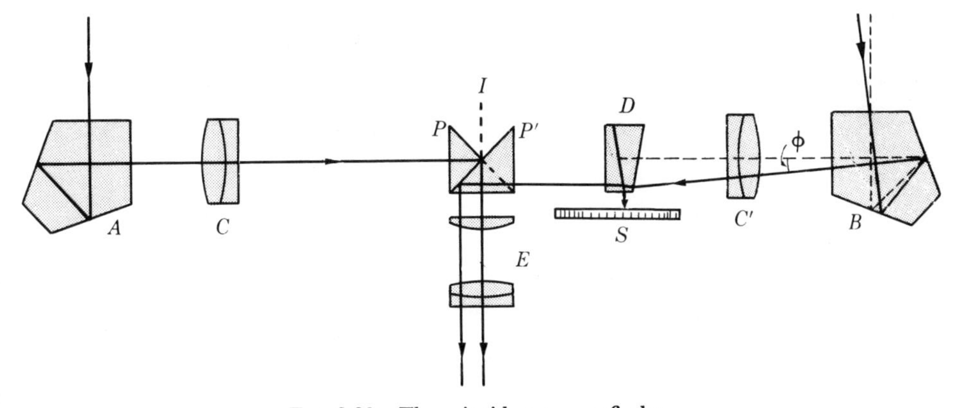

FIG. 6-29. The coincidence rangefinder.

6-14 The ultramicroscope. The ultramicroscope, as far as objective and ocular are concerned, is identical with the ordinary compound microscope; in fact, the term applies not so much to a different instrument as to a different method of using the compound microscope. The wave nature of light sets an ultimate limit to the smallest size particle of which a microscope can *form an image.* The ultramicroscope makes it possible to *observe the position* of particles of submicroscopic dimensions, although no image of the particles is formed and their shape, size, and so on, cannot be ascertained.

An ultramicroscope requires the use of some sort of *dark-field* illuminator, of which one type is illustrated in Fig. 6-30. Light proceeding upward from below the microscope stage is totally internally reflected from the surface of a *paraboloid condenser* as shown, so that an intense illuminating beam is traveling transversely across the field of view of the microscope. If a cell containing some submicroscopic particles in suspension is placed in the field of view, these particles "scatter" the light from the condenser in all directions, and appear as bright points of light against a dark background. The phenomenon of scattering, which we shall discuss later in Chap. 7, is also responsible for the luminous appearance of a beam of light passing through a cloud of smoke, or through a milky solution. Particles as small as 4×10^{-7} cm in diameter, or 1/100 of the wave length of visible light, may be "seen" by this means.

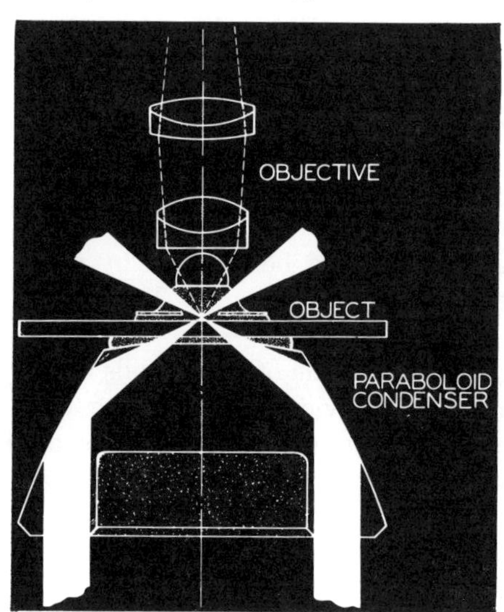

FIG. 6-30. Illuminating system of the ultramicroscope.

6-15 The prism spectrometer. Optical instruments may be grouped into two general classes, *image-forming* instruments and *analyzing* instruments. Instruments of the former class, such as those we have been discussing, serve to form the image of some given object; those of the latter class are used to determine the composition, intensity, or state of polarization of a beam of light. We shall consider next the prism spectrometer and some of its modifications. These are analyzing instruments, used primarily to discover what wave lengths are present in a given light beam.

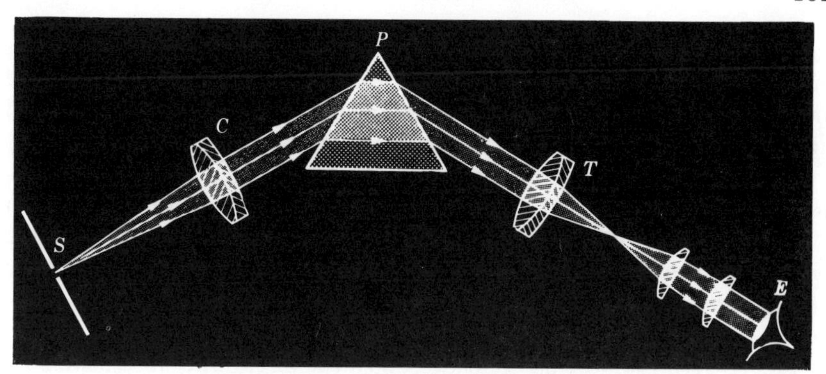

FIG. 6-31. Principle of the prism spectrometer.

The essential elements of a prism spectrometer are illustrated in Fig. 6-31. A narrow slit S, illuminated by the light to be analyzed, is located at the first focal point of the achromatic lens C, called the *collimator*. The parallel beam of light emerging from the collimator falls on the prism P, is deviated, and the emergent light is examined by the telescope T. Since, as we have seen, the index of refraction of optical materials varies with wave length, the various wave lengths present in the light are deviated by different angles.

The observer E sees a number of images of the slit side by side, each formed by light of a particular wave length. If the source emits light of all wave lengths, the images form a continuous succession, called a *continuous spectrum*. If the source emits only a few definite wave lengths, the images of the slit are separated from one another and appear as a series of bright lines, each the color of the light producing it. (See Fig. 11-1.) (This is the origin of the term "line-spectrum." If a small circular hole were used instead of a slit, the images would be a series of colored circles.)

In using a spectrometer, the prism is first removed, and the telescope T, whose ocular is provided with cross hairs, is rotated about a vertical axis until the image of the slit, formed by the collimator and telescope, coincides with the cross hairs. The optical axis of the telescope is now in the direction of the beam emerging from the collimator, and the angular position of the telescope is read from a graduated circle. The prism is next placed in position on a stand which may be rotated about a vertical axis coincident with the axis of rotation of the telescope. Care must be taken that the prism faces are parallel to this axis. The telescope is now swung until the deviated beam comes into the field of view and the prism is rotated slowly one way or another, meanwhile watching the deviated beam through the telescope, until a position is found such that the deviation of the line whose wave length is desired is a minimum. The cross hairs are then set on this line, and the position of the telescope is again read from the graduated circle. The difference between this reading and the first gives the angle of minimum deviation for this particular wave length.

To measure the angle of the prism, A, it is rotated into the position shown in Fig. 6-32, so that a portion of the light beam from the collimator is reflected from each face. The direction of each reflected beam is found by observing with the telescope the reflected images of the slit. The difference between the telescope settings is equal to twice the angle of the prism.

Eq. (2-11) may now be applied. Since the angles A and δ_m have been measured, the index of the prism may be computed for light of the wave length used. Conversely, from known values of the index, one can compute an unknown wave length.

A *spectroscope* is an instrument for visual observation of a spectrum,

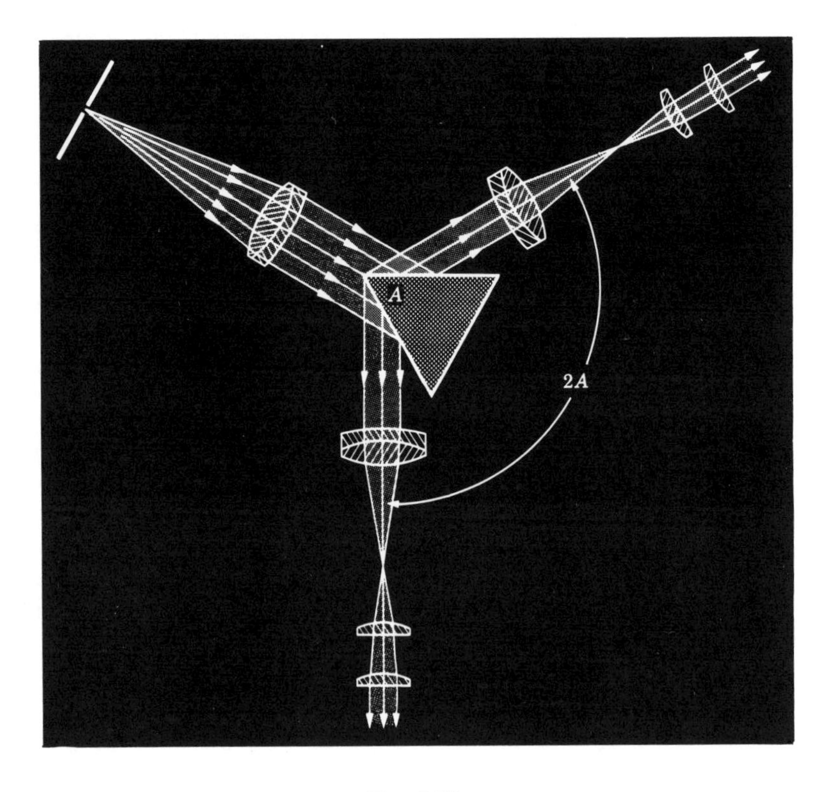

FIG. 6-32.

rather than for measurement of wave lengths or angles. A simple and compact form, to be held in the hand and pointed toward the source, consists of a slit, collimating lens, and a direct-vision prism, usually of the Amici type illustrated in Fig. 6-33. The outer prisms are of crown glass while the central one is flint. Paths of a few rays are traced through the system. The letters C, D, F, refer to the corresponding Fraunhofer lines.

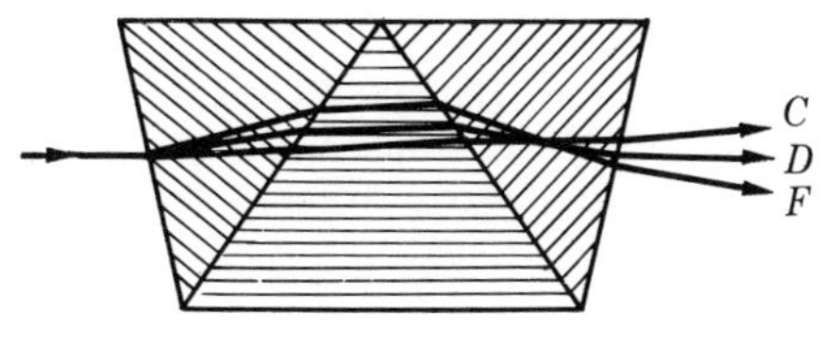

FIG. 6-33. Direct-vision spectroscope.

A *spectrograph* is essentially a spectrometer, but is constructed specifically for photographing a spectrum. The spectrometer in Fig. 6-31 could be used as a spectrograph by introducing a photographic plate in the plane of the image formed by the telescope objective, instead of examining this image with the ocular of the telescope. Practically all measurements of unknown wave lengths are made today by photographing, on the same plate with the unknown spectrum, a line spectrum of some substance, the wave lengths of which are already known. The known spectrum then provides a scale by means of which the unknown wave lengths may be measured.

Problems—Chapter 6

(1) Assume the eye to be filled with a homogeneous medium of index 1.330. (a) Compute the radius of curvature of the cornea, if the second focal point of the eye is at the retina, 25 mm behind the vertex of the cornea. (b) What is the height of the retinal image of an arrow 10 cm long, at a distance of 2 m from the eye?

(2) (a) What spectacles are required for reading purposes by a person whose near point is at 200 cm? (b) The far point of a myopic eye is at 30 cm. What spectacles are required for distant vision?

(3) (a) Where is the near point of an eye for which a spectacle lens of power +2 diopters is prescribed? (b) Where is the far point of an eye for which a spectacle lens of power −0.5 diopter is prescribed for distant vision?

(4) A person with normal vision has a range of accommodation from 25 cm to infinity. Over what range would he be able to see objects distinctly when wearing the spectacles of a friend whose correction is (a) + 4 diopters? (b) −3 diopters?

(5) The far point of a nearsighted individual is one meter in front of his eyes. (a) What power should his spectacles have, in order that he may see distinctly an object at infinity? (b) If, with these glasses, his near point is at 25 cm, where would it be without them?

(6) The focal length of a simple magnifier is 12.5 cm. (a) What is the angular magnification if the final image is formed at infinity? (b) What is the angular magnification if the final image is 25 cm in front of the eye?

(7) A compound microscope has an objective that produces a lateral magnification of 10×. What focal length ocular will produce an overall magnification of 100×?

(8) A certain microscope is provided with objectives of focal lengths 16 mm, 4 mm, and 1.9 mm, and with oculars of angular magnifications 5× and 10×. Each objective forms an image 160 mm beyond its second focal point. What is (a) the largest, (b) the least overall magnification obtainable?

(9) The focal length of the ocular of a certain microscope is 2.5 cm. The focal length of the objective is 16 mm. The distance between objective and ocular is 22.1 cm. The final image formed by the ocular is at infinity. Treat all lenses as thin. (a) What should be the distance from the objective to the object viewed? (b) What is the lateral magnification produced by the objective? (c) What is the overall magnification of the microscope?

(10) The image formed by a certain microscope objective lies 180 mm from its second focal point. The focal length of the objective is 9 mm. An ocular of focal length 50 mm is used to project a real image on a screen 1 m from the second focal point of the ocular. What is the lateral magnification between object and final image? Assume all lenses to be thin.

(11) A crude telescope is constructed of two spectacle lenses of focal lengths 100 cm and 20 cm respectively. (a) Find its angular magnification. (b) Find the height of the image formed by the objective of a building 200 ft high and distant one mile.

(12) The focal length of the objective of a telescope of the astronomical type is 30 cm. A normal magnification of 5× is desired. Treat all lenses as thin. (a) What should be the diameter of the objective? (b) What should be the focal length of the ocular? (c) How far behind the ocular is the exit pupil, when the telescope is focused on the moon? (d) If first sharply focused on the moon, how far and in what direction must the ocular be moved to focus on an object 20 m away?

(13) A large astronomical telescope has an objective of diameter 100 cm and focal length 2000 cm. The instrument is used to observe a distant object and the image formed by the ocular is at infinity. Assume the ocular to be a thin lens. (a) Draw a diagram of the telescope, showing the objective, its second focal point, the ocular and both its focal points, and the exit pupil, in the correct relative positions. (b) What is the angular magnification if the focal length of the ocular is 2 cm? (c) What is the normal magnification?

(14) A demonstration model of an astronomical telescope is to be made from spectacle lenses. A +1.5 diopter lens is selected for use as the objective, and the ocular is composed of a +5 diopter lens and a +10 diopter lens separated by 15 cm. The lenses are arranged in the order listed above, and the system is so adjusted that distant objects are imaged at infinity. (a) How should the lenses be spaced? (b) Where is the exit pupil? (c) What is the angular magnification of the telescope?

(15) A telescope is sighted on the image of a scale formed by reflection in a plane galvanometer mirror. Both the telescope objective and the scale are 1 m from the mirror, one slightly above the other. (a) What is the lateral magnification of the image of the scale formed by the objective, if the objective has a focal length of 50 cm? (b) What should be the angular magnification of the ocular if the scale is to be read as easily through the telescope as it would be by the unaided eye with the scale at the normal near point (25 cm)?

(16) A Huygens ocular is constructed of two thin lenses of focal lengths 10 cm and 5 cm, respectively, separated by 7.5 cm. The ocular is used in an astronomical telescope whose objective is 30 cm to the left of the front lens of the ocular. (a) Find the position of the exit pupil. (b) Find the focal length of the ocular.

(17) A Huygens ocular is constructed of two thin lenses of focal lengths 40 mm and 20 mm respectively, separated by 30 mm. (a) Locate the focal points of the ocular and find its focal length. (b) When the ocular is used in conjunction with an objective lens to form an astronomical-type telescope, where must the objective be located if the exit pupil is 20 mm from the second lens of the ocular?

(18) (a) How large is the exit pupil of a 7-power telescope having an objective lens 50 mm in diameter? (b) If the objective lens is 6 inches in focal length, what must be the focal length of the ocular? (c) What focal length ocular should be used with this objective lens in order to produce a telescope having normal magnification for daytime use?

(19) A Galilean telescope is to be constructed, using the same objective as in Prob. 11. What type and focal length lens should be used as an ocular, if the telescopes are to have the same magnification? Compare the lengths of the telescopes.

(20) The dimensions of a lantern slide are 3 inches × 4 inches. It is desired to project an image of the slide, enlarged to 6 ft × 8 ft, on a screen 30 ft from the projection lens. Treat all lenses as thin. (a) What should be the focal length of the projection lens? (b) Where should the slide be placed? (c) The diameter of the projection lens is 2 inches and the filament of the projection lamp may be considered a circle, 0.5 inch in diameter. What should be the focal length and diameter of the condensing lens?

(21) A projector for 1 inch × 1 inch lantern slides uses a projection lamp whose filament may be considered to be a circle 0.25 inch in diameter. The condensing lens, of focal length 1 inch, forms an image of the filament which just fills a projection lens 1.25 inches in diameter. Consider all lenses as thin. (a) Find the focal length of the projection lens. (b) How large an image of the slide will the projector produce on a screen located 25.5 ft from the slide? (c) What would be the effect on the image on the screen if half of the projection lens were covered by an opaque card?

(22) The focal length of a certain f/2.8 camera lens is 8 cm. (a) What is the diameter of the lens? (b) If the correct exposure of a certain scene is 1/200 sec at f/2.8, what would be the correct exposure at f/6.3?

(23) (a) An object at a distance x from a rangefinder of base length L (Fig. 6-34) is displaced a distance Δx. What is the corresponding change $\Delta\phi$ in the angle ϕ? Approximate finite changes by differentials, and recall that in practice ϕ is nearly 90°. (b) If the limit of accuracy with which the telescopes can be set on an object is 10 sec of arc, what is the corresponding error in the determination of the distance x if $L = 10$ ft and x is approximately 1000 yd? 10,000 yd? (c) What would these errors be if L were 100 ft?

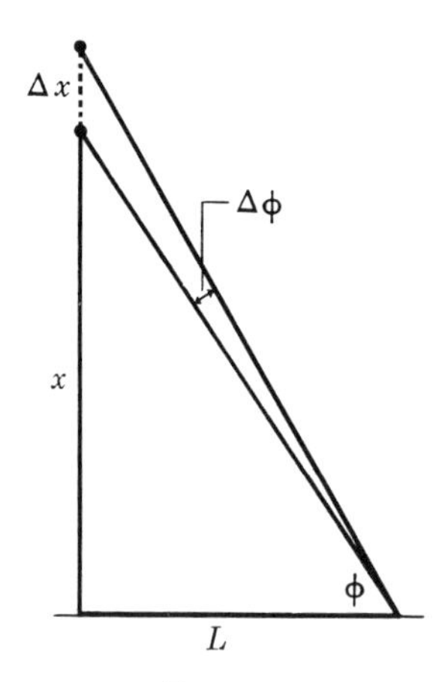

FIG. 6-34.

CHAPTER 7

POLARIZATION

7-1 Introduction. The preceding chapters have dealt, for the most part, with problems that can be treated by the methods of geometrical optics. The next four chapters will be devoted to *physical optics,* a part of the subject in which the wave nature of light plays a fundamental part. We shall begin with a discussion of *polarization.*

Fig. 1-2 is a diagram of the simplest type of electromagnetic wave, one in which the wave fronts are planes, while at any fixed point through which the wave train passes the electric and magnetic intensities oscillate along straight lines at right angles to one another, and to the direction of propagation. Fig. 7-1 shows a number of conventional methods of representing such a wave train advancing along the X-axis; in (a) by sine waves in the X-Z and X-Y planes, in (b) by E and H vectors along the Z- and Y-axes, and in (c) by an E vector alone, the H vector being

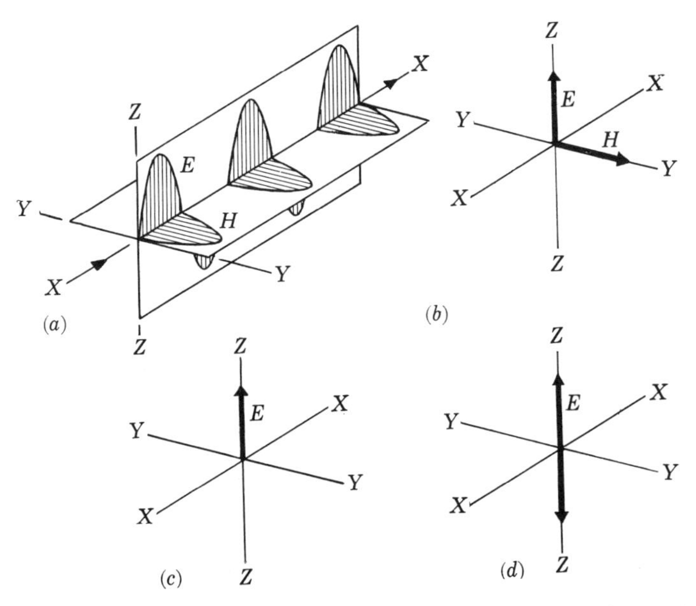

FIG. 7-1. Conventional method of representing a linearly polarized wave train.

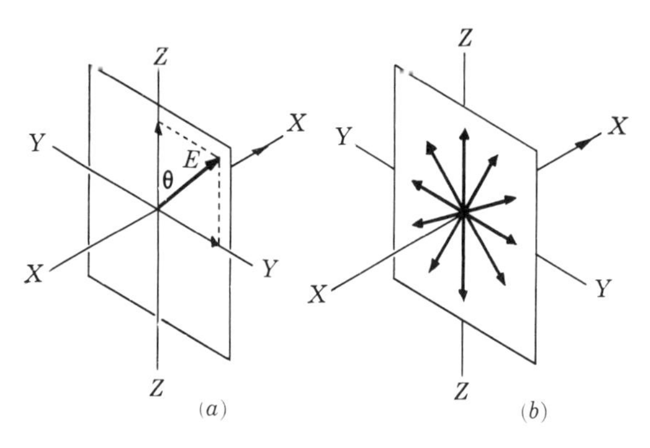

FIG. 7-2.

left to the reader's imagination. Another convention we shall often use is shown in (d), where the double arrow indicates the two equal and oppositely directed maximum values of E.

The wave in Fig. 7-1 is said to be *linearly polarized*, meaning that at any fixed point the tip of the E (or H) vector oscillates along a line. The wave is also called plane polarized, from the fact that each of the sine waves in Fig. 7-1 (a) lies in a plane. Other more complicated types of polarization will be discussed later in the chapter.

Both theory and experiment show that the waves from a simple radio or radar antenna (an oscillating dipole) are linearly polarized, with the electric vector in a plane containing the dipole. The waves from a source of light presumably originate in the molecules of the source and, if these radiate in the same way as does a dipole of finite size, the waves from any one molecule would also be linearly polarized. It is impossible, of course, to isolate a single molecule and study the wave train it emits. Every light source contains a tremendous number of molecules oriented in all possible directions. A beam of light waves from such a source, traveling along the X-axis toward the right as in Fig. 7-1, would then be expected to consist of a mixture of waves, in some of which E is oriented as in Fig. 7-1 (c), while in others the E vector (which must always lie in the Y-Z plane if the wave travels along the X-axis) makes some angle θ with the Z-axis, as in Fig. 7-2 (a). Since all values of the azimuth angle θ are equally probable, we would expect to find E vectors arranged symmetrically about the direction of propagation, as in Fig. 7-2 (b). Such is, in fact, found to be the case, and light of this sort is called *unpolarized* or *natural* light.

Although we cannot control the azimuths of the molecular dipoles in a light source, there are a number of methods for sorting out of a beam of natural light those wave trains or their components in which the electric vector oscillates in a particular azimuth. We consider first the processes of reflection and refraction at a boundary surface between two substances of different index. The problem was worked out in part in Sec. 2-2, where we derived the relation between the *angles* of incidence, reflection, and refraction. We now investigate the relative *quantities of light* in the incident, reflected, and refracted wave trains.

7-2 Reflection and refraction of linearly polarized light. In Fig. 7-3, a train of linearly polarized electromagnetic waves is incident on a plane surface bounding two substances of indices n and n', where we have assumed that $n' > n$. The wave is represented by a single ray, as in Fig. 7-1 (b). The boundary surface lies in the Y-Z plane. The origin of coordinates, O, is at the point of incidence and the normal to the surface at the point of incidence is the X-axis. The incident, reflected, and refracted rays all lie in the plane of incidence, the X-Y plane.

In order to predict the fraction of any type of incident light that is reflected at a surface, it is sufficient to analyze two cases only: when the

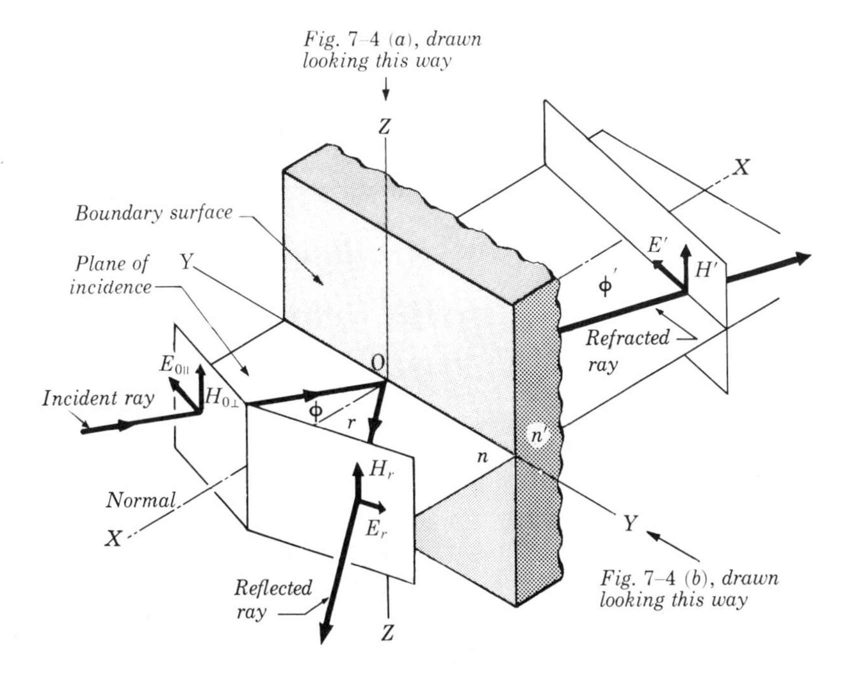

Fig. 7-3. Reflection and refraction of a linearly polarized electromagnetic wave, with the electric vector parallel to the plane of incidence. $(n' > n)$

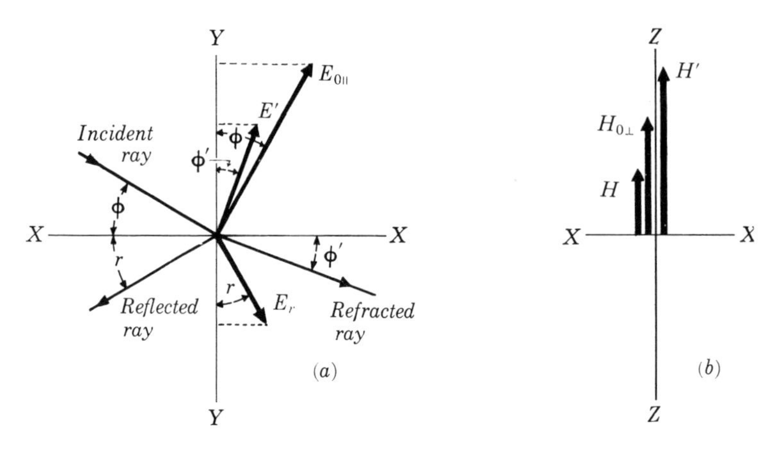

FIG. 7-4. The tangential components of the electric and magnetic intensities are the same on opposite sides of a boundary surface.

incident light is linearly polarized with the electric vector (1) perpendicular to, and (2) parallel to the plane of incidence. In Fig. 7-3, the electric vector is parallel to the plane of incidence and the magnetic vector is perpendicular to this plane. The electric and magnetic vectors in each wave train are in planes perpendicular to the respective rays.

The plane surfaces shown by light lines are reference planes only, to help in visualizing the three-dimensional nature of the problem.

The electric and magnetic fields in the three wave trains must satisfy the usual *boundary conditions* at the boundary surface, namely, the tangential components of E and H must have the same value at points on opposite sides of the surface. Of course, the electric and magnetic fields in all three sets of waves are varying both in space and in time. Let us represent the maximum values of the electric and magnetic intensities in the incident, reflected, and refracted waves respectively, by $E_{0\|}$ and $H_{0\perp}$, E_r and H_r, and E' and H'. Consider an instant when the fields in the three waves at the point of incidence O have their maximum values. Fig. 7-4 (a) shows the vectors $E_{0\|}$, E_r, and E' at this instant. The X-Y plane of Fig. 7-3 is the plane of the diagram in Fig. 7-4, and the Z-axis and all three magnetic vectors point toward the reader. Although the three vectors are drawn from a common origin, the vectors $E_{0\|}$ and E_r refer to fields just to the left of the Y-Z plane in the first medium, while the vector E' represents the field just to the right of the Y-Z plane in the second medium. The resultant tangential component of the electric intensity in the first medium is

$$E_{0\|} \cos \phi - E_r \cos r.$$

The tangential electric intensity in the second medium is

$$E' \cos \phi'.$$

Since the tangential components are equal,

$$E_{0\parallel} \cos \phi - E_r \cos r = E' \cos \phi'. \tag{7-1}$$

Consider now the boundary conditions imposed on H. In Fig. 7-4 (b), the X-Z plane in Fig. 7-3 is in the plane of the diagram and the Y-axis points toward the reader. The electric vectors (not shown) lie along the X-axis in this view. The magnetic vectors are slightly displaced from one another for clarity. Since they are already tangent to the boundary surface it is unnecessary to take components and

$$H_{0\perp} + H_r = H'. \tag{7-2}$$

From the general laws of electromagnetic waves, if the permittivity and permeability of the first and second media are represented respectively by ϵ and μ, ϵ' and μ',

$$\sqrt{\epsilon} E_{0\parallel} = \sqrt{\mu} H_{0\perp}, \quad \sqrt{\epsilon} E_r = \sqrt{\mu} H_r, \quad \sqrt{\epsilon'} E' = \sqrt{\mu'} H'. \tag{7-3}$$

But for all transparent dielectrics,

$$\mu = \mu' = \mu_0. \tag{7-4}$$

Also, the index of refraction n of a substance is related to its permittivity ϵ by the equation

$$\sqrt{\epsilon} = \sqrt{K_e \epsilon_0} = n\sqrt{\epsilon_0}, \quad \sqrt{\epsilon'} = \sqrt{K_e' \epsilon_0} = n'\sqrt{\epsilon_0}. \tag{7-5}$$

When the preceding three equations are combined with Eq. (7-2), we get

$$nE_{0\parallel} + nE_r = n'E'. \tag{7-6}$$

Eq. (7-6) expresses, in terms of electric intensities, the boundary conditions imposed on the magnetic intensities. Making use of the laws of reflection and refraction,

$$r = \phi,$$

$$n \sin \phi = n' \sin \phi', \tag{7-7}$$

we can now combine Eqs. (7-1) and (7-6) and express E_r and E' in terms of $E_{0\parallel}$ as follows,

$$E_r = \frac{\tan(\phi - \phi')}{\tan(\phi + \phi')} E_{0\|}, \tag{7-8}$$

$$E' = \frac{2 \sin \phi' \cos \phi.}{\sin(\phi + \phi') \cos(\phi - \phi')} E_{0\|}. \tag{7-9}$$

The relations between E and H given in Eq. (7-3) enable us to write also

$$H_r = \frac{\tan(\phi - \phi')}{\tan(\phi + \phi')} H_{0\perp}, \tag{7-10}$$

$$H' = \frac{\sin 2\phi}{\sin(\phi + \phi') \cos(\phi - \phi')} H_{0\perp}. \tag{7-11}$$

The four preceding equations enable one to compute, for any angle of incidence, the electric and magnetic intensities in the reflected and refracted waves when the incident wave is linearly polarized with the electric vector parallel to the plane of incidence. When the incident wave is linearly polarized with the electric vector perpendicular to the plane of incidence, reasoning exactly like that above can be applied. In this case the electric and magnetic fields are as shown in Fig. 7-5. We shall omit details and give the results only. These are,

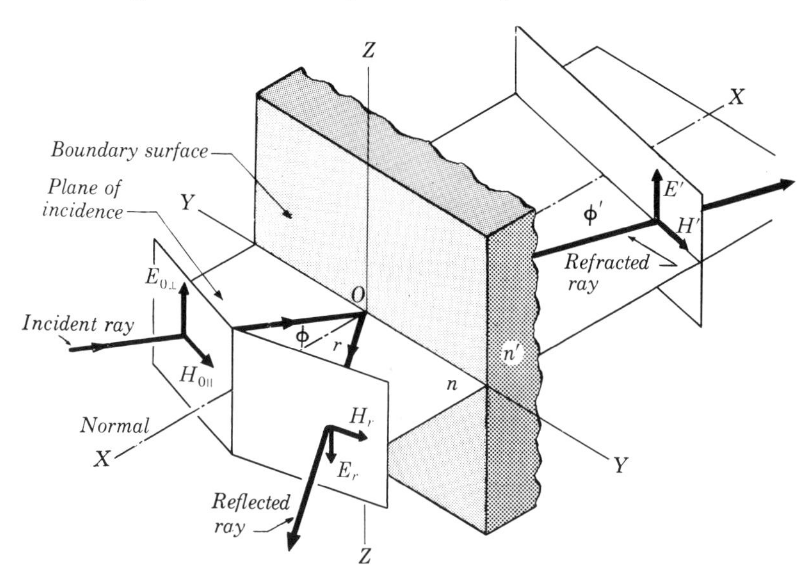

FIG. 7-5. Reflection and refraction of a linearly polarized electromagnetic wave, with the electric vector perpendicular to the plane of incidence. $(n' > n)$

$$E_r = -\frac{\sin(\phi - \phi')}{\sin(\phi + \phi')} E_{0\perp}, \tag{7-12}$$

$$E' = \frac{2 \sin \phi' \cos \phi}{\sin(\phi + \phi')} E_{0\perp}, \tag{7-13}$$

$$H_r = \frac{\sin(\phi - \phi')}{\sin(\phi + \phi')} H_{0\parallel}, \tag{7-14}$$

$$H' = \frac{\sin 2\phi}{\sin(\phi + \phi')} H_{0\parallel}. \tag{7-15}$$

Comparison of these equations with Eqs. (7-8) to (7-11) shows that the two cases are not identical. For example, it will be seen from Eqs. (7-8) and (7-10) that there is no reflected wave ($E_r = H_r = 0$) when $(\phi + \phi') = 90°$ (since $\tan 90° = \infty$) provided the incident wave is linearly polarized with the electric vector parallel to the plane of incidence, while there is always partial reflection of a wave polarized with the electric vector perpendicular to the plane of incidence.

Figs. 7-3 and 7-5 are drawn for the case in which $n' > n$ and, as a consequence, $\phi' < \phi$. If the reverse is true, as when light traveling in glass is reflected at a glass-air interface, then $\phi' > \phi$ and $\sin(\phi - \phi')$ and $\tan(\phi - \phi')$ are both negative. This means that the electric and magnetic vectors in the reflected waves are reversed, relative to their directions in the figures above. This point has a bearing on certain interference effects that will be described in Chap. 8.

7-3 Polarization by reflection. The equations derived in the preceding section give the *amplitudes* of the electric and magnetic fields in a train of reflected and refracted electromagnetic waves, in terms of the amplitude of the incident wave and the angles of incidence and refraction. The *energy* transported by such a wave, per unit time and per unit area (i.e., the Poynting vector) is proportional to the *square of the amplitude* of the wave. If the wave is a light wave, the quantity of light reflected or refracted (we shall give a more precise definition of "quantity of light" in Chap. 13) depends on the energy carried by the wave. Let I_0 and I_r represent the quantities of light in the incident and reflected beams. Then for linearly polarized light (or, more briefly, linear light) in which the electric vector is parallel to the plane of incidence as in Fig. 7-3, the ratio of reflected to incident light, or the *reflectance* r of the surface, is

$$r = \frac{I_r}{I_{0\parallel}} = \frac{E_r^2}{E_{0\parallel}^2} = \frac{\tan^2(\phi - \phi')}{\tan^2(\phi + \phi')}. \tag{7-16}$$

If the electric vector is perpendicular to the plane of incidence as in Fig. 7-5,

$$r = \frac{I_r}{I_{0\perp}} = \frac{E_r^2}{E_{0\perp}^2} = \frac{\sin^2(\phi - \phi')}{\sin^2(\phi + \phi')}. \tag{7-17}$$

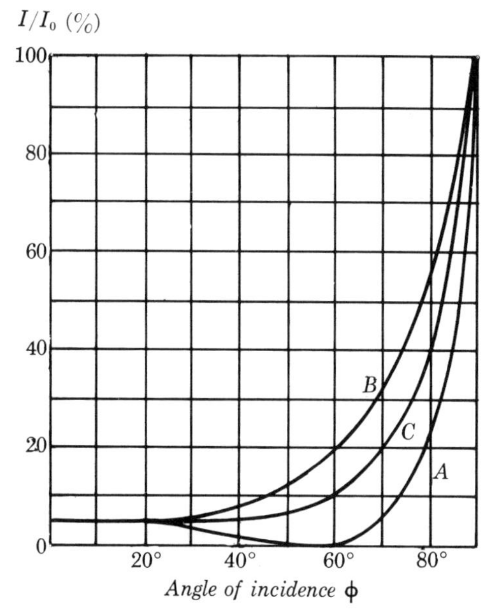

FIG. 7-6. Variation of reflectance with angle of incidence. Curve A: plane polarized light, E vector in incident light parallel to plane of incidence. Curve B: plane polarized light, E-vector perpendicular to plane of incidence. Curve C: natural or unpolarized light.

These equations were first derived by Fresnel and are called *Fresnel's formulae.* Curve A in Fig. 7-6 is a graph of Eq. (7-16), for the special case in which $n = 1$ and $n' = 1.523$. Curve B is a graph of Eq. (7-17).

If the incident light is linearly polarized with the electric vector making some angle other than 0° or 90° with the plane of incidence, it may be resolved into components parallel and perpendicular to the plane of incidence, as in Fig. 7-2 (a), and Eqs. (7-16) and (7-17) may be applied to find the fraction of each component reflected.

Natural or unpolarized light is a mixture of waves polarized in all possible azimuths. Each wave may be resolved into components as in Fig. 7-2 (a). By symmetry, the amount of each component will be equal. The reflectance of a surface for natural light is then given by one-half of Eq. (7-16) plus one-half of Eq. (7-17), or

$$r = \frac{I_r}{I_0} = \frac{1}{2}\frac{\tan^2(\phi - \phi')}{\tan^2(\phi + \phi')} + \frac{1}{2}\frac{\sin^2(\phi - \phi')}{\sin^2(\phi + \phi')}. \tag{7-18}$$

The corresponding curve is labelled C in Fig. 7-6.

For the special case in which the angle of incidence is zero and the first medium is air, Eq. (7-18), with the help of Snell's law, may be reduced to

$$r = \frac{I_r}{I_0} = \frac{(n-1)^2}{(n+1)^2}. \tag{7-19}$$

It follows that if $n = 1.5$, a typical value for glass, about 4% of the incident light is reflected at normal incidence.

The phenomenon of reflection may be used to separate or filter out linear light from a beam of natural light. It will be seen from Eq. (7-16), or from curve A in Fig. 7-6, that at one particular angle of incidence (about 57° for the material to which Fig. 7-6 applies) no light is reflected in which the electric vector is parallel to the plane of incidence. This angle, called the *polarizing angle* ϕ_P, is that at which $(\phi + \phi') = 90°$, since $\tan 90° = \infty$. The fact that $(\phi + \phi') = 90°$ means that at the polarizing angle the reflected and refracted rays are at right angles to one another.

The polarizing angle may be expressed in terms of indices of refraction as follows:

$$n \sin \phi_P = n' \sin \phi',$$

$$\phi_P + \phi' = 90°,$$

$$n \sin \phi_P = n' \sin (90° - \phi_P) = n' \cos \phi_P,$$

$$\frac{\sin \phi_P}{\cos \phi_P} = \frac{n'}{n},$$

$$\tan \phi_P = \frac{n'}{n}. \tag{7-20}$$

This relation is known as *Brewster's law*, after Sir David Brewster, who discovered it experimentally in 1812. If $n = 1$ and $n' = 1.5$, the polarizing angle is 56°.

In Fig. 7-7, a beam of natural light is incident at the polarizing angle on a reflecting surface. Each linearly polarized wave in the incident beam can be resolved into components with the electric vector parallel or perpendicular to the plane of incidence. At the polarizing angle, none of the components parallel to the plane of incidence is reflected. That is, all of the light which *is* reflected is linearly polarized with its electric vector perpendicular to the plane of incidence. From curves B and C in Fig. 7-6, it will be seen that at the polarizing angle about 15% of the "perpendicular" component, or about 7.5% of the incident light, is reflected. The transmitted light is not completely linearly polarized, but is a mixture of the "parallel" component, all of which is transmitted (since none is reflected) and the remaining 85% of the "perpendicular" component.

By allowing a beam of natural light to fall at the polarizing angle on a pile of glass plates, instead of on a single surface, more and more of the perpendicular component may be filtered out by repeated reflections. This is suggested in Fig. 7-7, where for simplicity only two plates are

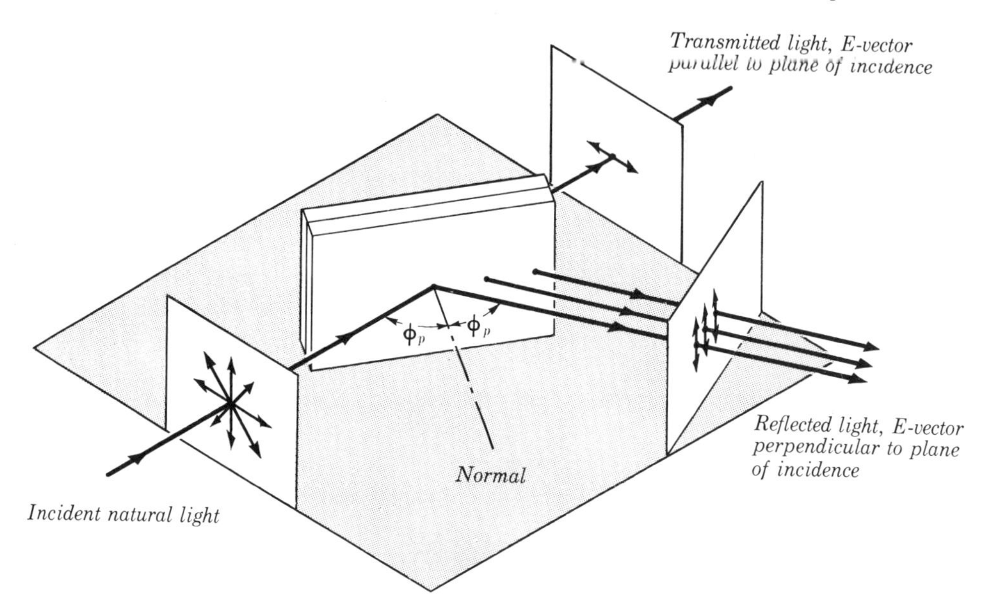

FIG. 7-7. Separation of natural light into two beams of linear light by reflection from a pile of plates.

shown. Eventually an almost complete separation of the light into two linearly polarized beams results. The transmitted beam consists almost entirely of waves in which the E-vector is parallel to the plane of incidence, while the reflected beam contains only waves in which the E-vector is perpendicular to this plane. Any device such as this, which transmits only waves in which the E-vector is in a single direction, is termed a *polarizer*.

Before leaving this question, it should be noted that the curves in Fig. 2-4 are graphs of Eq. (7-18), curve A applying when $n' > n$, and curve B when $n' < n$. Curve A in this figure is the same as curve C in Fig. 7-6. At the critical angle for total internal reflection, $\phi' = 90°$. But $\tan^2 (\phi - 90°) = \tan^2 (\phi + 90°)$, and $\sin^2 (\phi - 90°) = \sin^2 (\phi + 90°)$. Hence Eq. (7-18) reduces to

$$r = \frac{1}{2} + \frac{1}{2} = 1,$$

or, in other words, all of the incident light is reflected.

7-4 Double refraction. The progress of a wave train through a homogeneous isotropic medium such as glass may be determined by Huygens construction. The secondary wavelets in such a medium are spherical surfaces. There exist, however, many transparent crystalline substances which, while homogeneous, are optically *anisotropic*. That is, their optical properties are different in different directions. Crystals having this property are said to be *doubly refracting*, or to exhibit *birefringence*. *Two* sets of Huygens wavelets propagate from every wave front in such a crystal. In some crystals there is one particular direction, called the *optic axis*, in which the velocities of the two wavelets are equal. In other crystals the velocities are the same in two different directions. Crystals of the former type are called *uniaxial*, those of the latter, *biaxial*. Since all of the doubly refracting crystals used in optical instruments (chiefly quartz and calcite) are uniaxial, we shall consider only this type. One of the wavelets in a uniaxial crystal is spherical, while the other is an ellipsoid of revolution. The two wavelets are tangent in the direction of the optic axis.

In Fig. 7-8 (a), the origin of coordinates represents a point in a uniaxial crystal from which two Huygens wavelets are diverging. The Z-axis is in the direction of the optic axis. (The optic axis is a *direction* in the crystal, not just one line. Any other line parallel to the Z-axis is also an optic axis.) The diagram shows the intersections of the two wave surfaces with three mutually perpendicular planes. The intersections of all three planes with the spherical surface are circles with center at the origin. The Z-axis is the axis of revolution of the ellipsoidal surface. The intersection of this surface with any plane containing the Z-axis, such as the X-Z and Y-Z planes, is an ellipse whose minor axis is equal to the radius of the spherical surface. The intersection of the ellipsoidal surface with the X-Y plane is a circle of radius equal to the major axis of the ellipsoid.

The velocities of the wavelets, in any direction, are proportional to the lengths of the radius vectors drawn from the origin in that direction to the respective surfaces. The velocities are equal in the direction of the Z-axis or optic axis. The velocity of the ellipsoidal wavelet is the same in all directions in the X-Y plane. Fig. 7-8 is drawn for a crystal in which the velocity of the ellipsoidal wavelet is greater than that of the spherical wavelet, except in the direction of the optic axis. In some uniaxial crystals the velocity of the ellipsoidal wavelet is less than that of the spherical, again except along the optic axis. If the ellipsoids lie outside the spheres, as in Fig. 7-8, the crystal is said to have a negative birefringence. If the reverse is true, the birefringence is positive.

Parts (b), (c), and (d), of Fig. 7-8 are sections of the wave surfaces looking along the X-, Y-, and Z-axes respectively.

In Fig. 7-9 there is shown a cube of crystalline material having the properties represented by the wave surfaces in Fig. 7-8. The reference axes are oriented in the same way in both figures, so that the optic axis of

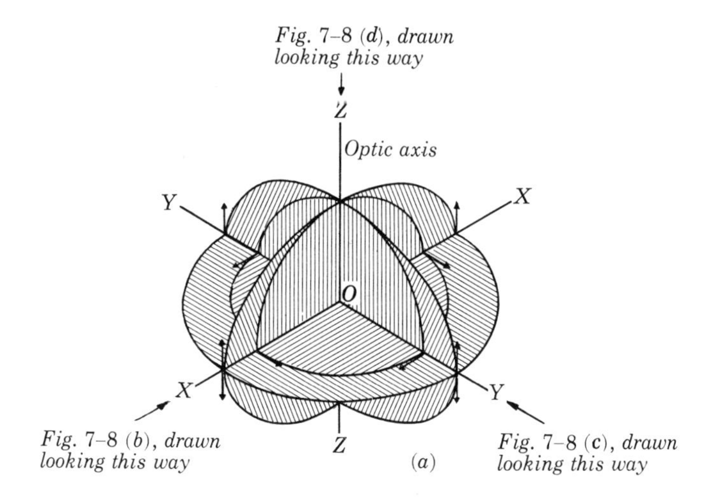

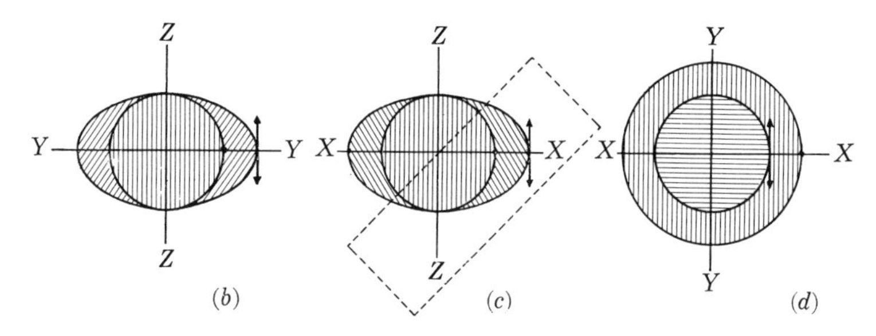

FIG. 7-8. Sections of Huygens' wavelets in a uniaxial crystal. The small arrows represent the direction of the electric vector.

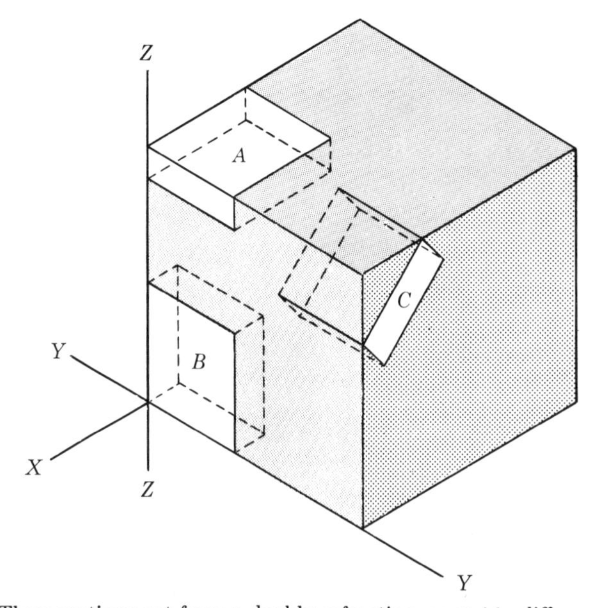

FIG. 7-9. Three sections cut from a doubly refracting crystal in different orientations.

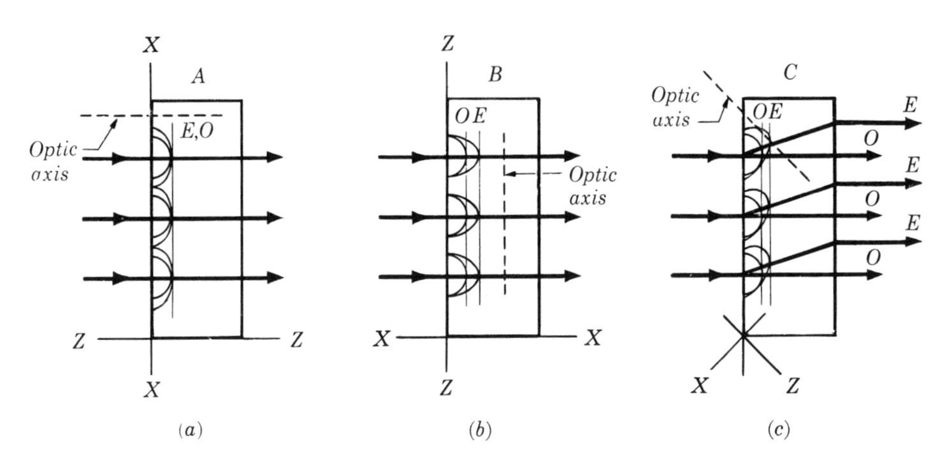

FIG. 7-10. Huygens' wavelets in a doubly refracting crystal when the optic axis is (a) perpendicular to the left face, (b) parallel to the left face, (c) at some arbitrary angle with the left face.

the crystal in Fig. 7-9 is parallel to the Z-axis. Suppose that we cut three sections from the crystal, as indicated by the letters A, B, and C, and orient them as shown in parts (a), (b), and (c) of Fig. 7-10. The optic axis is then perpendicular to the left face of section A, is parallel to the left face of section B, and makes an intermediate angle with the left face of section C. Let a beam of parallel light be incident normally on the left face of each section. The diagram shows how the Huygens wavelets are used to determine the progress of the wave fronts through the three sections.

The wavelets in section A are like those above (or below) the X-Y plane in Fig. 7-8, namely, a sphere and an ellipsoid, tangent in the direction of the optic axis. The envelopes of both sets of wavelets coincide, and a single wave front travels through the section with a velocity equal to that of the spherical wavelets.

The wavelets in section B are like those at the right of the Y-Z plane in Fig. 7-8. The envelopes of both sets are planes, but the wave front corresponding to the ellipsoidal wavelets travels with a greater velocity than that corresponding to the spherical wavelets.

In section C, where the optic axis makes an angle other than 0° or 90° with the surface, the envelope of the spherical wavelets is like that in the other sections. The envelope of the ellipsoidal wavelets, although it is a plane at right angles to the direction of the incident rays, is not tangent to the ellipsoids in the direction of the incident rays. Hence this wave front travels to the right and upward as it moves through the crystal, as indi-

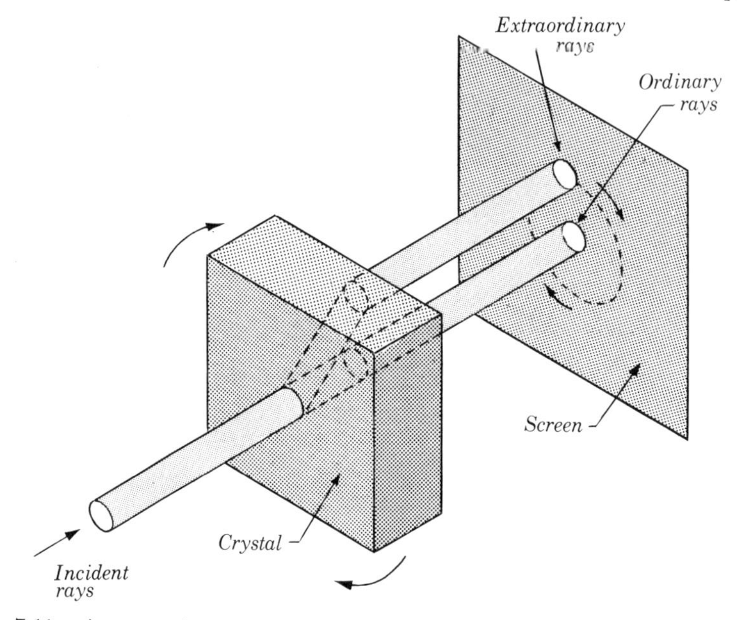

FIG. 7-11. A narrow beam of natural light can be split into two beams by a doubly refracting crystal.

cated by the rays sloping upward and to the right. In this case the rays are *not* normal to the wave front. A ray incident normally on the crystal is broken up into two rays in traversing the crystal. The rays corresponding to wave fronts tangent to the spherical wavelets are undeviated and are called *ordinary* rays. Rays corresponding to wave fronts tangent to the ellipsoidal wavelets are deviated even though the incident rays are normal to the surface, and are called *extraordinary* rays. If a narrow beam of parallel rays is incident on the crystal as in Fig. 7-11, two spots of light appear on a screen at the right. When the crystal is rotated about an axis parallel to the incident beam, the spot produced by the ordinary rays remains fixed, while the other revolves around it in a circle.

The index of refraction of a material is defined as the ratio of the velocity of light in empty space to the velocity in the material. Evidently a doubly refracting material has one index for the ordinary and another for the extraordinary ray. The latter index, however, does not have a unique value, since the velocity of the extraordinary ray is different in different directions. It is customary to specify the index for a direction at right angles to the optic axis (i.e., the X-Y plane in Fig. 7-8) in which the velocity is a maximum if the birefringence is negative and a minimum if the birefringence is positive. Some values of n_O and n_E, the indices for the ordinary and extraordinary rays, are listed in Table 7-1.

It will be seen that calcite has a negative and quartz a positive birefringence.

TABLE 7-1.

Indices of refraction of doubly refracting crystals (For light of wave length 589 mμ)

Material	n_O	n_E
Calcite	1.6583	1.4864
Quartz	1.544	1.553
Tourmaline	1.64	1.62
Ice	1.306	1.307

7-5 Polarization by double refraction. Experiment shows that the ordinary and extraordinary waves in a doubly refracting crystal are linearly polarized in mutually perpendicular directions, with the electric vector in the ordinary wave at right angles to the optic axis. Consequently, if some means can be found to separate one wave from the other, a doubly refracting crystal may be used as a polarizer. There are a number of ways in which this separation can be accomplished.

One method of separating the two components is by means of a *Nicol prism* or one of its modifications. The Nicol prism is a crystal of Iceland spar or calcite ($CaCO_3$), whose natural shape is shown by the full lines in Fig. 7-12 (a). To make a Nicol prism, the end faces of the crystal are cut

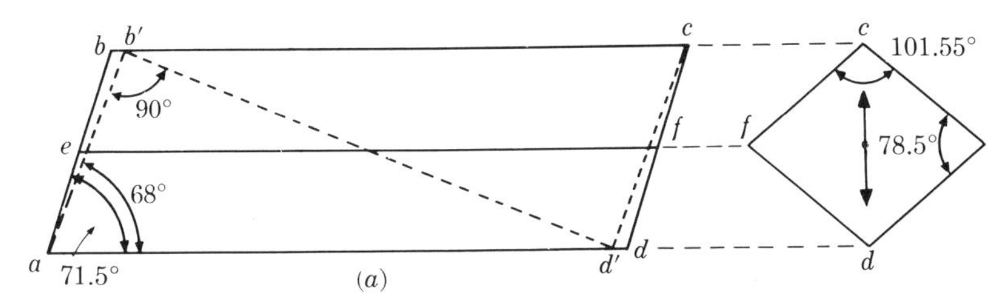

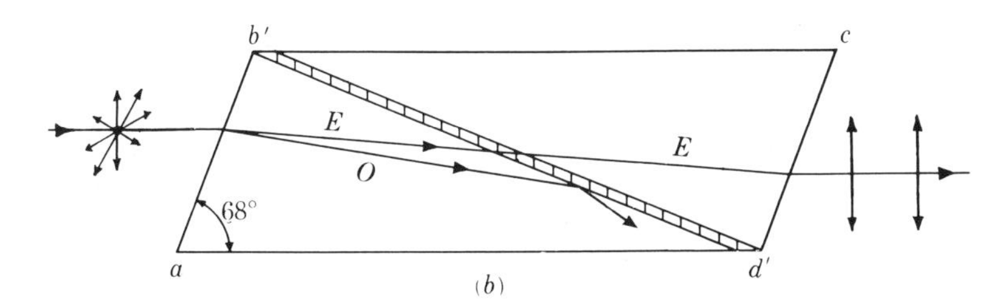

FIG. 7-12. (a) Natural crystal of Iceland spar. (b) A Nicol prism.

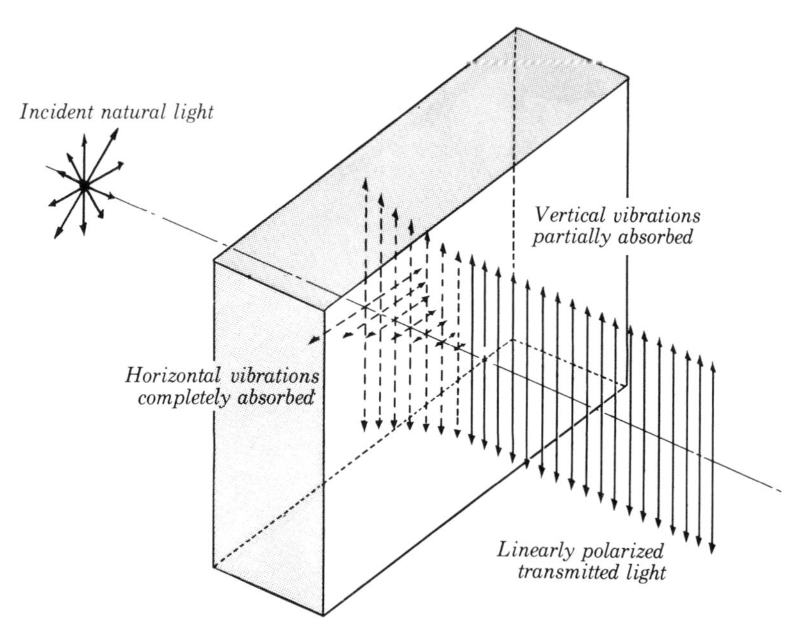

FIG. 7-13. Linearly polarized light transmitted by a dichroic crystal.

at a more obtuse angle, as shown by the dotted lines. The crystal is then cut along the shorter diagonal $b'd'$ and cemented together again with Canada balsam. The index of Canada balsam has such a value that the ordinary ray is totally reflected, while the extraordinary ray is transmitted as in Fig. 7-12 (b).

Certain doubly refracting crystals exhibit *dichroism,* that is, one of the polarized components is absorbed much more strongly than the other. Hence, if the crystal is cut of the proper thickness, one of the components is practically extinguished by absorption, while the other is transmitted in appreciable amount, as indicated in Fig. 7-13. Tourmaline is one example of such a dichroic crystal.

The angular and linear apertures of Nicol prisms are limited by the optical constants of calcite, and the scarcity of large pieces of optical quality. In 1934, Land developed a new type of dichroic polarizer, known as polaroid, which could be manufactured in thin sheets of large area. One type is made by preparing a suspension of dichroic crystal needles of herapathite (iodoquinine sulfate) in a volatile viscous medium, and then subjecting the suspension to a uniform flow process. This results in the orientation of the crystals, which are of microscopic size, parallel to the streamlines in the flow process. The solvent is then allowed to evaporate. A more recent type is manufactured by subjecting a sheet of polyvinyl

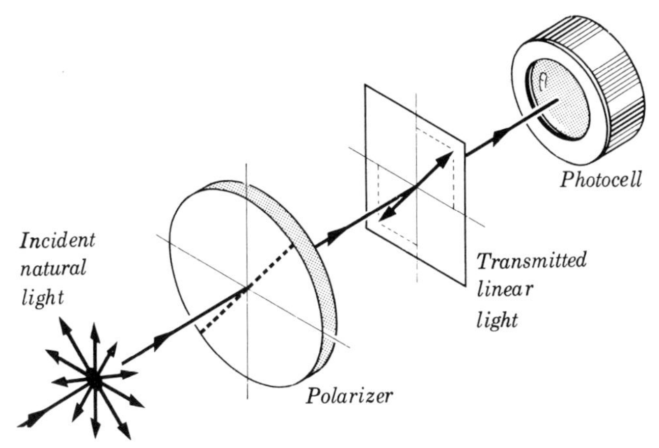

FIG. 7-14. The intensity of the transmitted linear light is the same at all azimuths of the polarizer.

alcohol, a rubberlike material, to a large tensile strain. This orients the molecules parallel to the direction of the strain and renders the material doubly refracting. When stained with iodine, the material becomes dichroic.

7-6 Percentage polarization. Malus' law. When light is incident on a polarizer as in Fig. 7-14, linear light only is transmitted. The polarizer may be a pile of plates, a Nicol prism, or a sheet of polaroid. It is represented as a polaroid disk in Fig. 7-14. The dotted line across the polarizer indicates the direction of the electric vector in the transmitted light, that is, it corresponds to the vertical direction in Fig. 7-13. The transmitted light falls on a photocell, and the current in a microammeter connected to the cell is proportional to the quantity of light incident on it.

If the incident light is unpolarized, then as the polarizer is rotated about the incident ray as an axis, the reading of the microammeter remains constant. The polarizer transmits the components of the incident waves in which the E-vector is parallel to the transmission direction of the polarizer, and by symmetry the components are equal for all azimuths.

If there is any variation in the meter reading as the polarizer is rotated, the incident light is not natural light and is said to be *partially* polarized. (But just what kind of light it is cannot be determined from this experiment alone.) Suppose the meter reading does vary. Let I_{max} and I_{min} represent the maximum and minimum values of the quantity of light incident on the photocell, or the maximum and minimum meter readings, since the two are proportional. The *percentage polarization* of the incident light is defined as

$$\text{Percent polarization} = \frac{I_{max} - I_{min}}{I_{max} + I_{min}} \times 100. \tag{7-21}$$

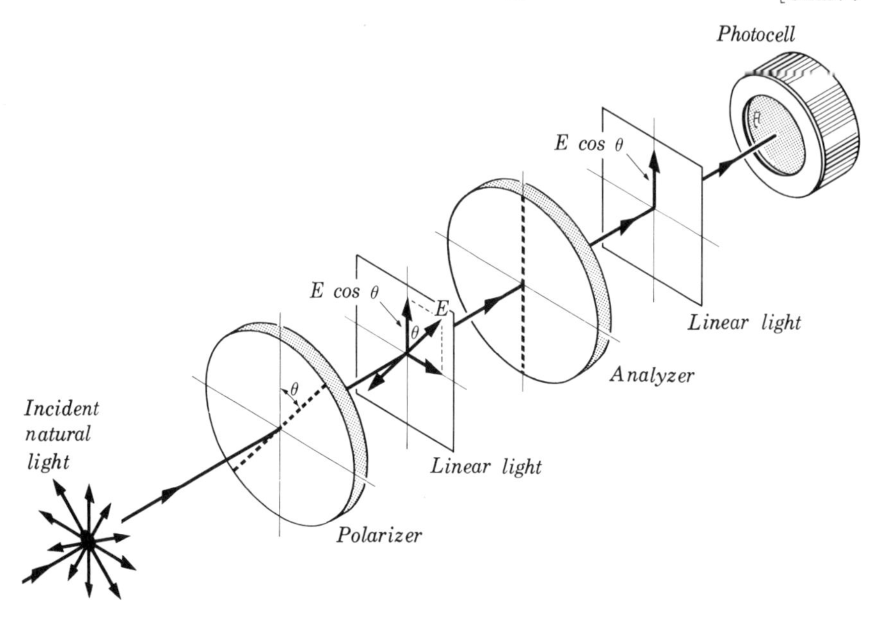

FIG. 7-15. The analyzer transmits only that component of the linear light parallel to its transmission direction.

Suppose now that a second polaroid is inserted in the light between polarizer and photocell, as in Fig. 7-15. Let the transmission direction of the second polaroid, or *analyzer*, be vertical, and let that of the polarizer make an angle θ with the vertical. The linear light transmitted by the polarizer may be resolved into two components as shown, one parallel and the other perpendicular to the transmission direction of the analyzer. Evidently only the parallel component, of amplitude $E \cos \theta$, will be transmitted by the analyzer. The transmitted light is a maximum when $\theta = 0$, and is zero when $\theta = 90°$, or when polarizer and analyzer are *crossed*. At intermediate angles, since the quantity of light is proportional to the square of the electric intensity, we have

$$I = I_{max} \cos^2\theta, \tag{7-22}$$

where I_{max} is the maximum amount of light transmitted and I is the amount transmitted at the angle θ. This relation, which was discovered experimentally by Etienne Louis Malus in 1809, is called *Malus' law.*

Evidently, the significance of the angle θ, in general, is the angle between the transmission directions of polarizer and analyzer. If either the analyzer or the polarizer is rotated, the amplitude of the transmitted beam varies in the same way with the angle between them.

Polaroid is now widely used in "sun glasses" where, from the standpoint of its polarizing properties, it plays the role of the analyzer in Fig. 7-15. We have seen in Sec. 7-3 that when unpolarized light is reflected, there is a preferential reflection for light polarized perpendicular to the plane of incidence. When sunlight is reflected from a horizontal surface, the plane of incidence is vertical. Hence in the reflected light there is a preponderance of light polarized in the horizontal direction, the proportion being greater the nearer the angle of incidence is to the polarizing angle. The transmission direction of the polaroid in the sun glasses is vertical, so none of the horizontally polarized light is transmitted.

Apart from this polarizing feature, these glasses serve the same purpose as any dark glasses absorbing 50% of the incident light, since even in an unpolarized beam, half the light can be considered as polarized horizontally and half vertically. Only the vertically polarized light is transmitted.

7-7 Retardation plates. Circular and elliptical light. A section cut from a doubly refracting crystal like that lettered B in Fig. 7-9 is called a *retardation plate.* Fig. 7-10 (b) shows how two waves travel through the plate with different velocities when light is incident normally on the face of the section. The action of a retardation plate is illustrated in more detail in Fig. 7-16.

The polarizer in part (b) of Fig. 7-16 transmits a beam of linear light, the E-vector in which is shown making an angle of 45° with the vertical. The beam can be resolved into components, in one of which the E-vector is horizontal, while in the other it is vertical. These components are shown in part (c) of Fig. 7-16 by the convention of Fig. 7-1 (c) and also in part (d) of Fig. 7-16 by the convention of Fig. 7-1 (a). This linear light is incident normally on the left face of the retardation plate, of which part (e) of Fig. 7-16 is a schematic cut-away view. The axes on the plate are lettered to correspond to those in Figs. 7-8 and 7-9. Instead of showing the Huygens wavelets in the plate (as in Fig. 7-10 (b)), the waves in the plate are indicated by the convention of Fig. 7-1 (a).

The optic axis of the plate is parallel to the Z-axis. The electric vector in the ordinary wave is always perpendicular to the optic axis, and in the extraordinary wave it is parallel to the optic axis. If the azimuth of the polarizer were set so that the electric vector in the light transmitted by it were vertical, only an extraordinary wave would travel through the plate. If the electric vector in the transmitted light were horizontal, only an ordinary wave would be set up. When the polarizer makes some angle with the vertical, as in Fig. 7-16, both types of wave travel through the plate. Neglecting the small loss of light by reflection, the amplitudes of the ex-

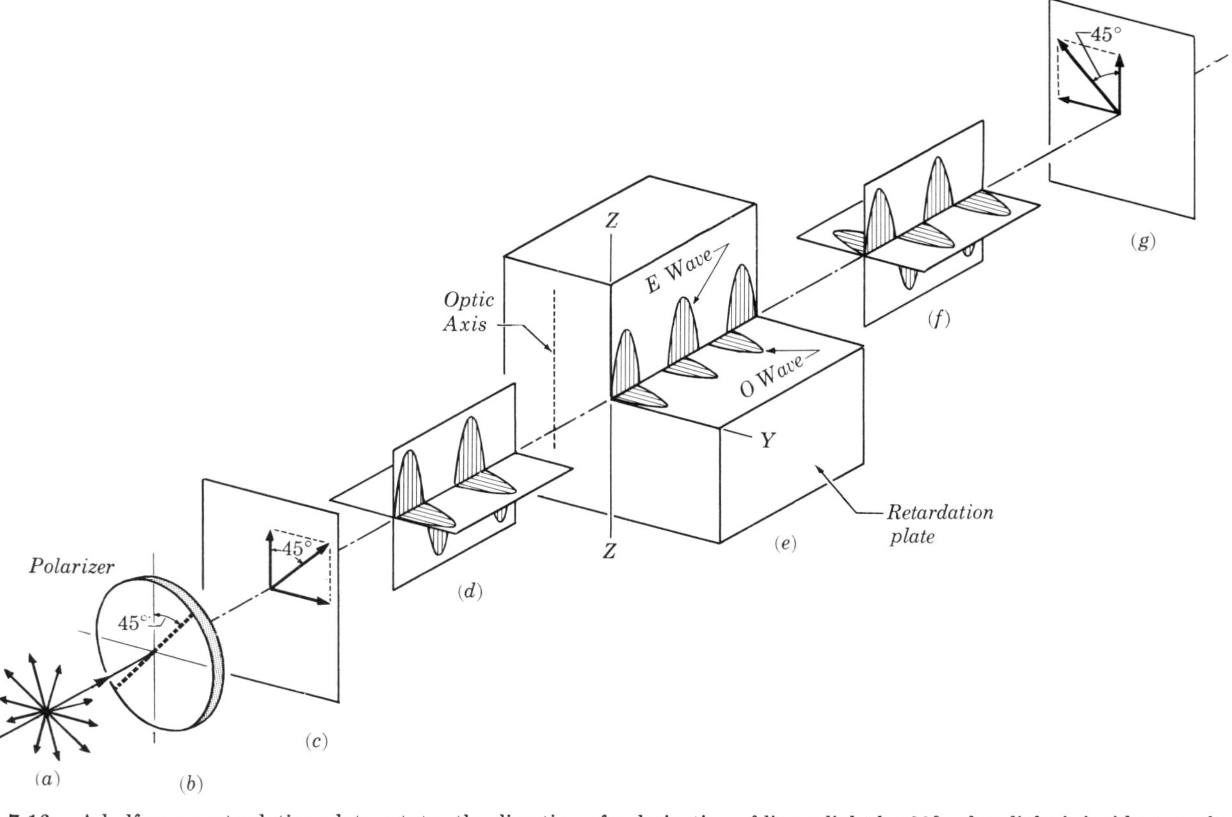

FIG. 7-16. A half-wave retardation plate rotates the direction of polarization of linear light by 90°, when light is incident on the plate at an angle of 45° with the optic axis.

traordinary and ordinary waves in the plate are equal to the amplitudes of the vertical and horizontal components of the light transmitted by the polarizer. For the special case shown in Fig. 7-16, in which the transmitted light makes an angle of 45° with the optic axis, the horizontal and vertical components are equal and hence the extraordinary and ordinary waves in the plate are of equal amplitude.

The velocity of the extraordinary wave, for the type of crystal assumed, is greater than that of the ordinary wave. Therefore the wave length of the former in the crystal is greater than that of the latter. As a consequence, the ordinary wave falls behind the extraordinary wave and hence on emergence from the plate the phase relation between the waves will have altered. After emergence, both waves resume their original velocities and wave lengths, so there is no *further* change in phase. Since the plate causes one wave to lag behind the other, it is called a *retardation plate.*

It should be evident that the phase difference between the emergent waves depends on the difference between the indices n_O and n_E and on the thickness of the plate. If these quantities are such that in traveling through the plate one wave drops behind the other by just one-quarter of a wave length, we have a *quarter-wave* plate; if the lag is one-half a wave length, we have a *half-wave* plate. These are the two most useful retardation plates.

In Fig. 7-16, three full O-waves and two and a half E-waves are shown within the plate. (This implies that the ratio of indices for the two waves is 6/5, much greater than in any known crystal. The wave lengths in real crystals are much more nearly equal than in the diagram.) The O-wave is therefore just one-half a wave length behind the E-wave on emergence. In other words, the plate in Fig. 7-16 is a half-wave plate.

Parts (f) and (g) of Fig. 7-16 show the two waves on emergence. It will be seen that they combine to form a linearly polarized wave, but with the direction of the electric vector at right angles to that in the incident wave. In other words, *a half-wave plate rotates the azimuth of a beam of linear light by 90°*, provided the light is incident on the plate with the electric vector making an angle of 45° with the optic axis.

Consider now the effect of a quarter-wave plate. Fig. 7-17 (a) corresponds to part (f) of Fig. 7-16, the preceding portion of the optical setup being the same except that the retardation plate is only half as thick as that in Fig. 7-16. The O-wave therefore drops behind by only one-quarter of a wave length. The nature of the resultant wave is shown in Fig. 7-17 (c), which has been drawn as if it were a wire model because much of

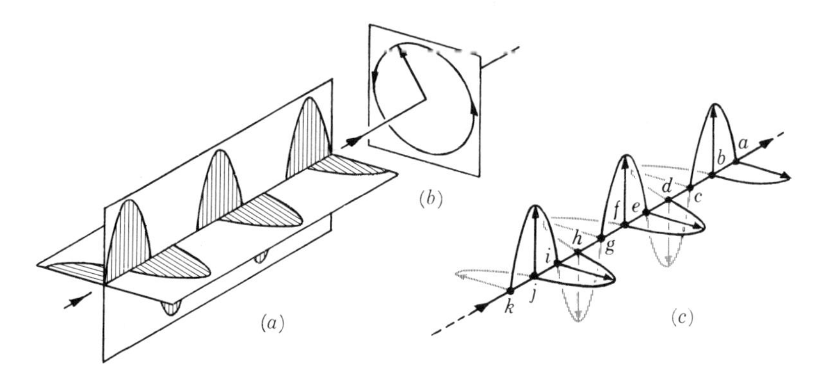

FIG. 7-17. A quarter-wave retardation plate converts linear light to circular light; when linear light is incident on the plate at an angle of 45° with the optic axis.

the figure is invisible if the reference planes are considered opaque. The resultant electric vector has been constructed at a number of lettered points.

At points *a*, *c*, *e*, and *g*, the magnitude of the vertical component is zero and that of the horizontal component is a positive or negative maximum. The resultant at these points, shown by an arrow, is therefore horizontal and equal to the maximum of the horizontal wave. At points *b*, *d*, and *f*, the magnitude of the horizontal component is zero and the resultant equals the maximum of the vertical wave, equal to that of the horizontal wave. At a point halfway between *a* and *b*, the magnitude of each component is 0.707 times the maximum. The magnitude of the resultant is the square root of the sums of the squares of these components; that is, it is equal to the maximum value of either wave and hence equal to the resultant at points *a* and *b*. Its direction makes an angle of 45° with the horizontal and vertical planes. To avoid further confusion in the diagram, the resultant at this point has not been drawn but is left to the reader's imagination. At a point halfway between *b* and *c*, the same construction shows that again the magnitude of the resultant is the same, and that it makes an angle of 45° with both reference planes, but on the opposite side of the vertical plane. It should be evident without further analysis that the magnitude of the resultant electric intensity is the same at all points, but that it rotates around the direction of propagation, making one revolution as the wave advances one wave length. If we consider a plane perpendicular to the ray, through some fixed point such as point *a*, then as the wave advances, the electric field in this plane, instead of oscillating in magnitude along a fixed direction as in Fig. 1-2, remains *constant* in magnitude but rotates in direction. An electromagnetic wave of this sort is called *circularly polarized*, and if it is a light wave we speak of it as *circular light*. Thus *a quarter-wave plate converts linear light to circular light*, when linear light is incident on the plate at an angle of 45° with the optic axis. A conventional method for representing circular light is shown in part (b) of Fig. 7-17.

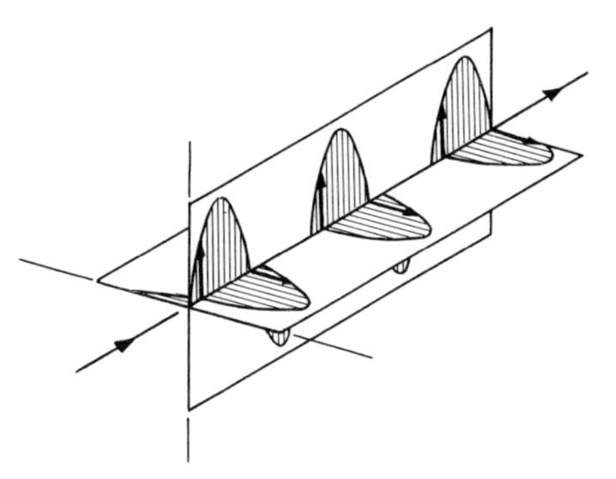

Fig. 7-18. The resultant of two beams of linear light of the same frequency, in mutually perpendicular planes, is, in general, elliptical light.

Retardations of one-quarter and one-half wave lengths are evidently special cases. We now consider the general case of some arbitrary phase relation, as in Fig. 7-18. When the horizontal and vertical components are added vectorially it will be found that while the resultant rotates around the direction of propagation as with circular light, its magnitude does not remain constant. The projection of the tip of the resultant on a plane at right angles to the direction of propagation is an ellipse. The light is said to be *elliptically polarized* and for brevity is called elliptical light. Circular light and linear light are special cases of elliptical light, the former resulting when the phase difference between the components is 90° or 270°, the latter when the phase relation is 0° or 180°.

The phase difference between the waves emerging from a retardation plate can be computed as follows. Let t represent the crystal thickness (in the direction of propagation), n_E the index for the extraordinary ray, and n_O the index for the ordinary ray. If λ is the wave length in vacuum, the wave lengths of the extraordinary and ordinary waves in the crystal are

$$\lambda_E = \frac{\lambda}{n_E}, \quad \lambda_O = \frac{\lambda}{n_O}.$$

The number of waves of each type in the crystal is

$$\frac{t}{\lambda_E} = \frac{tn_E}{\lambda} \text{ (extraordinary)},$$

$$\frac{t}{\lambda_O} = \frac{tn_O}{\lambda} \text{ (ordinary)}.$$

The difference between these expressions is the number of waves that one wave train lags behind the other on emergence, since the waves are necessarily in phase at the left face of the crystal. Each wave corresponds to a phase angle of 2π radians, so the phase difference between the emergent waves is

$$\phi_E - \phi_O = \frac{2\pi t}{\lambda}(n_E - n_O). \tag{7-23}$$

Example. What minimum thickness of crystalline quartz is required for a quarter wave plate?

The indices of refraction of quartz, for light of wave length 589 $m\mu$, are

$$n_E = 1.553, \quad n_O = 1.544.$$

From Eq. (7-23), the thickness required for a given phase difference of the emergent waves is

$$t = \frac{\lambda\,(\phi_E - \phi_O)}{2\pi\,(n_E - n_O)}.$$

For a quarter wave plate, $\phi_E - \phi_O = \pi/2$, so

$$t = \frac{589 \times 10^{-7} \times \pi/2}{2\pi(1.553 - 1.544)}$$
$$= 16 \times 10^{-4} \text{ cm}.$$

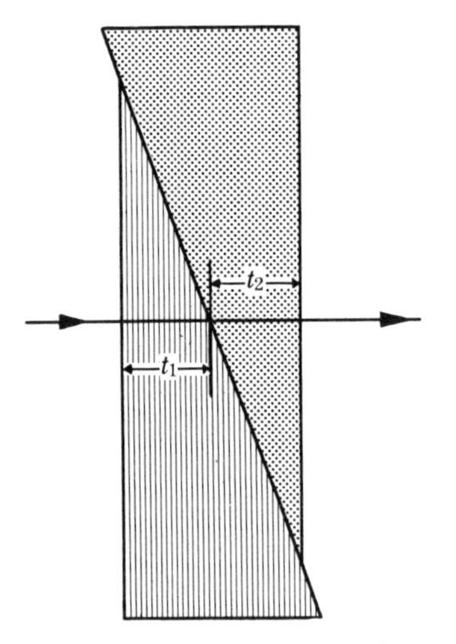

FIG. 7-19. Principle of the Babinet compensator.

In the analysis of polarized light it is often desirable to be able to produce a phase difference of any arbitrary amount between two linearly polarized waves in mutually perpendicular planes. A single crystal such as a quarter-wave plate can, of course, produce only one particular phase difference, dependent on the thickness of the crystal and the wave length of the light. The *Babinet compensator* is a device by means of which the phase difference is continuously variable. It consists of two wedge-shaped sections of crystalline quartz as in Fig. 7-19. The optic axis is lengthwise in one section and transverse in the other. Thus a wave that travels through the first crystal as an ordinary wave, travels through the second as an extraordinary wave and vice versa. If t_1 and t_2 are the thicknesses of the crystals at any point, the phase difference produced by the first crystal is $\frac{2\pi t_1}{\lambda}(n_E - n_O)$ and that produced by the second is $-\frac{2\pi t_2}{\lambda}(n_E - n_O)$. The resultant phase difference is the sum of these, or

$$\frac{2\pi}{\lambda}(t_1 - t_2)(n_E - n_O).$$

The crystals are mounted so that one can be slid over the other. The difference $(t_1 - t_2)$, and hence the phase difference, can be made to have any desired value.

7-8 Transmission of elliptically polarized light by an analyzer. Fig. 7-18 illustrates how an elliptically polarized wave can be compounded from two linearly polarized waves of equal amplitude, at right angles to one another, and differing in phase. Conversely, a given elliptically polarized wave can be resolved into two component linearly polarized waves of this sort. But it is also possible to resolve the elliptically polarized wave into two linearly polarized waves of *unequal* amplitude, polarized in planes parallel to the major and minor axes of the ellipse and differing in phase by $\pi/2$ radians as in Fig. 7-20. This viewpoint is more useful for our present problem.

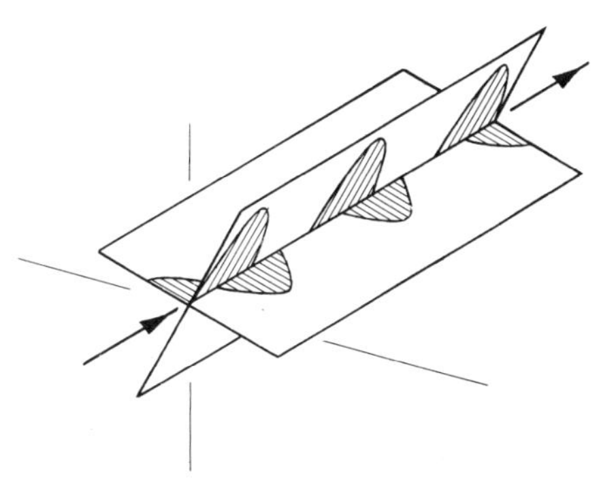

FIG. 7-20. An elliptically polarized wave can be resolved into two linearly polarized waves of unequal amplitudes, differing in phase by $\pi/2$ radians.

Suppose a beam of elliptically polarized light is incident on an analyzer and the transmission direction of the analyzer makes an angle θ with the major axis of the ellipse. (Fig. 7-21.) Let E_1 and E_2 represent the amplitudes of the E-vectors in the linearly polarized component waves, parallel and perpendicular to the major axis. Only the component of each of these components parallel to the transmission direction will be transmitted. The problem is somewhat more complicated by the fact that the components are out of phase by $\pi/2$ radians.

If $\theta = 0$, only the vibration of amplitude E_1, parallel to the transmission direction, is transmitted. As the analyzer is rotated and θ increases, the amplitude of the transmitted beam decreases and becomes equal to E_2

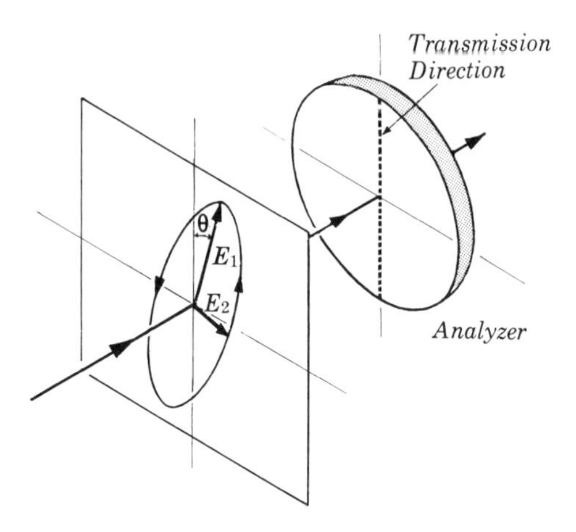

FIG. 7-21. Transmission of elliptically polarized light by an analyzer.

when $\theta = 90°$. Thus the amplitude fluctuates between a maximum of E_1 and a minimum of E_2 and the beam is never completely extinguished as with linear light. In the special case where $E_1 = E_2$ the light is circularly polarized and the amplitude of the transmitted light is independent of the azimuth of the analyzer. In this respect circular light behaves like natural or unpolarized light, that is, circular light cannot be detected by an analyzer. However, if a quarter-wave plate, or a Babinet compensator adjusted to give a phase difference of $\pi/2$ radians, is inserted in the beam, the circular light becomes linear light and the analyzer shows maxima, and minima of zero intensity, as it is rotated. What would be the effect of inserting a quarter-wave plate in a beam of unpolarized light?

7-9 Optical stress analysis. When a polarizer and an analyzer are mounted in the "crossed" position, i.e. with their transmission directions at right angles to one another, no light is transmitted through the combination. But if a doubly refracting crystal is inserted between polarizer and analyzer, the light after passing through the crystal is, in general, elliptically polarized and some light will be transmitted by the analyzer. Thus the field of view, dark in the absence of the crystal, becomes light when the crystal is inserted.

Some substances, such as glass, celluloid, and bakelite, while not normally doubly refracting, become so when subjected to mechanical stress. From a study of the specimen between "crossed Nicols" much information regarding the stresses can be obtained. Improperly annealed glass, for example, may be internally stressed to an extent which might cause it later to develop cracks. It is evidently important that optical glass should be free from such a condition before it is subjected to expensive grinding and polishing. Hence such glass is always examined between crossed Nicols before grinding operations are begun.

The double refraction produced by stress is the basis of the science of *photoelasticity*. The stresses in opaque engineering materials such as girders, boiler plates, gear teeth, etc., can be analyzed by constructing a transparent model of the object, usually of bakelite, and examining it between crossed Nicols. Very complicated stress distributions such as those around a hole or a gear tooth, which it would be practically impossible to analyze mathematically, may thus be studied by optical methods. Fig. 7-22 is a photograph of a photoelastic model under stress.

Liquids are not normally doubly refracting, but some become so when an electric field is established within them. This phenomenon is known

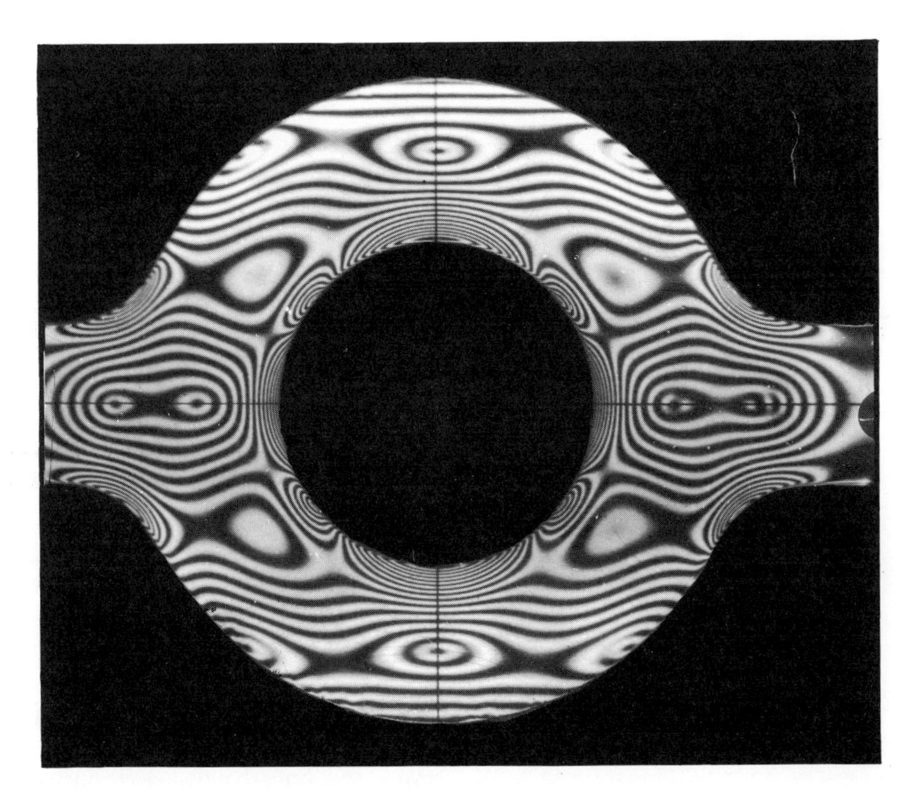

Fig. 7-22. Photoelastic stress analysis.
(Courtesy Dr. W. M. Murray, M.I.T.)

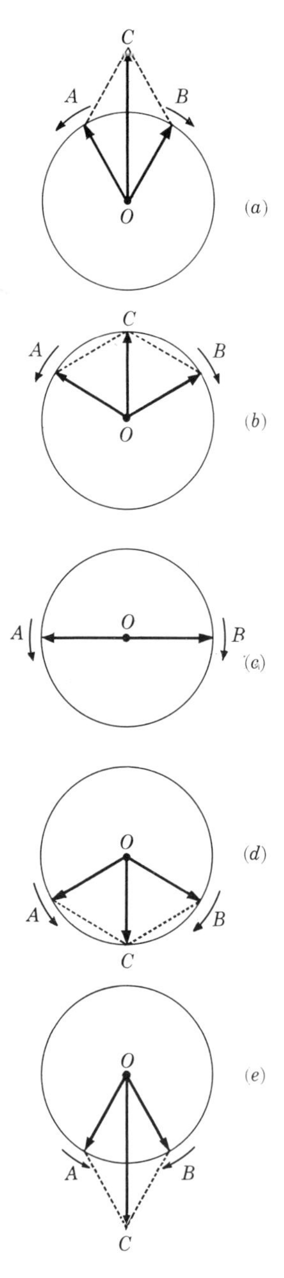

FIG. 7-23. A beam of linear light can be resolved into two beams of circular light.

as the *Kerr effect.* The existence of the Kerr effect makes it possible to construct an electrically controlled "light valve." A cell with transparent walls contains the liquid between a pair of parallel plates. The cell is inserted between crossed Nicols. Light is transmitted when an electric field is set up between the plates and is cut off when the field is removed. This is the method used by Anderson to modulate the intensity of a light beam in his measurements of the velocity of light. (See Sec. 1-6.)

7-10 Optical activity. It has been useful in the cases previously discussed to consider a linearly polarized wave train as the resultant of two linearly polarized waves, vibrating in mutually perpendicular planes. A linearly polarized wave may also be looked upon as the resultant of two *circularly* polarized waves whose E-vectors rotate in opposite directions. Thus in Fig. 7-23, which represents an "end-on" view of a wave train advancing toward the reader, the vectors OA and OB, rotating in opposite directions, represent the circularly polarized waves. Their resultant OC corresponds to a linearly polarized wave whose vibration direction is vertical.

When a beam of linearly polarized light is sent through certain types of crystals, the crystal separates the linearly polarized beam into two components, circularly polarized in opposite directions. These components travel through the crystal with different velocities and on emergence from the crystal one component will have rotated through a greater angle than the other. If, for example, the electric vector in the incident wave is vertical, as in Fig. 7-23 (a), the vectors OA and OB are both initially vertical. Suppose that after passing through the crystal one vector, say OA, has made some integral number of revolutions while the other has turned through a slightly smaller angle. The situation on emergence is then as in Fig. 7-24. The resultant OC lies in a plane making an angle α with the vibration plane of the incident wave. This phenomenon is called *rotation of the plane of polarization*, and substances which exhibit the effect are called *optically active*. Those which rotate the plane of polarization to the right, looking along the advancing beam, are called dextrorotatory or right-handed; those which rotate it to the left, laevorotatory or left-handed.

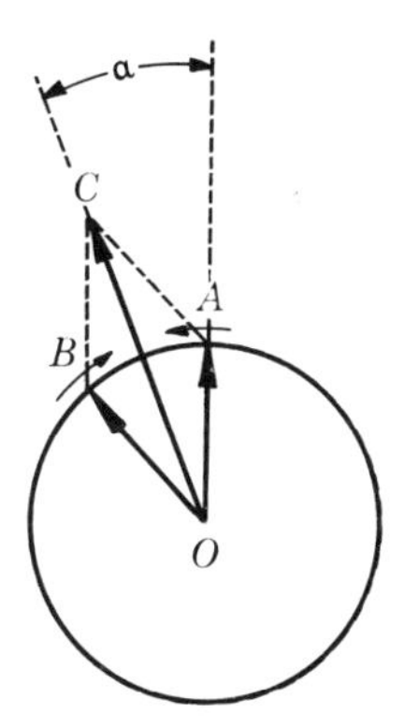

FIG. 7-24.

Optical activity may be due to an asymmetry of the molecules of a substance, or it may be a property of a crystal as a whole. For example, solutions of cane sugar are dextrorotatory, indicating that the optical activity is a property of the sugar molecule. The rotation of the plane of polarization by a sugar solution is used commercially as a method of determining the proportion of cane sugar in a given sample. Crystalline quartz is also optically active, some natural crystals being right-handed and others left-handed. Here the optical activity is a consequence of the crystalline structure, since it disappears when the quartz is melted and allowed to resolidify into a glassy noncrystalline state called fused quartz.

7-11 The scattering of light. The sky is blue. Sunsets are red. Skylight is largely linearly polarized, as can readily be verified by looking at the sky directly overhead through a polarizing plate. One and the same phenomenon is responsible for all three of the effects noted above.

In Fig. 7-25 sunlight (unpolarized) comes from the left along the X-axis and passes over an observer looking vertically upward along the Z-axis. One of the molecules of the earth's atmosphere is located at point O. We know that these molecules are aggregates of electrically charged particles.

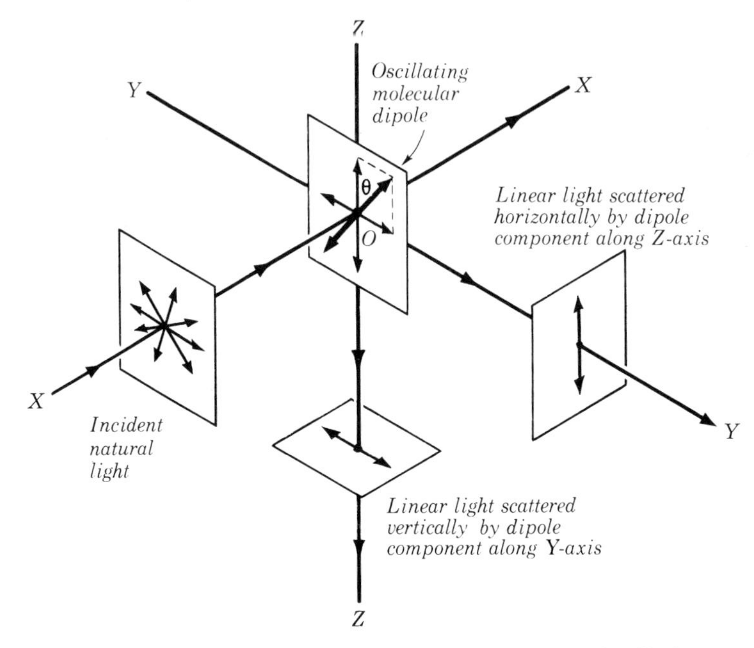

FIG. 7-25. Scattering of light by an oscillating molecular dipole.

The electric field in the beam of sunlight exerts a force on the positive charges in one direction, and on the negative charges in the opposite direction. Since the charges are not bound in a perfectly rigid structure, a small relative displacement of the charges results. However, since the field in the light wave is continually reversing in direction, the molecular charges are displaced first in one direction and then in the opposite direction. These directions necessarily lie in the Y-Z plane in Fig. 7-25, since the electric field in the wave is at right angles to the direction of propagation. There is no field, and hence no vibration, in the direction of the X-axis.

An arbitrary component of the incident light, vibrating at an angle θ with the Z-axis, sets the electric charges in the molecule vibrating in the same direction, as indicated by the heavy line through point O. In the usual way, we can resolve this vibration into two, one along the Y- and the other along the Z-axis. The result, then, is that each component in the incident light produces the equivalent of two molecular "antennas," oscillating with the frequency of the incident light, and lying along the Y- and Z-axes.

It is well known that an antenna does not radiate in the direction of its own length. Hence the antenna along the Z-axis sends no light to the observer directly below it. It does, of course, send out light in other directions. The only light reaching the observer comes from the component of vibration along the Y-axis and, as is the case with the waves from any antenna, this light is linearly polarized with the electric field parallel to the antenna. The vectors on the Z-axis below point O show the direction of the electric vector in the light reaching the observer. The vectors on the Y-axis give the direction of the electric vector in the light radiated in this direction by the antenna vibrating along the Z-axis.

This absorption and reradiation of energy by the molecules is called *scattering*. The energy of the scattered light is abstracted from the original beam, which becomes weakened in the process.

The vibration of the charges in the molecule is a *forced* vibration, like the vibration of a mass on a spring when the upper end of the spring is moved up and down with simple harmonic motion. The amplitude of the forced vibrations is greater, the closer the driving frequency approaches the natural frequency of vibration of the spring-mass system. Now the natural frequency of the electric charges in a molecule is the same as that of a wave length in the ultraviolet. The frequencies of the waves in visible light are less than the natural frequency, but the higher their frequency, or the shorter their wave length, the closer is the driving frequency to the natural frequency, the greater the amplitude of vibration, and the greater the intensity of the scattered light. In other words, blue light is scattered more than red, with the result that the hue of the scattered light is blue.

Toward evening, when sunlight has to travel a large distance through the earth's atmosphere to reach a point over or nearly over an observer, a large proportion of the blue light is removed from it by scattering. White light minus blue light is yellow or red in hue. Thus when sunlight, with the blue component removed, is incident on a cloud, the light reflected from the cloud to the observer has the yellow or red hue so commonly seen at sunset.

From the explanation above, it follows that if the earth had no atmosphere we would receive no skylight at the earth's surface and the sky would appear as black in the daytime as it does at night. This conclusion is borne out by observations at high altitudes, where there is less atmosphere above the observer.

It is not too difficult a matter to compute the way in which the amount of scattered light should vary with wave length, and the problem affords an opportunity to bring to bear on a problem in optics a number of diverse relations from the fields of electricity and mechanics. The theory was first worked out by Lord Rayleigh.

Our problem is first to compute the amplitude of forced oscillations of the electric charges in a molecule set up by the oscillating field in a light wave, and second to find the power radiated by this oscillating doublet. From the principles of mechanics, we know that when a systen of mass m, capable of oscillating with a natural frequency f_0, is acted on by a sinusoidally varying driving force of frequency f and maximum value F_m, the amplitude A of the forced oscillations is

$$A = \frac{F_m}{4\pi^2 m}\,\frac{1}{f_0^2 - f^2}, \tag{7-24}$$

and the frequency of oscillation, f, equals that of the driving force.

To find the natural frequency of oscillation of the electric charges in a molecule, we choose as the simplest molecular model a Thomson molecule, consisting of a sphere of positive charge in which the electrons are embedded. If the molecule contains Z electrons, the total positive charge is Ze, where e is the electronic charge. For the purposes of this calculation we can think of all the electrons as combined in a single negative charge of magnitude Ze and mass Zm, where m is the electronic mass (see Fig. 7-26). The equilibrium position of the electrons is at the center of the sphere of positive charge. It is an elementary problem in electrostatics to show that when the electrons are displaced a distance x from the center, the force pulling them back is

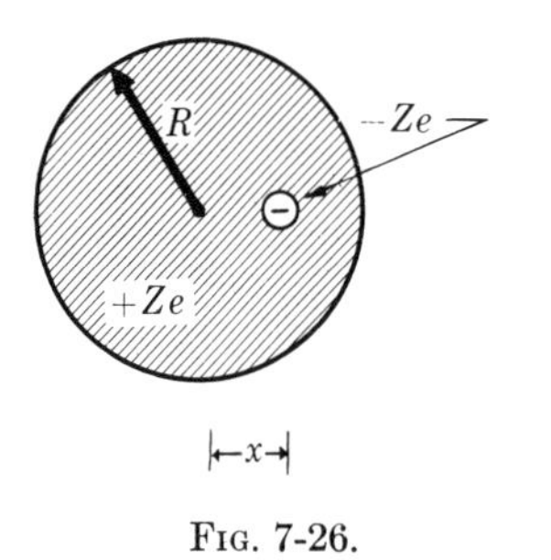

FIG. 7-26.

$$F = -\frac{1}{4\pi\epsilon_0}\,\frac{Z^2e^2}{R^3}\,x, \tag{7-25}$$

where R is the radius of the sphere of positive charge and we are using rationalized mks units. This is a linear restoring force, of the form

$$F = -kx.$$

If the electrons are displaced from the center and released, they execute simple harmonic motion of natural frequency

$$f_0 = \frac{1}{2\pi}\sqrt{\frac{k}{m}} = \frac{1}{2\pi}\sqrt{\frac{1}{4\pi\epsilon_0}\,\frac{Z^2e^2}{R^3 Zm}}. \tag{7-26}$$

Let us calculate the order of magnitude of this natural frequency. Take as our Thomson molecule an atom of oxygen, which has eight electrons. Then

$$Z = 8,$$

$$e = 1.6 \times 10^{-19} \text{ coulomb},$$

$$m = 9.1 \times 10^{-31} \text{ kgm},$$

$$\frac{1}{4\pi\epsilon_0} = 9 \times 10^9 \text{ coul}^2/\text{newton-m}^2.$$

From many lines of evidence, the radius R of the atom can be taken as

$$R = 10^{-8} \text{ cm} = 10^{-10} \text{ m}.$$

When these numerical values are inserted in Eq. (7-26) we get

$$f_0 = 7.2 \times 10^{15} \text{ vibrations/sec}.$$

This is equal to the frequency of a "light" wave of wave length

$$\lambda_0 = \frac{c}{f_0} = 0.42 \times 10^{-7} \text{ m} = 42 \text{ m}\mu,$$

which is in the ultraviolet. In the visible spectrum, where wave lengths are of the order of 500 mμ = 5×10^{-7} m, the frequency f is

$$f = \frac{c}{\lambda} = 0.6 \times 10^{15} \text{ cycles/sec}.$$

Hence the square of the natural frequency, f_0^2, is much larger than the square of the frequency of a wave of visible light, f^2, and we may neglect the latter in the denominator of Eq. (7-24) if we are concerned only with the scattering of visible light.

The maximum force F_m exerted on the charge Ze by the electric field of the light wave is

$$F_m = ZeE,$$

where E is the maximum electric intensity in the wave. When we insert this expression for F_m in Eq. (7-24), and neglect the term f^2, we get for the amplitude of forced oscillations of frequency f,

$$A = \frac{ZeE}{4\pi^2 m f_0^2}. \tag{7-27}$$

This completes the first part of the problem.

Electromagnetic theory predicts that the average power radiated by an oscillating electric doublet is

$$P_{av} = \frac{4\pi^3}{3\epsilon_0}\frac{p_m^2 f^4}{c^3}, \tag{7-28}$$

where p_m is the maximum electric moment of the doublet and f is the frequency of oscillation. The maximum electric moment of a molecule consisting of positive and negative charges of magnitude Ze, oscillating with an amplitude A, is

$$p_m = ZeA.$$

Inserting this expression for p_m in Eq. (7-28), together with the value of A from Eq. (7-27), we get for the average power radiated by a single molecule

$$P_{av} = \frac{1}{12\pi\epsilon_0}\frac{Z^4e^4E^2}{m^2c^3}\frac{f^4}{f_0^4}. \tag{7-29}$$

If there are n molecules per unit volume, the average power radiated (i.e., scattered) per unit volume is n times as great. It is customary to express this power as a fraction of the power per unit area, or the Poynting flux, in the incident wave. The latter is

$$S_{av} = \frac{1}{2}c\epsilon_0 E^2,$$

and the ratio of the scattered power to the incident flux is

$$\frac{nP_{av}}{S_{av}} = \frac{1}{6\pi\epsilon_0^2}\frac{nZ^4e^4}{m^2c^4}\frac{f^4}{f_0^4}, \tag{7-30}$$

or in terms of wave lengths it is

$$\frac{1}{6\pi\epsilon_0^2}\frac{nZ^4e^4}{m^2c^4}\frac{\lambda_0^4}{\lambda^4}. \tag{7-31}$$

The scattering is therefore inversely proportional to the 4th power of the wave length, so that the shorter blue waves are scattered more than the longer red waves. Either Eq. (7-30) or Eq. (7-31) is known as the *Rayleigh scattering formula.*

Problems—Chapter 7

(1) (a) At what angle above the horizontal must the sun be in order that sunlight reflected from the surface of a calm body of water shall be completely linearly polarized? (b) Is the plane of the E-vector in the reflected light horizontal or vertical?

(2) A parallel beam of "natural" light is incident at an angle of 58° on a plane glass surface. The reflected beam is completely linearly polarized. (a) What is the angle of refraction of the transmitted beam? (b) What is the refractive index of the glass?

(3) Light is incident on a water surface (index 4/3) at such an angle that the reflected light (ray 1 in Fig. 7-27) is completely linearly polarized. (a) What is the angle of incidence? (b) A block of glass (index 3/2) having a flat upper surface is immersed in the water as indicated in Fig. 7-27. The light reflected from the surface of the glass (ray 2) is completely linearly polarized. Find the angle between the surface of the water and the surface of the glass.

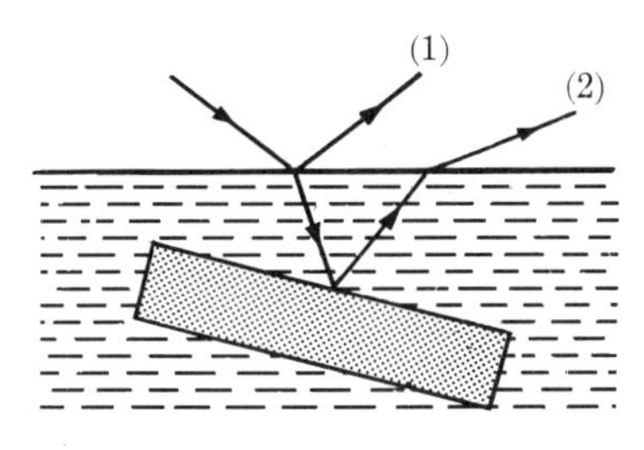

FIG. 7-27.

(4) Linearly polarized light is incident on the surface of a glass plate of index 1.732 at an angle of incidence of 60°. The electric vector in the incident light makes an angle of 30° with the plane of incidence. What fraction of the incident light is reflected at the first surface?

(5) A beam of natural light is incident on the surface of a piece of glass of index 1.523 at an angle of incidence of 70°. (a) What fraction of the incident light is reflected? (b) In the reflected beam, what is the ratio of the component of the E-vector in the plane of incidence to the component of the E-vector at right angles to the plane of incidence?

(6) A polarizer and an analyzer are oriented so that the maximum amount of light is transmitted. To what fraction of its maximum value is the intensity of the transmitted light reduced when the analyzer is rotated through (a) 30°, (b) 45°, (c) 60°?

(7) How must a polarizer and an analyzer be oriented so that a beam of natural light is reduced to (a) $\frac{1}{2}$, (b) $\frac{1}{4}$, (c) $\frac{1}{8}$, of its original intensity?

(8) (a) Construct a diagram like that of Fig. 7-17, showing the state of polarization of the light transmitted by a quarter-wave plate when the electric vector in the incident linear light makes an angle of 60° with the optic axis. (b) Describe the state of polarization.

(9) (a) A beam of circularly polarized light is passed normally through a quarter-wave plate. What is the state of polarization of the light after it emerges from the plate? (b) A beam of circularly polarized light is passed normally through an eighth-wave plate. What is the state of polarization of the light after it emerges from the plate?

(10) A parallel beam of linearly polarized light of wave length 589 mμ (in vacuum) is incident on a calcite crystal as in Fig. 7-10 (b). (a) Find the wave lengths of the ordinary and extraordinary waves in the crystal. (b) What minimum thickness of crystal is necessary to produce phase differences of $\pi/4$ radians, $\pi/2$ radians, and π radians between the emergent waves? (c) If the electric vector in the incident light makes an angle of 45° with the optic axis, what is the state of polarization of the emergent light in the three cases in part (b)?

(11) A sheet of cellophane is a half-wave plate for light whose wave length is 400 mμ. (a) Assuming that the variations in the indices of refraction with wave length can be neglected, how would the sheet behave with respect to light whose wave length is 800 mμ? (b) White light is incident on this sheet after passing through a Nicol prism so oriented that the E-vector of the light incident normally on the sheet makes an angle of 45° with the optic axis of the cellophane. If the light transmitted by the sheet of cellophane is then examined by another Nicol prism oriented like the first, what will be the color of the transmitted light (i.e., which will be transmitted to a greater extent, the short wave length end of the visible spectrum or the long wave length end)?

(12) Assume the values of n_O and n_E for quartz to be independent of wave length. A certain quartz crystal is a quarter-wave plate for light of wave length 800 mμ (in vacuum). What is the state of polarization of the transmitted light when linearly polarized light of wave length 400 mμ (in vacuum) is incident on the crystal, the plane of polarization making an angle of 45° with the optic axis?

(13) Suppose the thickness t_1 of one of the quartz wedges of a Babinet compensator (see Fig. 7-19) is 2 mm. Find the required thickness t_2 of the other wedge, if it is desired to produce a phase difference of $2\pi/3$ radians between the ordinary and extraordinary waves for light of wave length 500 mμ in vacuum.

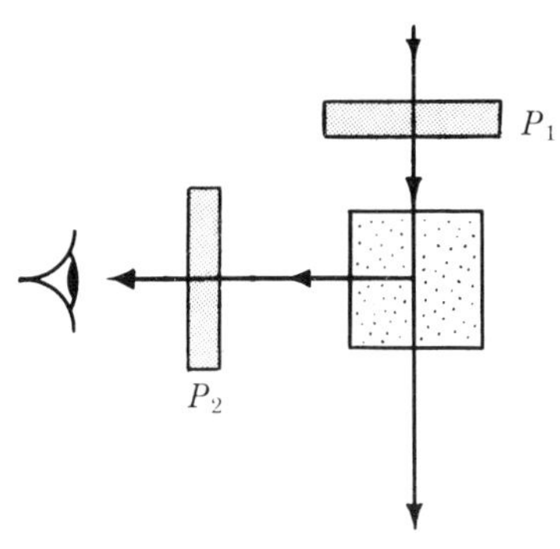

FIG. 7-28.

(14) A beam of light, after passing through the polarizing plate P_1 in Fig. 7-28, traverses a cell containing a scattering medium. The cell is observed at right angles through another polarizing plate, P_2. Originally, the plates are oriented until the brightness of the field seen by the observer is a maximum. (a) Plate P_2 is rotated through 90°. Is extinction produced? (b) Plate P_1 is now rotated through 90°. Is the field bright or dark? (c) Plate P_2 is then restored to its original position. Is the field bright or dark?

CHAPTER 8

INTERFERENCE

8-1 Interference in thin films. At any point where two or more trains of waves cross one another they are said to *interfere.* This does not mean that any wave train is impeded by the presence of the others, but refers to the combined effect of them all at the point in question. The *principle of superposition* states that the resultant displacement at any point and at any instant may be found by adding the instantaneous displacements that would be produced at the point by the individual wave trains if each were present alone. The term "displacement" as used here is a general one. If one is considering surface ripples on a liquid, the displacement means the actual displacement of the surface above or below its normal level. If the waves are sound waves, the term refers to the excess or deficiency of pressure. If the waves are electromagnetic, the displacement means the magnitude of the electric or magnetic field intensity.

The brilliant colors that are often seen when light is reflected from a soap bubble or from a thin layer of oil floating on water are produced by interference effects between the two trains of light waves reflected at opposite surfaces of the thin films of soap solution or of oil.

In Fig. 8-1, part (a) represents a train of plane waves, incident on a thin film of a material whose index is greater than that of air. For simplicity, it has been assumed that the waves are linearly polarized with the electric vector parallel to the plane of incidence, as in Fig. 7-3, but the results are the same whatever the state of polarization of the incident

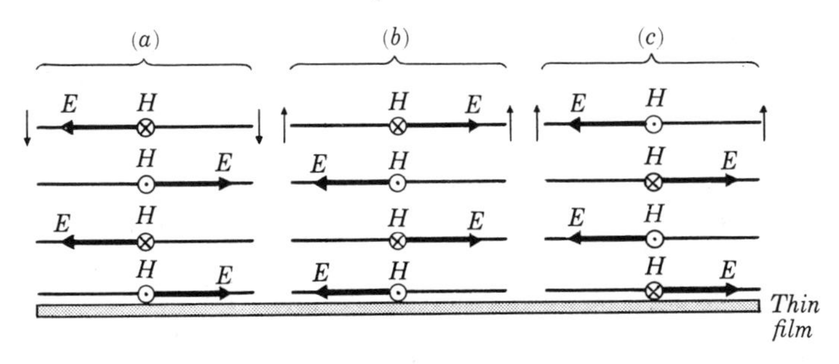

FIG. 8-1. Electric and magnetic vectors in a train of waves (a) incident on; (b) and (c) reflected from the surfaces of a thin film.

beam. The horizontal lines are wave fronts spaced one-half a wave length apart. The direction of the electric vector in the wave fronts is shown by an arrow and the direction of the magnetic vector by the usual convention of dots and crosses. The incident wave train is in part reflected at the first surface of the film, and in part transmitted. Of the portion transmitted, a part is reflected at the second surface and emerges through the first surface. The wave train reflected from the first surface is shown in part (b) and that reflected at the second surface in part (c). Actually, of course, both reflected wave trains travel back along the same path as the incident train, and parts (b) and (c) should be moved to the left so as to overlap part (a). The separation in the diagram is schematic only.

The lowest wave front in beam (a) is just making contact with the upper surface, and the lowest wave front in beam (b) is the reflected wave front that originates at the upper surface at this instant. The film thickness is assumed to be a sufficiently small fraction of the wave length so that the lowest wave front in beam (c), reflected from the lower surface, can be considered to coincide with the corresponding wave fronts in beams (a) and (b).

Reference to Figs. 7-3 and 7-4 shows that for the wave reflected from the first surface, where the index of the reflecting medium is greater than that of the medium in which the wave is traveling, the direction of the electric vector is reversed, relative to its direction in the incident wave, while that of the magnetic vector is not. The electric vector is said to be reflected with *reversal of phase*. The wave front reflected at the second surface is reflected from a medium whose index is less than that of the medium in which the wave is traveling. Diagrams corresponding to Figs. 7-3 and 7-4, for the case in which the index of the second medium is less than that of the first, were not included in Chap. 7, but it was explained in Sec. 7-2 that in this case the magnetic vector is reflected with reversal of phase while the electric vector is not. It will be seen from an examination of Fig. 8-1 that in wave (b) the electric vector in the lowest wave front is opposite in direction to that in the lowest wave front in beam (a), while the magnetic vectors are in the same direction. In wave (c), the magnetic vector is opposite in direction to that in the lowest wave front in beam (a), while the electric vectors are in the same direction.

A comparison of the *reflected* beams, (b) and (c), shows that both the electric and magnetic vectors in the two beams are in opposite directions. Furthermore, the amplitudes of the respective vectors are very nearly the same in both beams, since from Eq. (7-19) the reflectance of both surfaces of the film is the same and, since the reflectance is small, the amplitude of the wave incident on the second surface is only slightly smaller than that

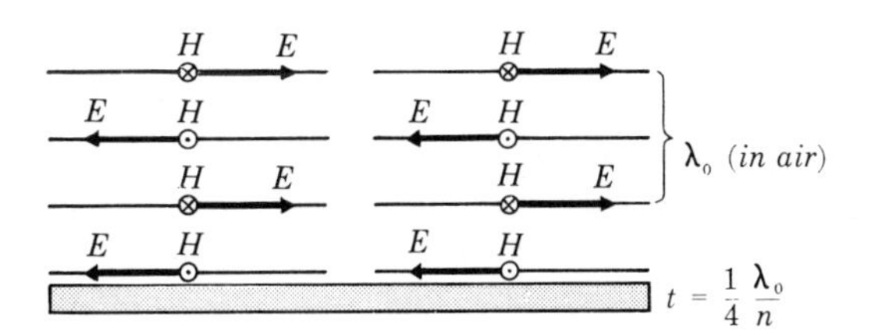

Fig. 8-2. Waves reflected from the surface of a film of thickness ¼ wave length.

of the original wave. (A more complete analysis would have to take into account the effect of additional reflections within the film.) By the principle of superposition, the resultant electric and magnetic intensities in the reflected beam are both very nearly zero at every point and at every instant. In other words, no light is reflected from a film of thickness small compared with the wave length of the light.

Suppose now that the film thickness is gradually increased. Let λ_0 be the wave length of the incident light in air, and let n be the index of the film. The wave length in the film is then λ_0/n. When the thickness is equal to one-quarter of the wave length in the film, or $\frac{1}{4}\frac{\lambda_0}{n}$, the wave train reflected from the second surface drops behind that reflected from the first surface by one-half a wave length (since the light must travel through the film and back). The reflected wave trains are then as shown in Fig. 8-2. It will be seen that both the electric and magnetic vectors in the reflected waves are in the same direction at all points, so that the amplitudes of both in the reflected light are twice as great as in either beam. In other words, light is strongly reflected from a film of thickness $\frac{1}{4}\frac{\lambda_0}{n}$, when incident normally on the film.

If the thickness is increased to $\frac{1}{2}\frac{\lambda_0}{n}$, the wave reflected from the second surface drops behind the other by one wave length and destructive interference again results. When the thickness is $\frac{3}{4}\frac{\lambda_0}{n}$, there is strong reflection again, and so on. Hence there is no reflection (or very little reflection) from a film of thickness zero, $\frac{1}{2}\frac{\lambda_0}{n}$, $\frac{2}{2}\frac{\lambda_0}{n}$, etc., and strong reflection from a film of thickness $\frac{1}{4}\frac{\lambda_0}{n}$, $\frac{3}{4}\frac{\lambda_0}{n}$, $\frac{5}{4}\frac{\lambda_0}{n}$, etc.

If the film is in the shape of a thin wedge of narrow angle, as in Fig. 8-3, and is viewed by reflected monochromatic light, it appears to be crossed

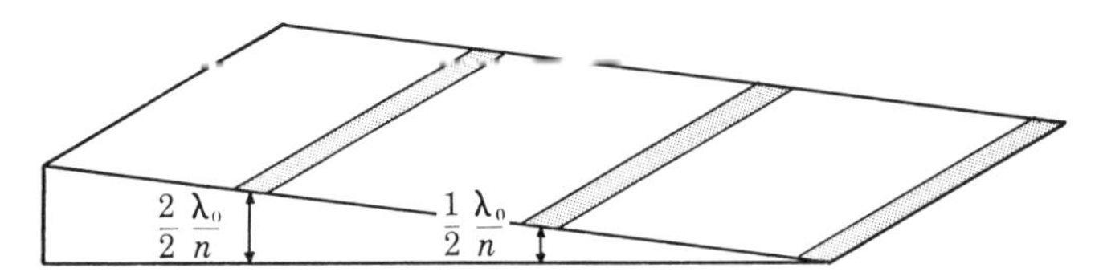

FIG. 8-3. Destructive interference results at thickness of zero, one-half a wave length, one wave length, etc.

by parallel bright bands of the color of the light used, separated by dark bands. At the apex, the film is dark. At a distance from the apex such that the film thickness is one-quarter of a wave length, it is bright. Where the thickness equals one-half a wave length it is dark, and so on. If the film is illuminated first by blue, then by red light, the spacing of the red bands is greater than that of the blue, as is to be expected from the greater wave length of the red light. The bands produced by intermediate wave lengths occupy intermediate positions. If the film is illuminated by white light, its color at any point is that due to the mixture of those colors which may be reflected at that point, while the colors for which the thickness is such as to result in destructive interference are absent. Just those colors which are absent in the reflected light, however, are found to predominate in the transmitted light. At any point, the color of the film by reflected light is complementary to its color by transmitted light.

The frontispiece is reproduced from a Kodachrome photograph of the light reflected from a thin film of soap solution. The film is formed on a glass ring and mounted in a vertical plane. As a result of drainage of the liquid in a downward direction, the thickness of the film increases from top to bottom. The black section at the top indicates that the thickness in this region is less than one-quarter of the wave length of light.

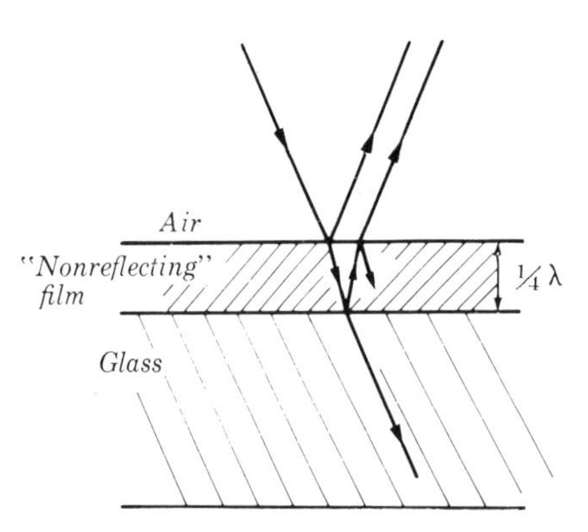

FIG. 8-4. Destructive interference between light reflected from the surfaces of a "nonreflecting" film.

8-2 Nonreflecting films. The phenomenon of interference is utilized in the production of so-called "nonreflecting" glass. A thin layer or film of transparent material is deposited on the surface of the glass as in Fig. 8-4. If the index of this material is properly chosen at some value intermediate between that of air and the glass, equal quantities of light are reflected from its outer surface, and from the boundary surface between it and the glass. Furthermore, since in both reflections the light is reflected from a medium more dense than that in which it is traveling, the same phase change occurs in each reflection. It follows that if the film thickness is one-quarter wave length (normal incidence is assumed), the light reflected from the first surface is 180° out of phase with that reflected from the second, and complete destructive interference results.

The thickness can, of course, be one-quarter wave length for one particular wave length only. This is usually chosen in the yellow-green portion of the spectrum where the eye is most sensitive. Some reflection then takes place at both longer and shorter wave lengths and the reflected light has a purple hue. The overall reflection from a lens or prism surface can be reduced in this way from 4 or 5 percent to a fraction of one percent. The treatment is highly effective in reducing the loss of light by reflection in instruments such as rangefinders and periscopes which have a large number of air-glass surfaces.

8-3 Newton's rings. If the convex surface of a lens is placed in contact with a plane glass plate, as in Fig. 8-5, a thin film of air is formed between the two surfaces. The thickness of this film is very small at the point of contact, gradually increasing as one proceeds outward. The loci of points of equal thickness are circles concentric with the point of contact. Such a film is found to exhibit interference colors, produced in the same way as the colors in a thin soap film. The interference bands are circular, concentric with the point of contact. When viewed by reflected light, the center of the pattern is black, as is a thin soap film. Note that in this case the *magnetic* vector is reversed in phase in the light reflected from the upper surface of the film (which here is of smaller index than that of the medium in which the light is traveling before reflection), and that the phase of the *electric* vector is reversed in the light reflected from the lower surface. When viewed by transmitted light, the center of the pattern is bright. If white light is used, the color of the light reflected from the film at any point is complementary to the color transmitted.

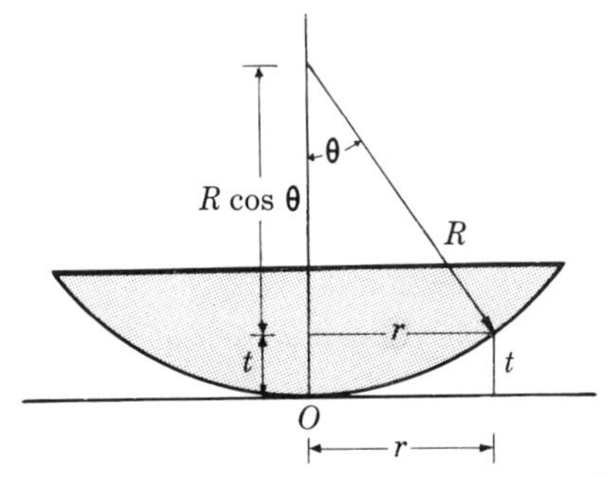

FIG. 8-5. Air film between a convex and a plane surface.

The thickness of the film at a distance r from the point of contact O

may be found as follows. From the construction in Fig. 8-5, it will be seen that

$$t = R - R\cos\theta = R(1 - \cos\theta).$$

If θ is small, $\cos\theta$ may be written

$$\cos\theta = 1 - \frac{\theta^2}{2},$$

or

$$1 - \cos\theta = \frac{\theta^2}{2}.$$

Furthermore,

$$\theta = \frac{r}{R} \text{ (approximately).}$$

Substituting these values, we find

$$t = \frac{r^2}{2R} \text{ (approximately).}$$

If light travels through the film vertically, destructive interference will occur wherever the thickness t is equal to $\frac{1}{2}\lambda_0$, $\frac{2}{2}\lambda_0$, etc.

Hence by measuring the radius of a bright or dark ring, the wave length of the light producing it may be calculated or, conversely, if the wave length is known, one can find the thickness of the air film.

These interference patterns were studied by Newton, and are called after him, *Newton's rings.* Newton did not attempt to account for them by a wave theory, but in terms of a theory of his own that a reflecting surface was subject to "fits of easy transmission" and "fits of easy reflection." Newton's "Opticks," which contains a complete account of all of his experiments in this field, should be read by every serious student of physics.

Fig. 8-6 is a photograph of Newton's rings, formed by the air film between a convex and a plane surface.

The surface of an optical part which is being ground to some desired curvature may be compared with that of another surface, known to be correct, by bringing the two in contact and observing the interference fringes. For example, if a plane surface (an "optical flat") is desired, a glass plate whose lower surface is accurately plane is placed over the surface to be tested. If both surfaces are accurately plane, the entire area of contact will be dark, or, if contact between them is made at one edge, a series of straight interference fringes, parallel to the line of contact, will be observed. If the surface being ground is not plane, the interference fringes are curved. By noting the shape and separation of the fringes, the departure of the surface from the desired form may be determined.

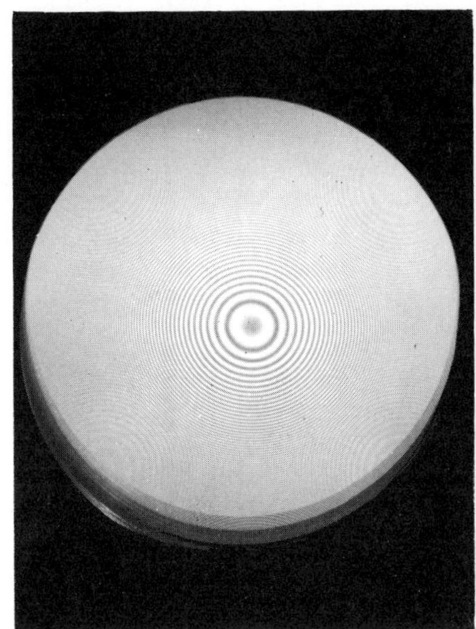

FIG. 8-6. Newton's rings formed by interference in the air film between a convex and a plane surface.
(Courtesy of Bausch & Lomb Optical Co.)

FIG. 8-7. The surface of a telescope objective under inspection during manufacture.
(Courtesy of Bausch & Lomb Optical Co.)

Fig. 8-7 is a photograph made at one stage of the process of manufacturing a telescope objective. The lower, larger diameter, thicker disk is the master. The smaller upper disk is the objective under test. The "contour lines" are Newton's interference fringes, and each one indicates an additional departure of the specimen from the master of one-half a wave length of light. That is, at 10 lines from the center spot the space between the specimen and master is 5 wave lengths, or about 0.0002 inch. This specimen is very poor.

8-4 Standing waves. It will be recalled that a direct and a reflected wave train, traveling in opposite directions in a vibrating string or an organ pipe, give rise to a system of standing waves in the string or the pipe. These standing waves, whose amplitude at every point is found by applying the principle of superposition, are fundamentally an interference phenomenon. The "interference pattern," in the examples mentioned, is such that the "medium" (the string or the air column) is divided into a

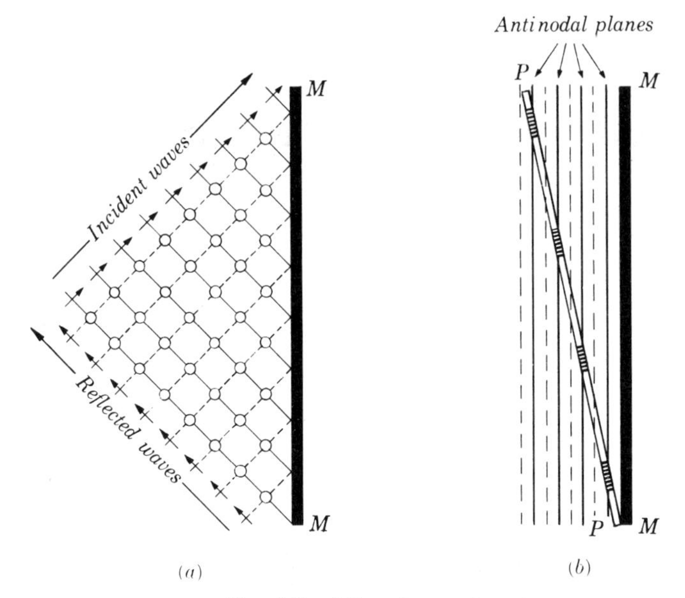

FIG. 8-8. Wiener's experiment.

number of vibrating segments, separated by points which remain at rest, called *nodes*. The regions of maximum amplitude of vibration, intermediate between the nodes, are called loops or *antinodes*. There is no longer any appearance of *traveling* waves in the medium, the nodes and antinodes remaining in the same position. If the displacement at the ends of the medium, as fixed by the boundary conditions, is zero, then the ends are necessarily nodes. This is the case in a closed organ pipe, or a string held rigidly at both ends. If either end is free, as the open end of an organ pipe, that end is a displacement antinode.

All of these standing wave phenomena may be produced by interference between trains of direct and reflected light waves. The boundary conditions which the light waves must fulfill are determined by the nature of the reflecting surface. If it is a metallic reflector, then we can see quite readily what these conditions must be, for our light wave is an electromagnetic wave, and we have seen that within a conducting material the electric field must be zero. This implies that the resultant E-vector of the incident and reflected waves must be zero at the reflecting surface, or that this surface must be a node.

Suppose that plane waves of monochromatic light, represented by the full lines in Fig. 8-8 (a), are allowed to fall on a polished silver surface MM at an angle of incidence of 45°. The electric vector in the reflected light undergoes a phase change of π radians and the reflected waves are shown by dashed lines. At the points marked by small circles the two wave trains meet crest to trough, and at all points of planes through these circles, parallel to MM, the electric intensity is always zero. These planes are then *nodal planes*, and are shown by dotted lines in Fig. 8-8 (b). In planes halfway between the nodal planes, the two wave trains meet crest to crest, and trough to trough. These are the *antinodal* planes and in them the electric intensity is a maximum. They are shown by full lines in Fig. 8-8 (b).

The existence of the standing wave pattern was shown experimentally by Wiener in 1890. An extremely thin film of photographic emulsion, PP in Fig. 8-8 (b), was inclined at a small angle with the reflecting surface. After development, the film was found to be crossed by a series of dark lines (indicated by shading in Fig. 8-8 (b)), located along the lines of intersection of the film with the antinodal planes. No blackening resulted along the lines of intersection of the film with the nodal planes.

Wiener also showed that in order to obtain these results, the light should be plane polarized with the electric vector perpendicular to the plane of incidence (the plane of the diagram in Fig. 8-8). The reader can readily verify that only if this is the case will the E-vectors in the incident and reflected waves point in opposite directions.

The Wiener experiment was repeated in 1932 by Fry and Ives at the Bell Telephone Laboratories, using the photoelectric effect in a thin metal film on a quartz wedge instead of a photographic emulsion. Their conclusions bear out the theory in all respects.

8-5 The Lippmann process of color photography. A method of color photography invented by Lippmann in 1881 makes use of the phenomenon of standing light waves. A specially prepared photographic plate, Fig. 8-9, having an extremely fine grain emulsion, is covered on the emulsion side with a layer of mercury to form a reflecting surface. It is then exposed in a camera, but with the glass instead of the emulsion side of the plate toward the light.

Suppose light of various wave lengths, corresponding to the various colors of the object photographed, is incident on the plate as shown. A standing wave pattern is set up at each point, the separation of the antinodes corresponding to the particular color incident on the plate at that point. When developed, the plate contains a number of layers of silver located at the antinodes. If now the plate is illuminated with white light and viewed by reflection, the silver layers will reflect strongly, at each point of the image, only that color (or colors) by which they were origi-

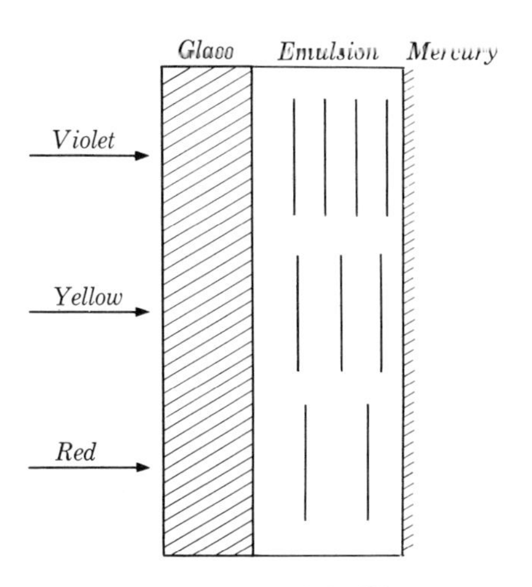

FIG. 8-9. Principle of the Lippmann method of color photography.

nally formed. Each silver layer, by itself, is so thin as to reflect only a small amount of light. But as the white light passes through the film and is partially reflected at each layer, all light of the same wave length as that by which the layers were formed, reflected from any one layer, is in phase with that reflected from every other layer. These beams then combine so as to reenforce one another's effects, giving rise to an intense reflection for this particular color. Other colors, for which the path differences between the reflected beams is not exactly one wave length, will be reflected with much less intensity.

It appears contradictory at first that, although each silver layer reflects equal (or very nearly equal) amounts of all colors, the one particular color whose wave length is twice that of the separation of the layers should predominate to a great extent in the reflected light. An example may make this point clearer. It will be recalled that the energy of a harmonic vibration, or the intensity of a sinusoidal wave, is proportional to the square of its amplitude. The resultant intensity of a number of wave trains, if there is no definite phase relation between them, is found by *adding the intensities* of the individual wave trains. But if the wave trains are all in phase, the resultant intensity is found by *adding the amplitudes* of the individual waves, and squaring to obtain the resultant intensity. A number of wave trains whose phase relations are quite at random are called *noncoherent,* while if a definite phase relation does exist between them, they are called *coherent.*

Assume for instance that we have three wave trains to combine, each of amplitude 2 (arbitrary units). If the phase relation between them is random, we first compute the intensity of each wave, which is 2^2, or 4, and add the intensities, giving 12 (arbitrary units) as the resultant intensity. If the three are in phase, however, they combine to produce a wave of *amplitude* 6, or *intensity* $6^2 = 36$. In general, given n wave trains of amplitude A, their resultant intensity, if noncoherent, is nA^2, while if coherent, the intensity is $(nA)^2$. The intensity in the latter case is n times that in the former. If n is very large, as would be the case in the Lippmann color photograph, the coherent reflection far exceeds the noncoherent.

Lippmann color photographs are extremely brilliant, but the plates are difficult to prepare, and of course no prints can be made from them.

8-6 The Michelson interferometer. The phenomenon of interference is of very practical value in connection with precise measurements of length. The *Michelson interferometer*, which makes possible the measurement of a distance in terms of the wave length of light waves, is illustrated in Fig. 8-10. Light from a small source S is made parallel by a lens L, and falls on a glass plate A, inclined at 45° to the direction of the incident beam. The front surface of this plate is "half-silvered" that is, covered with an extremely thin coating of silver, so that approximately one-half of the beam is reflected, while the remainder is transmitted through the plate. The reflected and transmitted beams strike the plane mirrors B and D respectively, and are returned to the plate A. At the half-silvered surface a part of the beam from B is transmitted, and a part of the beam from D reflected. Since both are derived from the same beam originally incident on A, they are capable of producing interference effects. The light reflected from B passes once through the plate A, while that from D passes through it three times. In order that each beam may travel the same distance in glass, another glass plate C, of the same thickness as A, is inserted in the path of the upper beam, which ensures that each beam passes through the same thickness of glass.

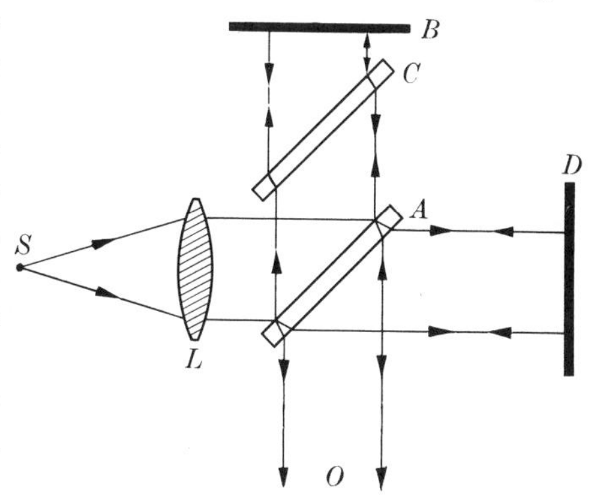

FIG. 8-10. The Michelson interferometer.

An observer O sees the surface of mirror B through the half-silvered plate A, and sees the surface of mirror D reflected in A. If the distances from A to the two mirrors are exactly equal, and B and D are exactly at right angles to one another and at 45° with A, the image of D coincides with the surface of B. If these adjustments are not made exactly, then in effect a thin air film exists between the surface of B and the image of D, and the interference fringes in this thin film are what are observed.

The mirror B is fixed in position, and the mirror D is mounted on a screw so that it may be moved along a line perpendicular to its face. As D is displaced, the system of fringes moves laterally across the field of view, a displacement of the mirror of one-half a wave length causing each fringe to move from its original position to that formerly occupied by the next adjacent fringe. By counting the number of fringes which pass a fixed reference point, the distance moved by the mirror can be measured to within a small fraction of the wave length of the light used.

In seeking for some permanent standard of length, so that the standard meter might be reproduced if it should ever be destroyed, it was decided that the wave length of the light emitted by some chosen chemical element in an electrical discharge would be a standard of the most fundamental and unchanging sort. After some searching, one of the wave lengths emitted by cadmium vapor, lying in the red portion of the spectrum, was chosen as the most suitable. Making use of an interferometer of the type described above, Michelson compared the standard meter with the wave length of these red cadmium waves. The value found is,

$$1 \text{ meter} = 1{,}553{,}164.13 \text{ wave lengths},$$

or the wave length of these waves is

$$\lambda = 643.84696 \text{ m}\mu.$$

The precision of this result is better than one part in a million, and it represents one of the most precise physical measurements ever attempted.

8-7 Double slit interference. Young's experiment. One of the earliest demonstrations of the fact that light can produce interference effects was performed in 1800 by the English scientist Thomas Young. The experiment was a crucial one at the time, since it added further evidence to the growing belief in the wave nature of light. A corpuscular theory was quite inadequate to account for the effects observed.

Young's apparatus is shown in Fig. 8-11. Light from a source at the left, not shown, falls on a narrow slit S_1. Two more slits, S_2 and S_3, are parallel to S_1 and equidistant from it. When a screen is placed at the right of these slits a number of alternate bright and dark bands are observed on it, parallel to the slits. If either of the slits S_2 or S_3 is covered, the dark lines disappear and the screen is illuminated in a broad band. A photograph of these interference fringes is reproduced in Fig. 9-2 (b). The diffuse bands of light in Fig. 9-2 (a) show the appearance of the screen when only a single slit S_2 or S_3 is used. A corpuscular theory cannot account for the fact that a point on the screen, bright when only one slit is exposed, becomes dark when both slits are exposed. The explanation in terms of the wave theory follows. A more complete analysis will be found in Sec. 9-3.

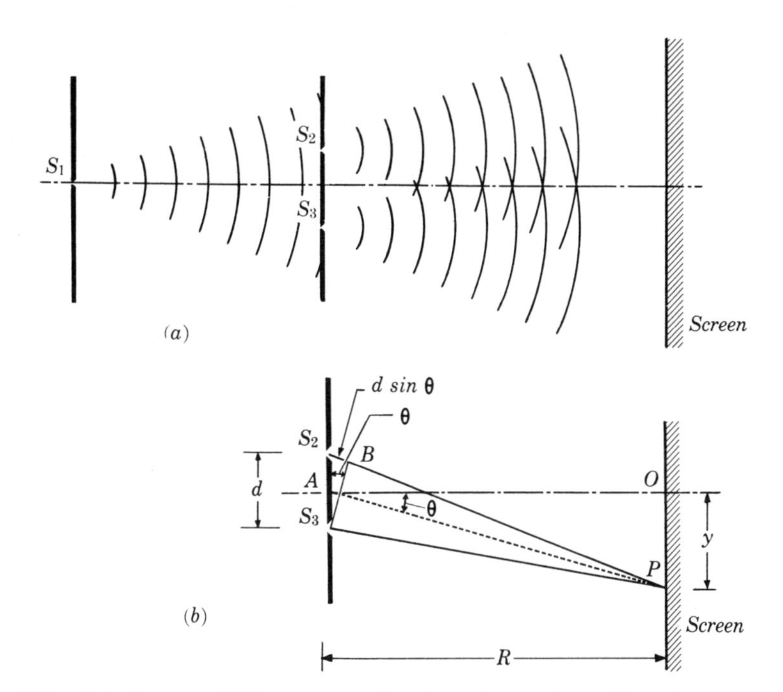

FIG. 8-11. Interference of light waves passing through two slits. Young's experiment.

According to Huygens' principle, cylindrical wavelets spread out from slit S_1 and reach slits S_2 and S_3 at the same instant. A train of Huygens wavelets diverges from both of these slits. Let d represent the distance between the slits, and consider a point P on a screen, in a direction making an angle θ with the axis of the system (Fig. 8-11 (b).) With P as a center, and PS_3 as radius, strike an arc intersecting PS_2 at B. If the distance R from slits to screen is large in comparison with the distance d between the slits, the arc S_3B can be considered a straight line at right angles to PS_3, PA, and PS_2. Then the triangle BS_2S_3 is a right triangle, similar to POA, and the distance S_2B equals $d \sin \theta$. This latter distance is the difference in path length between the waves reaching P from the two slits. The waves spreading out from S_2 and S_3 necessarily start out in phase, but they will not be in phase at P because of this difference in length of path. If λ is the wave length, the number of waves in the distance $d \sin \theta$ is $d \sin \theta/\lambda$, and since the phase of a wave increases by 2π radians in each wave length, the phase difference ϕ between the waves reaching P is

$$\phi = 2\pi \frac{d \sin \theta}{\lambda}.$$

If point P coincides with point O, then θ is zero and the phase difference ϕ is zero. The waves therefore reach O in phase, their amplitudes add, and there is a bright line at the center of the interference pattern. As P moves farther and farther out from the center, the angle θ and the phase difference ϕ both increase. When $\phi = \pi$ radians, or when

$$\sin\theta = \frac{\lambda}{2d},$$

the waves are exactly out of phase and there is a dark line on the screen. When $\phi = 2\pi$ radians, or

$$\sin\theta = \frac{\lambda}{d}, \tag{8-1}$$

there is another bright line, and so on. Evidently bright lines are observed when

$$\sin\theta = \frac{\lambda}{d}, \frac{2\lambda}{d}, \frac{3\lambda}{d}, \text{ etc.,}$$

and dark lines when

$$\sin\theta = \frac{\lambda}{2d}, \frac{3\lambda}{2d}, \frac{5\lambda}{2d}, \text{ etc.}$$

Hence by measuring the angle θ for a dark or bright line in the interference pattern, and the distance d between the slits, the wave length of the light can be computed.

8-8 Interference—many slits. An extremely useful optical device known as a *grating* consists, in one of its forms, of a large number of equidistant narrow slits side by side. Light waves spreading out from the slits interfere with one another in the same way as the waves from the two slits in Young's experiment. The first gratings were constructed by Joseph Fraunhofer in 1821, and consisted of a number of fine wires stretched across a frame, with spaces of a few hundredths or tenths of a millimeter between the wires. Gratings are now made by ruling equidistant lines with a diamond cutting tool on a glass or metal surface, the latter type being used in reflection rather than transmission. The spacing of the lines is of the order of a few ten-thousandths of an inch, and there may be as many as 50,000 lines in a grating.

The theory and applications of the grating are discussed more fully in the next chapter. At this point we shall give only an elementary analysis, using the same method as that in the preceding section. In Fig. 8-12, a slit S_1 is illuminated from the left by monochromatic light. Since we wish to have wavelets start out in phase from *all* of the slits in the grating, a

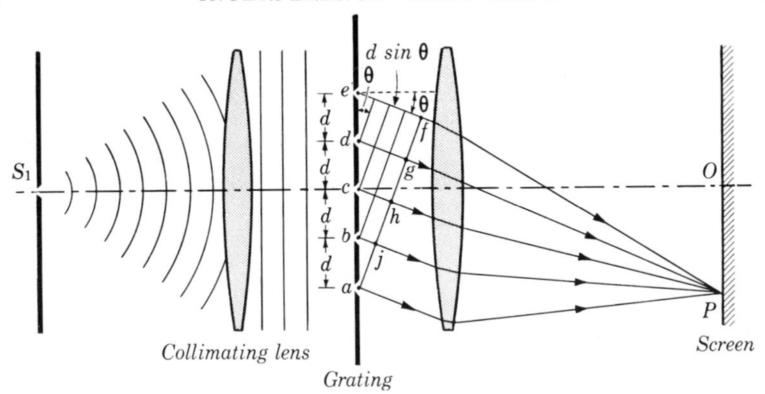

FIG. 8-12. Interference between light waves passing through a large number of slits Principle of the grating.

collimating lens is inserted between slit and grating, with the slit in its first focal plane. The wave fronts emerging from the lens are then planes at right angles to the axis of the system.

One might place a screen at the right of the grating in Fig. 8-12 and observe an interference pattern as in Fig. 8-11. In practice, however, the grating is almost always followed by a second lens as in Fig. 8-12, with a screen or photographic film in its second focal plane. Since a lens brings parallel rays to a focus in its second focal plane, the lens forms on the screen a reduced image of the pattern that would appear on a screen at infinity.

The wavelets diverging from the slits in the grating start out in phase, but travel along different paths in reaching the point P. Let us consider those portions of the wavelets that leave the grating in an arbitrary direction making an angle θ with the axis of the system. Construct the line af at right angles to this direction. We have shown in Sec. 3-1 that the number of waves is the same in all rays from a plane through af to the image point P. Hence the relative phases of the wavelets remain unchanged after they pass through this plane, and it suffices to consider only their relative phases at a, j, h, g, and f, where the rays from the slits intersect the line af. It is evident from the diagram that the distance bj equals $d \sin \theta$, ch equals $2d \sin \theta$, and so on. If the angle θ has such a value that bj equals just one wave length, then ch equals two wave lengths, and so on. At this particular angle the phase differences between the wavelets reaching P are 2π, 4π, etc., so that the amplitudes all add and the screen is bright along a line through P, parallel to the slits. That is, P lies on a bright line if

$$\sin \theta = \frac{\lambda}{d}, \quad \frac{2\lambda}{d}, \quad \frac{3\lambda}{d}, \text{ etc.}$$

Notice that these angles are exactly the same as those at which bright lines appear in the interference pattern of two slits. In what way, then, does the interference pattern of a large number of slits differ from that of two slits? The difference lies not in the *positions* of the maxima, but in the distribution of light on the screen between the maxima. The simple theory given above does not enable one to calculate this distribution, but we shall show in Sec. 9-3 that as the number of slits is increased the maxima become much *brighter* and much *narrower*. With only two slits, the brightness decreases gradually from that at a maximum to zero at a minimum, while with many slits it falls practically to zero at an extremely small angle away from a maximum and remains very nearly zero until the next maximum is reached. The increased sharpness of the line enables the angle θ to be determined with much greater precision and hence makes possible a more accurate measurement of wave length. The nature of the interference patterns of two, three, four, and five slits is shown in Fig. 9-2.

Problems—Chapter 8

(1) (a) Find the thickness of a soap film ($n = 1.33$) for a strong first-order reflection of yellow light, $\lambda = 600$ mμ (in vacuum). Assume normal incidence. (b) What is the wave length of the light in the film?

(2) Will an exceedingly thin *air* film between two glass plates appear bright or dark by reflected light? Why?

(3) A glass plate 0.40 micron thick is illuminated by a beam of white light normal to the plate. The index of refraction of the glass is 1.50. What wave lengths within the limits of the visible spectrum ($\lambda = 40 \times 10^{-6}$ cm to $\lambda = 70 \times 10^{-6}$ cm) will be intensified in the reflected beam?

(4) A continuous spectrum is projected normally on a glass plate coated with a uniform film of lacquer. As seen by reflection, the spectrum appears to have dark bands centered at 600 mμ and 428.6 mμ. The index of refraction of the glass is 1.600 and that of the lacquer is 1.500. Find the thickness of the lacquer film.

(5) Two rectangular pieces of plane glass are laid one upon the other on a table. A thin strip of paper is placed between them at one edge so that a very thin wedge of air is formed. The plates are illuminated by a beam of sodium light at normal incidence. Bright and dark interference bands are formed, there being ten of each per centimeter length of wedge measured normal to the edges in contact. Find the angle of the wedge.

(6) Monochromatic light is incident normally on a thin wedge-shaped film of transparent plastic of refractive index 1.40. The angle of the wedge is 10^{-4} radian. Interference fringes are observed, with a separation of 0.25 cm between adjacent bright fringes. Compute the wave length (in air) of the incident light.

(7) A wedge-shaped vertical soap film (index 4/3), 2.75 cm $\times$ 2.75 cm, is illuminated normally by red light of wave length 600 mμ (in vacuum). The upper edge of the film is observed to be black when viewed by reflected light. Six horizontal bright bands appear to traverse the film, the center of the sixth bright band coinciding with the bottom of the film. Find the angle of the wedge.

(8) The radius of curvature of the convex surface of a plano-convex lens is 30 cm. The lens is placed convex side down on a plane glass plate, and illuminated from above with red light of wave length 650 mμ. Find the diameter of the third bright ring in the interference pattern.

(9) Newton's rings are observed when a plano-convex lens is placed convex side down on a plane glass surface and the system is illuminated from above by monochromatic light. The radius of the first bright ring is 1 mm. (a) If the radius of the convex surface is 4 m, what is the wave length of the light used? (b) If the space between the lens and the flat glass surface is filled with water, what is the radius of the first bright ring?

(10) (a) Measure the rings in Fig. 8-6 and determine the radius of curvature of the convex surface. Assume a wave length of 600 mμ. (b) If the space between the surfaces in Fig. 8-6 were filled with water, what would be the diameter of the first dark ring? (c) Assuming the surface of the telescope objective and the test glass to be in contact at the dark spot at the center of Fig. 8-7, by how many wave lengths are they separated at the left edge of the lens?

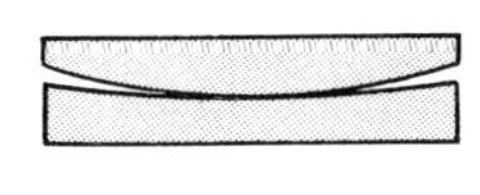

FIG. 8-13.

(11) The radius of curvature of the convex surface of a plano-convex lens is 200 cm. The lens is placed convex side down on the concave surface of a plano-concave lens as shown in Fig. 8-13. The radius of curvature of the concave surface is 400 cm. The lenses are illuminated from above with red light of wave length 625 mμ. Find the diameter of the third bright ring in the interference pattern seen by reflected light.

(12) The surfaces of a prism of index 1.52 are to be made "nonreflecting" by coating them with a thin layer of transparent material of index 1.30. The thickness of the layer is such that at a wave length of 550 mμ (in vacuum), light reflected from the first surface is 180° out of phase with that reflected from the second surface. (a) Find the thickness of the layer. (b) What is the phase difference for violet light of wave length 400 mμ? For red light of wave length 700 mμ? Assume normal incidence.

(13) A Michelson interferometer has mirrors (B and D in Fig. 8-10) which are 4 cm $\times$ 4 cm. An observer sees 20 vertical fringes across mirror B when the system is illuminated by sodium light of wave length 589 mμ. Find the angle between mirrors B and D.

(14) Find the distance between the silver layers in a Lippmann color photograph if formed by light of wave length 500 mμ (in vacuum). Take 1.53 as the index of the gelatine film.

(15) Let the distance d between the slits in Fig. 8-11 be 0.1 mm and the perpendicular distance to the screen be 50 cm. Compute the distance on the screen between the central maximum and the first maximum at either side, for violet light, $\lambda = 400$ mμ, and red light, $\lambda = 700$ mμ.

CHAPTER 9

DIFFRACTION

9-1 Diffraction. According to geometrical optics, if an opaque object is placed between a point light source and a screen as in Fig. 9-1, the edges of the object will cast a sharp shadow on the screen (the penumbra will be negligible if the source is sufficiently small). No light will reach the screen at points within the geometrical shadow, while outside the shadow the screen will be uniformly illuminated. The photograph reproduced in Fig. 9-8 was made by placing a razor blade halfway between a pinhole illuminated by monochromatic light and a photographic film, so that the film made a record of the shadow cast by the blade. Fig. 9-7 is an enlargement of a region near the shadow of an edge of the blade. A small amount of light has "bent" around the edge, into the geometrical shadow, which is bordered by alternate bright and dark bands. In the first bright band, just outside the geometrical shadow, the illumination is actually greater than in the region of uniform illumination at the extreme left.

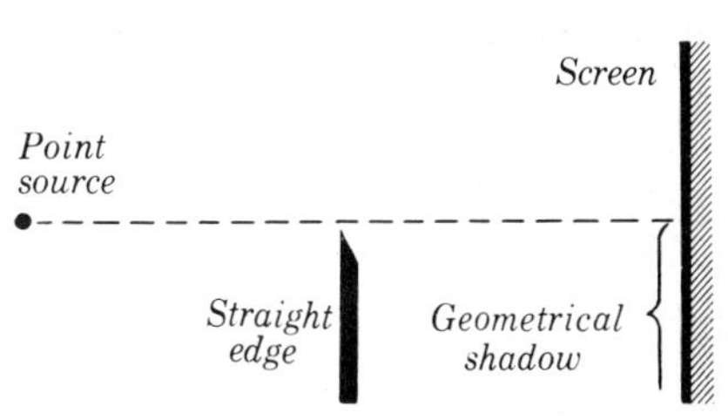

Fig. 9-1. Geometrical shadow of a straight edge.

This simple experiment serves to give some idea of the true complexity of what is often considered the most elementary of optical phenomena, the shadow cast by a small source of light.

The reason that a pattern like that of Fig. 9-8 is not commonly observed in the shadow of an object is merely that most light sources are not point sources. If a shadow of a razor blade is cast by a frosted bulb incandescent lamp, for example, the light from every point of the surface of the lamp forms its own pattern of bright and dark lines, but these overlap to such an extent that no individual pattern can be observed.

The term *diffraction* is applied to problems such as this, in which one is concerned with *the resultant effect produced by a limited portion of a wave front.* Since in most diffraction problems some light is found within the region of geometrical shadow, diffraction is sometimes defined as "the bending of light around an obstacle." It should be emphasized, however, that the process by which diffraction effects are produced is going on continuously in the propagation of every wave front. Only if a part of the wave front is cut off by some obstacle are the effects commonly called "diffraction effects" observed. But since every optical instrument does in fact make use of only a limited portion of a wave front (a telescope, for example, utilizes only that portion of a wave front admitted by the objective lens) it is evident that a clear comprehension of the nature of diffraction is essential for a complete understanding of practically all optical phenomena.

The main features observed in diffraction effects can be predicted with the help of Huygens' principle, according to which every point of a wave front can be considered the source of a secondary wavelet which spreads out in all directions. The shape of the wave front at a later time is found by constructing the envelope of the secondary wavelets. As proposed by Huygens, this principle was a useful graphical method for finding the shape of the new wave front and not a great deal of physical reality was attached to the wavelets themselves. Later on, when the wave nature of light came to be more fully understood, it was realized that Huygens' method had a deeper significance than was at first supposed.

In Fig. 9-10, O represents a point source of electromagnetic waves. From the known laws of propagation of these waves we can find the energy reaching some other point P. An alternate method, however, is to break the problem up into two steps. We surround the point O with an arbitrary closed surface S, and compute the amplitude and phase of the waves from O at every infinitesimal element of area dS of this surface. Each of these elements is then treated as a small source sending out secondary wave trains. The waves from all elements pass through the point P, and the disturbance at this point is found by combining these wave trains with proper amplitude and phase. The problem is therefore essentially one in interference, except that instead of dealing with two wave trains, we now have an infinite number of infinitesimal wave trains.

The complete theory, which we cannot go into here, shows that the amplitude of the secondary wave train originating at any element is proportional to the area of that element and inversely proportional to the distance of the element from the source O. (The "amplitude" can be taken as either that of the E-vector or the H-vector in the wave.) As each secondary wave advances, its amplitude decreases in inverse proportion to the distance traveled. Also, the amplitude of the secondary wave is a maximum in the direction radially outward from the source O and falls to zero in the opposite direction. This latter property of the secondary waves removes one of the difficulties with the simpler form of Huygens' principle, namely, that if the wavelets spread out in *all* directions from each infinitesimal source they should combine to give a wave traveling

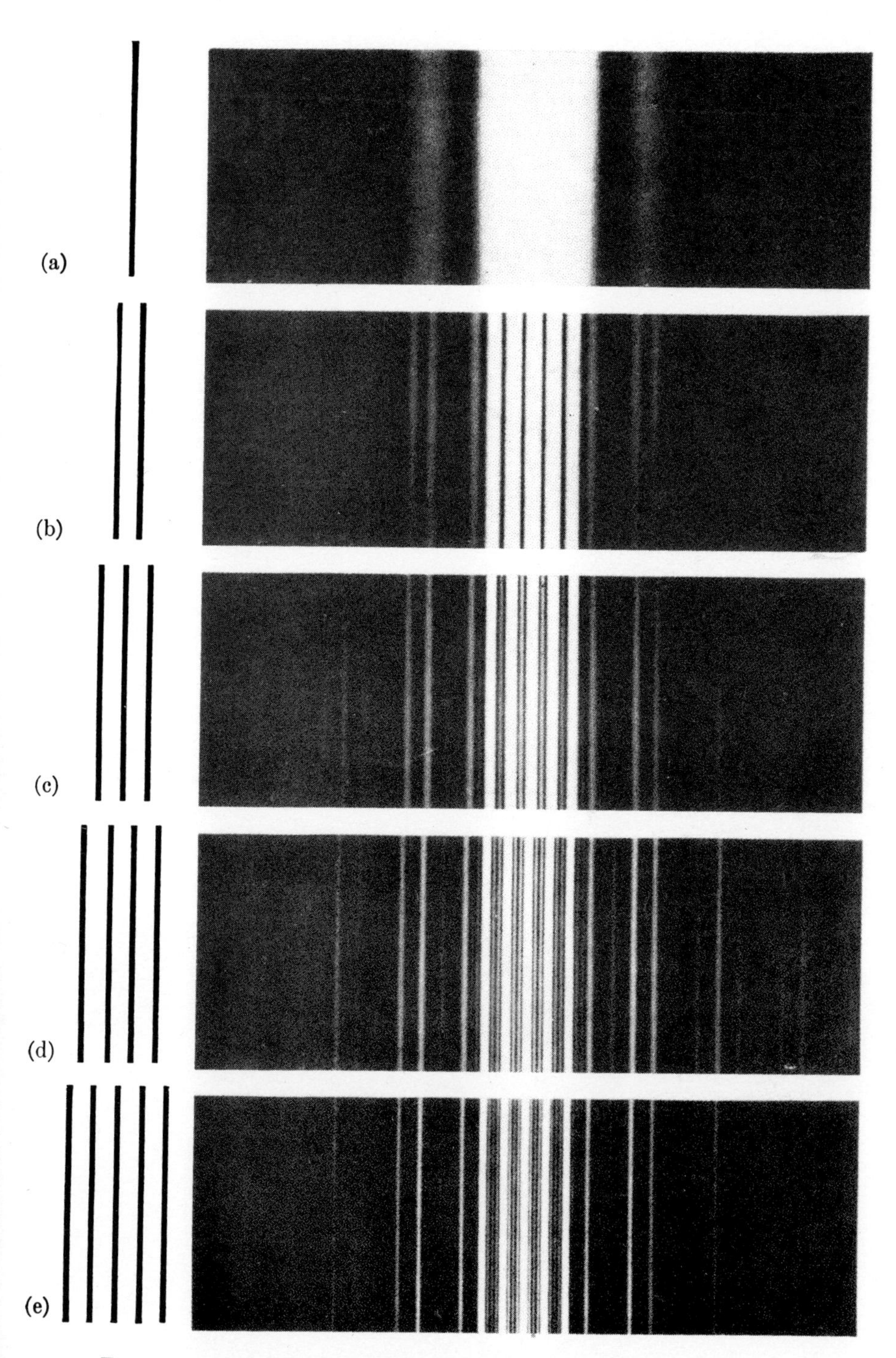

FIG. 9–2. (a) Fraunhofer diffraction pattern of a single slit. (b), (c), (d), (e) Interference patterns of 2, 3, 4, and 5 slits respectively. The slit systems are indicated at the left.

FIG. 9–3. Fraunhofer diffraction pattern of a single slit.

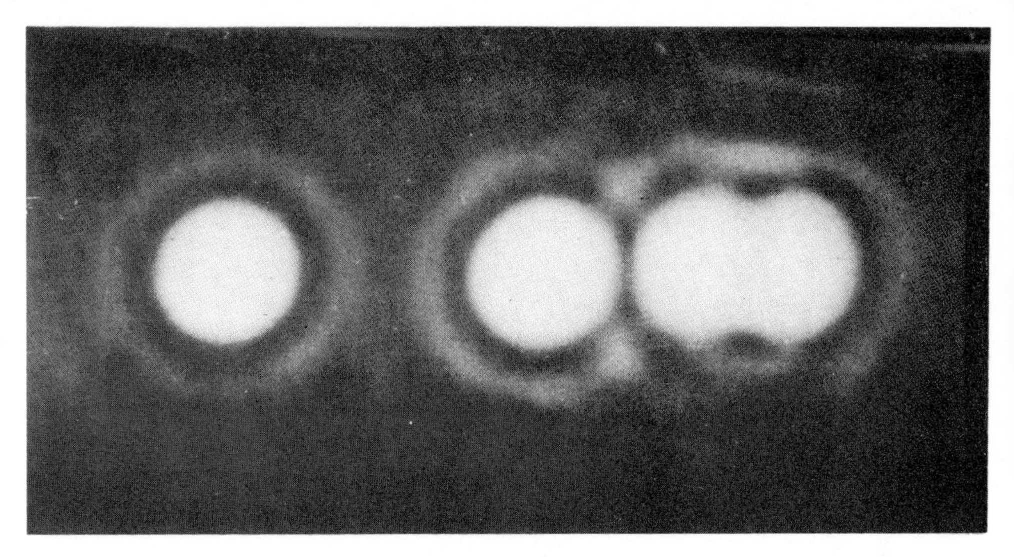

(a)

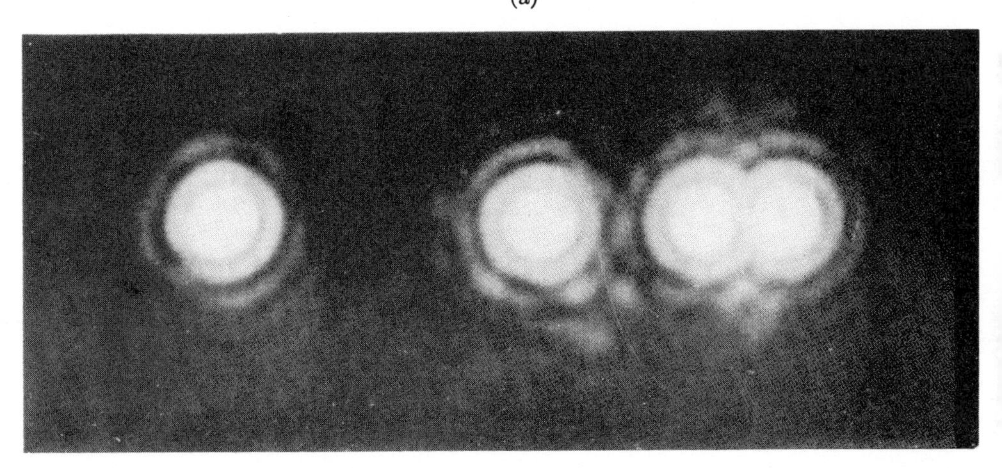

(b)

(c)

Fig. 9–4. Fraunhofer diffraction patterns of four "point" sources, with a circular opening in front of the lens. In (a), the opening is so small that the patterns at the right are just resolved by Rayleigh's criterion. Increasing the aperture decreases the size of the diffraction patterns as in (b) and (c).

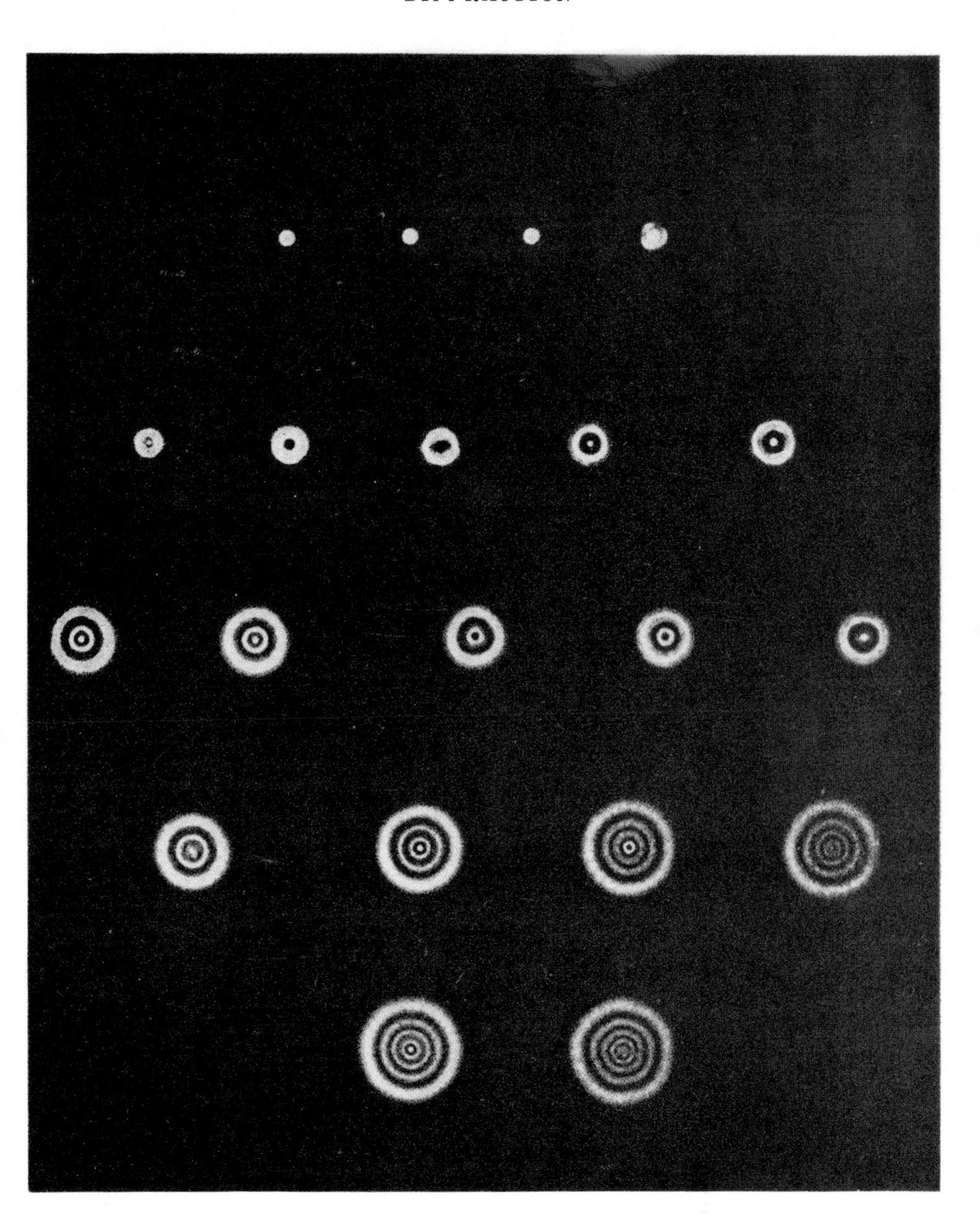

FIG. 9–5. Fresnel diffraction patterns of circular holes of various sizes.

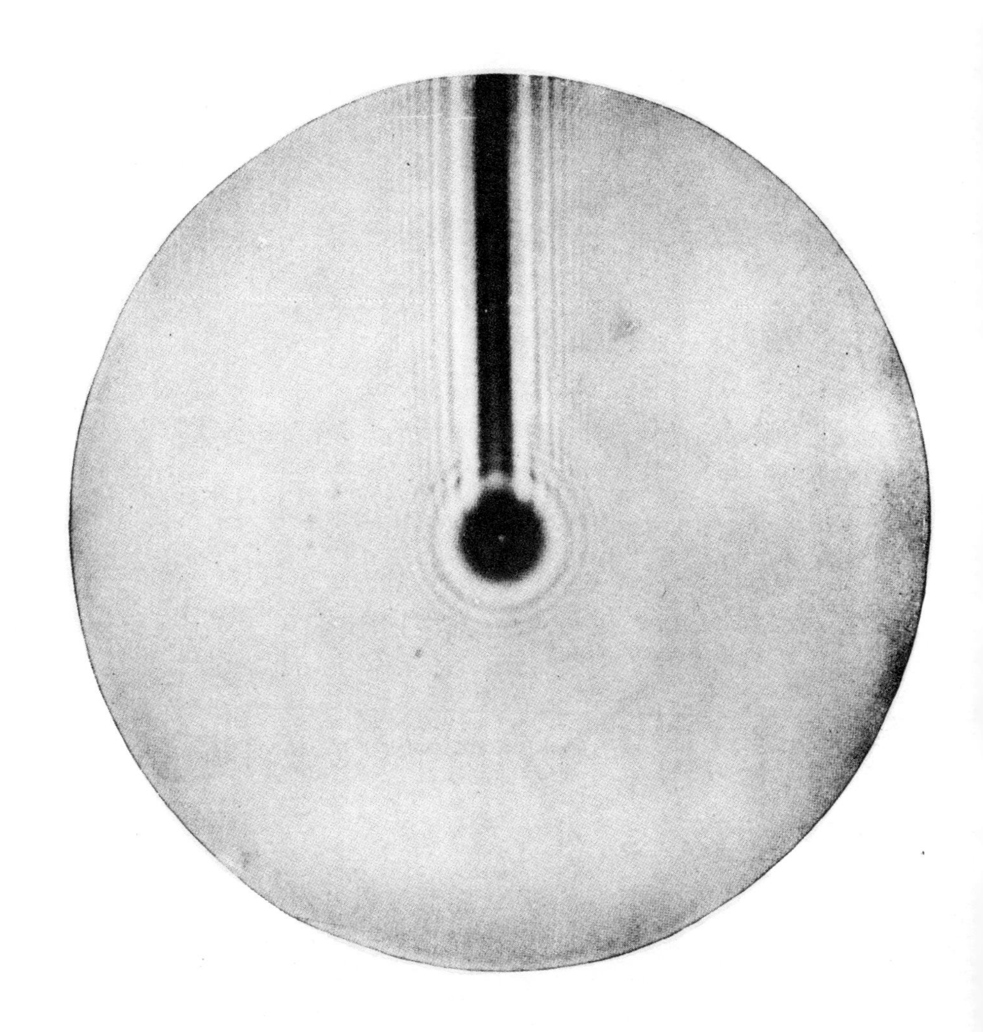

FIG. 9–6. Shadow of a ball bearing.

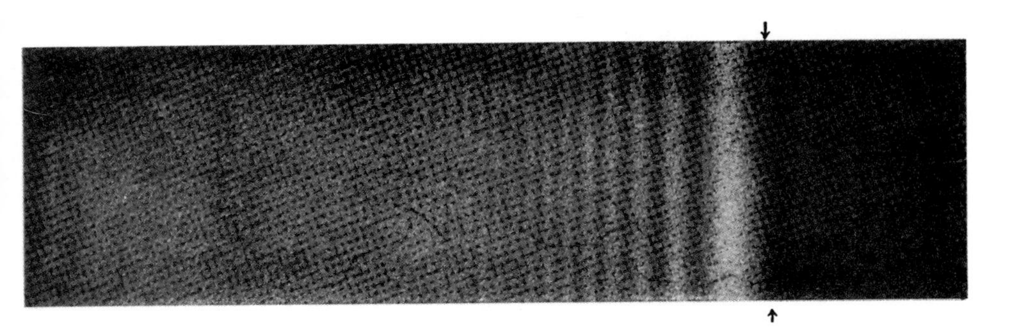

Fig. 9–7. Shadow of a straight edge.

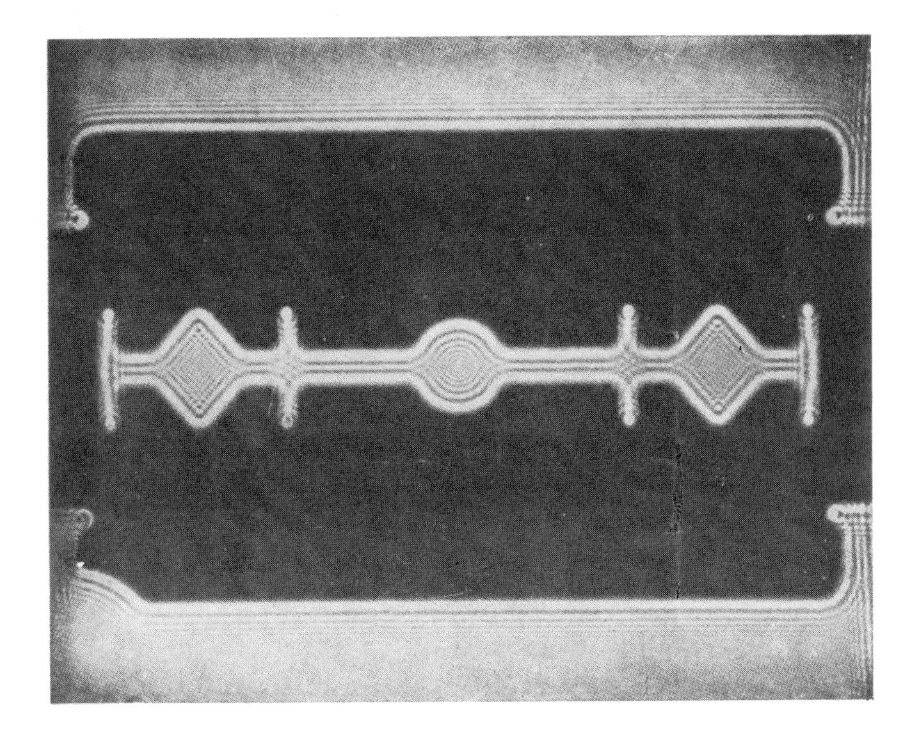

Fig. 9–8. Shadow of a razor blade.

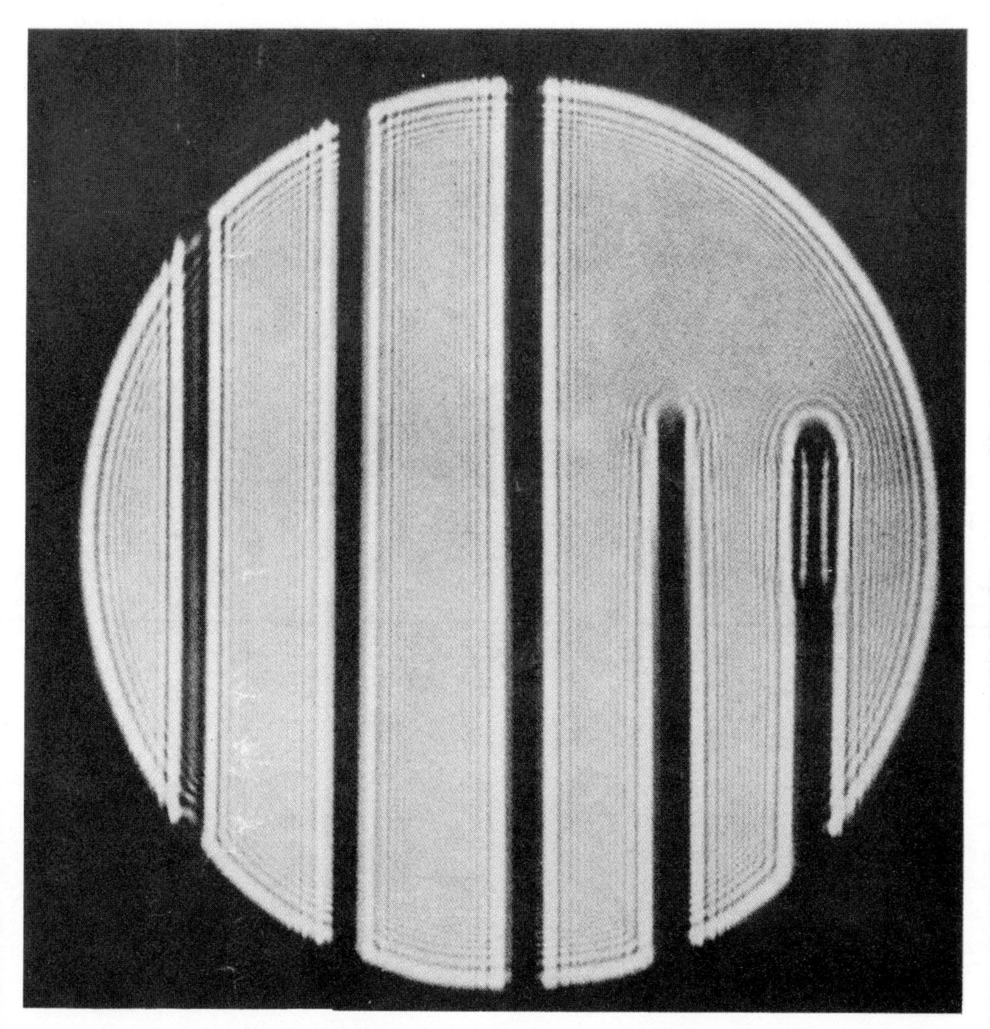

FIG. 9–9. Shadows of needles of various sizes.

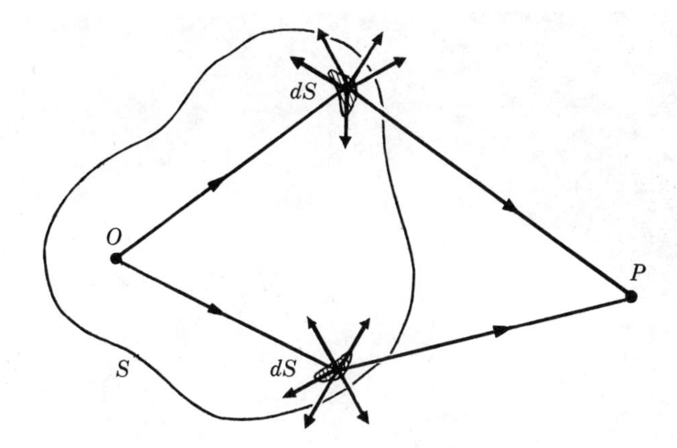

FIG. 9-10. Each element dS of a closed surface S surrounding the point source O, can be considered a secondary source.

backward as well as forward. Since the amplitude at the rear of the waves is zero, there will evidently be no "back" wave.

The extension of Huygens' principle, then, is first that each element of a surface through which a train of electromagnetic waves is passing sends out a *continuous train* of waves instead of the single spherical pulses of the simple theory and, second, that the amplitude at any point is found by combining these waves according to the principles of interference, instead of graphically by drawing the envelope of the secondary wavelets.

At first sight, it seems as if we had simply gone out of our way to make the problem more complicated, since the energy reaching point P from the source O, in Fig. 9-10, could have been computed directly. It develops, however, that if there is an obstacle between O and P as in Fig. 9-1, so that a portion of the waves from O is obstructed, the effect at P of the unobstructed portion of the waves can, in many instances, be computed much more readily with the help of the generalized form of Huygens' principle.

9-2 Diffraction by a slit. We consider first the diffraction of light by a narrow slit. In the drawing of Fig. 9-11, a beam of parallel monochromatic light is incident from the left on an opaque plate in which there is a narrow horizontal slit. According to the principles of geometrical optics, the transmitted beam has the same cross section as the slit and a screen in the path of the beam is illuminated uniformly over an area of the same size and shape as the slit, as in part (a) of Fig. 9-11. Actually, what one observes on the screen is the diffraction pattern shown in part (b). The

FIG. 9-11. (a) Geometrical "shadow" of a slit. (b) Diffraction pattern of a slit. The slit width has been greatly exaggerated.

beam spreads out vertically after passing through the slit, and the diffraction pattern consists of a central bright band, which may be much wider than the slit width, bordered by alternating dark bands and bright bands of decreasing intensity. A diffraction pattern of this nature can readily be observed by looking at a point source such as a distant street light through a narrow slit formed between two fingers in front of the eye. The retina of the eye then corresponds to the screen in Fig. 9-11.

The width of the slit in Fig. 9-11 has been greatly exaggerated for clarity. To obtain a vertical spread having the proportions shown in the figure, the slit width would have to be of the order of 5 wave lengths of light. (See page 232.) There is also a small transverse spreading of the beam, which has been omitted in the diagram for simplicity.

Fig. 9-3 is an enlargement of a photograph made by placing a photographic film in the plane of the screen in a setup like that in Fig. 9-11.

Let us now apply the extended form of Huygens' principle, described in Sec. 9-1, to compute the distribution of light on the screen. The infinitesimal elements of area dS (see Fig. 9-10) are obtained by subdividing the wave front passing through the slit into narrow strips, parallel to the long edges of the slit. A section through the slit is shown in Fig. 9-12 (a). The division of the wave front into narrow strips, which are seen end on, is indicated by the short lines across the wave front. From each of these strips, secondary wavelets spread out in all directions as shown.

In Fig. 9-12 (b), a screen is placed at the right of the slit and P is one point on a line in the screen, the line being parallel to the long edges of the slit and perpendicular to the plane of the diagram. The light reaching a point on the line is calculated by applying the principle of superposition to all the wavelets arriving at the point, from all the elementary strips of the wave front. Because of the varying distances to the point, and the varying angles with the original direction of the light, the amplitudes and phases of the wavelets at the point will be different.

The problem is greatly simplified when the screen is sufficiently distant, or the slit sufficiently narrow, so that all rays from the slit to a point on the screen can be considered parallel, as in Fig. 9-12 (c). The former case, where the screen is relatively close to the slit (or the slit is relatively wide) is referred to as *Fresnel* diffraction, the latter as *Fraunhofer* diffraction. There is, of course, no difference in the nature of the diffraction process in the two cases, and Fresnel diffraction merges gradually into Fraunhofer diffraction as the screen is moved away from the slit, or as the slit width is decreased.

Fraunhofer diffraction occurs also if a lens is placed just beyond the slit as in Fig. 9-12 (d), since the lens brings to a focus in its second focal plane all light traveling in a specified direction. That is, the lens forms in its focal plane a reduced image of the pattern that would appear on an infinitely distant screen in the absence of the lens.

The photograph in Fig. 9-3 is a Fraunhofer diffraction pattern.

When a narrow slit is placed in front of a lens as in Fig. 9-13, the image of a distant *axial* point source P_2 is spread out into the diffraction pattern P_2'. Point sources P_1 and P_3, not on the lens axis, are imaged as shown at P_1' and P_3'. If there are point sources all along the line P_1P_3, the images of these points merge to give a diffraction pattern like that in Fig. 9-3. The usual way of providing a line source is to place a slit along the line P_1P_3, illuminate this slit from the left, and then in effect move the slit to infinity by inserting between it and the lens in the diagram a second lens called a *collimating* lens, with the illuminated slit in its first focal plane.

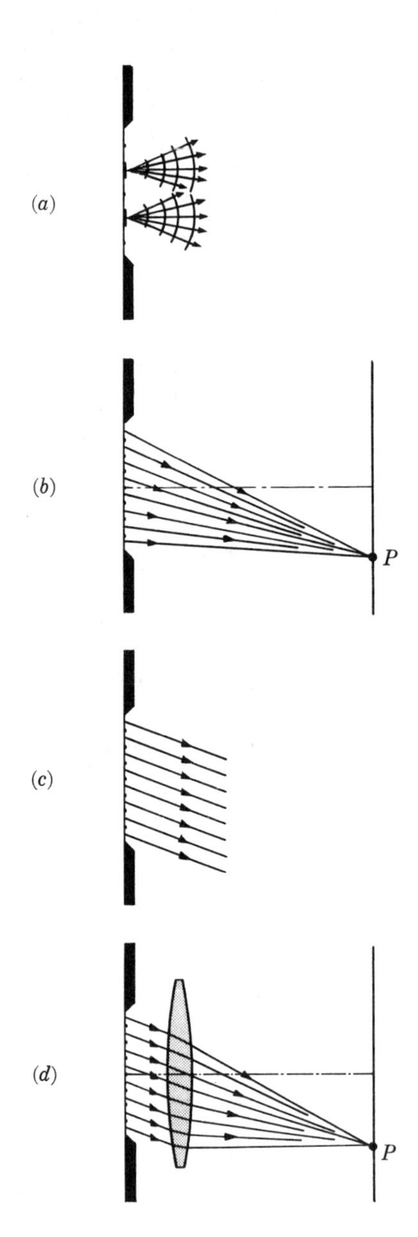

FIG. 9-12. Diffraction by a slit.

Because of its relative simplicity, we shall consider only Fraunhofer diffraction at this point. Fresnel diffraction is discussed in Sec. 9-8. Some of the important aspects of Fraunhofer diffraction by a slit can be deduced very easily. Consider the two infinitesimal strips in the wave front passing through a slit, one just below the upper edge of the slit and the other just below its center line, as in Fig. 9-14, and the wavelets from these strips that travel in a direction making an angle α

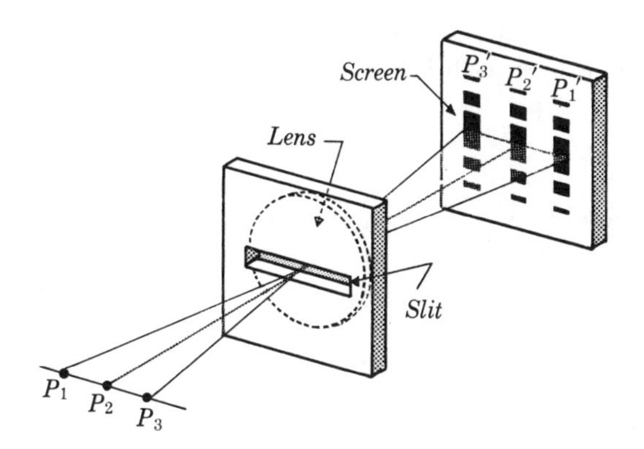

FIG. 9-13.

with the direction of the incident light. The two wave trains start out from the plane of the wave front in phase, but the upper one has to travel a greater distance than the lower before reaching the screen. This additional distance, from Fig. 9-14, is

$$\frac{D}{2} \sin \alpha,$$

where D is the slit width. We have shown in Sec. 3-1 that when parallel rays are imaged by a lens, there are the same number of waves in all rays from points in a plane at right angles to the rays, to the image. There-

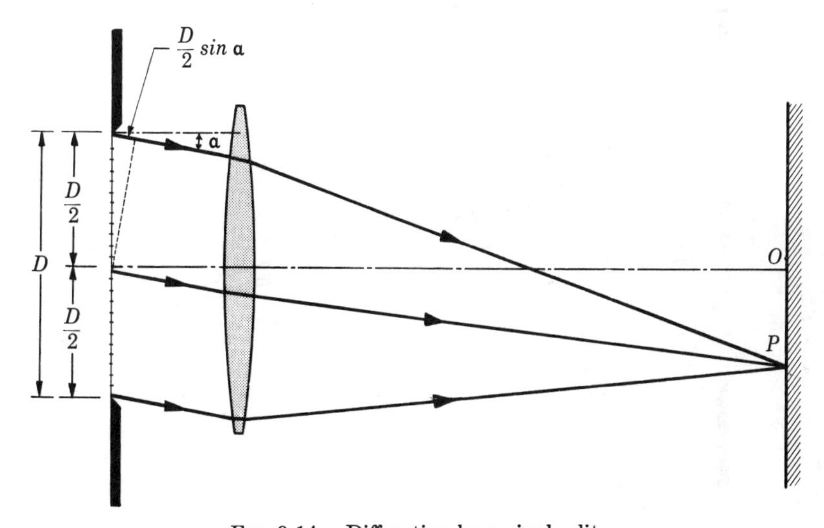

FIG. 9-14. Diffraction by a single slit.

fore the only phase difference between the waves arriving at P is that arising from the excess path length of the upper ray,

$$\frac{D}{2} \sin \alpha,$$

For those points on the screen that lie on a line through O, opposite the center of the slit, the angle α is zero and the path difference is zero. The wavelets from *all* strips on the transmitted wave front therefore reach points on this line in phase with one another, their amplitudes add, and the center of the diffraction pattern is bright. As we consider points farther and farther out from the center, the angle α increases and the path difference increases. When the path difference has become equal to one-half a wave length, the wavelets from the two strips in Fig. 9-14 reach the screen out of phase and complete destructive interference results. The path difference between the wavelets from the two strips next below those in Fig. 9-14 is also one-half wave length, so that these two wavelets cancel one another also. Proceeding in this way, it is seen that the light from each element in the upper half of the slit is cancelled by the light from the corresponding element in the lower half. Hence no light reaches the screen at points along a line, the direction to which lies at an angle α above or below the original direction of the light, provided that

$$\frac{D}{2} \sin \alpha = \frac{\lambda}{2},$$

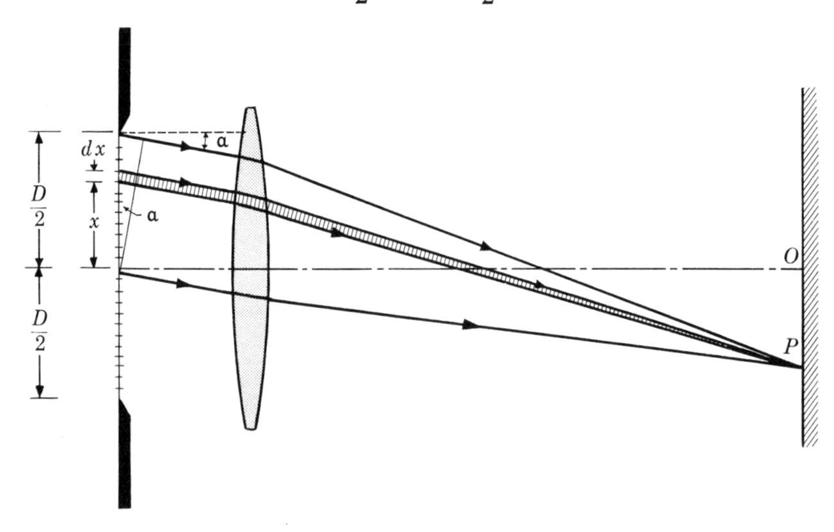

FIG. 9-15.

or

$$\sin \alpha = \frac{\lambda}{D}. \tag{9-1}$$

Similar reasoning shows that the screen is again dark when

$$\sin \alpha = \frac{2\lambda}{D}, \frac{3\lambda}{D}, \text{ etc.}$$

The preceding simple analysis gives the angular positions of the lines of zero intensity in the diffraction pattern, but does not tell how the intensity varies from point to point. A more complete discussion follows.

Let us as before subdivide the transmitted wave front into infinitesimal strips, all of equal width dx as in Fig. 9-15. At point P, the amplitude dE_m of the electric vector in the wavelet spreading out from any one strip is proportional to the width of the strip, dx, and we shall write it as

$$dE_m = kdx,$$

where k is a proportionality constant. Consider first the wavelet originating at the central strip in the wave front. If f is the frequency of the wave, the instantaneous electric intensity dE at point P due to this wavelet can be written

$$dE = dE_m \sin 2\pi ft = kdx \sin 2\pi ft.$$

The wavelet originating at a strip at a height x above the center line travels a greater distance than the wavelet originating at the center, by an amount $x \sin \alpha$. Since the phase of the wave train increases by 2π radians in a distance of one wave length, the phase difference ϕ at point P, between this wavelet and the wavelet from the central strip, is

$$\phi = 2\pi \frac{x \sin \alpha}{\lambda}.$$

Hence the electric intensity at P due to this wavelet is

$$\begin{aligned} dE &= dE_m \sin (2\pi ft - \phi) \\ &= kdx \sin \left(2\pi ft - 2\pi \frac{x \sin \alpha}{\lambda}\right). \end{aligned}$$

The resultant electric intensity at P due to the wavelets from all of the strips can now be found by integrating over the width of the slit. This gives

$$E = \int dE = \int_{-\frac{D}{2}}^{+\frac{D}{2}} k \sin\left(2\pi ft - 2\pi \frac{x \sin \alpha}{\lambda}\right) dx.$$

The integral is of the form

$$\int a \sin (b - cx)\, dx,$$

where a, b, and c are constants. The integration is straightforward and yields

$$E = \left[kD \frac{\sin\left(\pi D \frac{\sin \alpha}{\lambda}\right)}{\pi D \frac{\sin \alpha}{\lambda}} \right] \sin 2\pi ft. \tag{9-2}$$

Since E varies with $\sin 2\pi ft$, the phase of the resultant is the same as the phase of the central wavelet. We shall need to make use of this fact later.

The term in brackets in Eq. (9-2) is the amplitude of the resultant oscillating electric field at point P. Let us represent this by E_m.

$$E_m = kD \frac{\sin\left(\pi D \frac{\sin \alpha}{\lambda}\right)}{\pi D \frac{\sin \alpha}{\lambda}}. \tag{9-3}$$

For brevity, we define a new quantity z by the equation

$$z = \pi D \frac{\sin \alpha}{\lambda}. \tag{9-4}$$

Then the amplitude at any point is

$$E_m = kD \frac{\sin z}{z}. \tag{9-5}$$

The amplitude is evidently not the same at all points, because although k and D do not change from point to point, the term $(\sin z/z)$ does, since z depends on the angle α through Eq. (9-4). The quantity of light incident at any point on the screen is proportional to the square of the electric vector, so if one is interested only in the *relative* amounts of light reaching different points, it suffices to consider the relative magnitudes of the quantity

$$\left(\frac{\sin z}{z}\right)^2$$

at different points, or at different angles α.

Fig. 9-16 is a graph of $(\sin z/z)^2$, as a function of z and of $\sin \alpha$, shown above a drawing of the diffraction pattern to which it applies. At the center of the pattern,

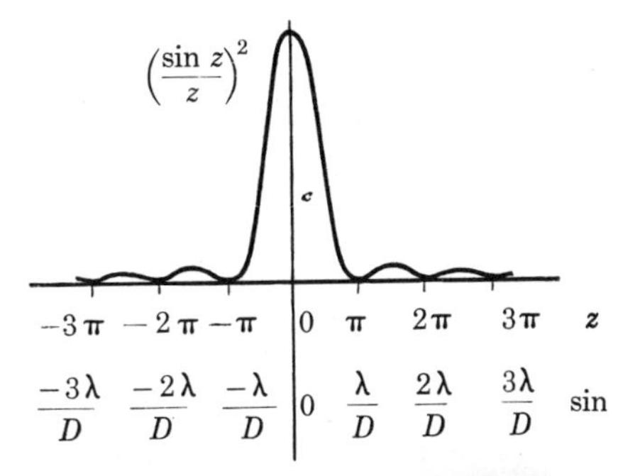

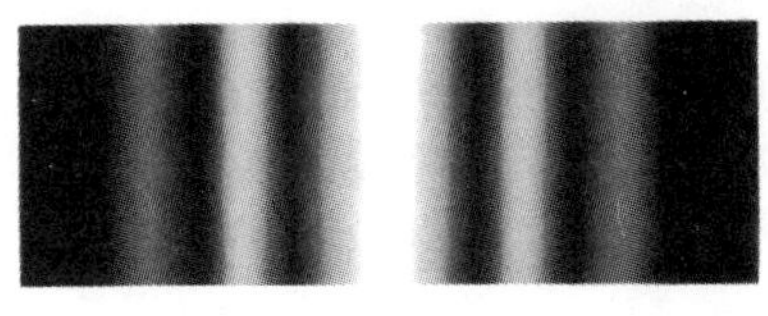

FIG. 9-16. Graph of intensity distribution in the Fraunhofer diffraction pattern of a narrow slit.

$$\alpha = 0,$$

$$\sin \alpha = 0,$$

$$z = \pi D \frac{\sin \alpha}{\lambda} = 0,$$

$$\sin z = 0,$$

and

$$\left(\frac{\sin z}{z}\right)^2 = \left(\frac{\sin 0}{0}\right)^2 = 1,$$

since $\sin z = z$ when z is very small.

As we consider points farther out from the center, z and $\sin z$ both increase from zero, since $\sin \alpha$ increases. However, z increases at a faster rate than $\sin z$, so the magnitude of $(\sin z/z)^2$ decreases from its value of 1 at the center. It becomes equal to zero when $z = \pi$ radians, since at this angle $\sin z = 0$ while z is finite. But

$$z = \pi D \frac{\sin \alpha}{\lambda},$$

so when $z = \pi$,

$$\sin \alpha = \frac{\lambda}{D}.$$

That is, the more complete theory predicts the first minimum at exactly the same angle as the simple theory on page 229.

Beyond the first minimum $(\sin z/z)^2$ increases to a secondary maximum and again becomes zero when $z = 2\pi$, or when

$$\sin \alpha = \frac{2\lambda}{D},$$

again in agreement with the simple theory. The secondary maximum occurs nearly, though not exactly, halfway between the first and second minima, that is, when $z = 3\pi/2$. (See problem 4 on page 255.) At this angle,

$$\left(\frac{\sin z}{z}\right)^2 = \left(\frac{\sin \frac{3\pi}{2}}{\frac{3\pi}{2}}\right)^2 = \left(\frac{1}{4.71}\right)^2 = \frac{1}{22}\cdot$$

The intensity at the mid-line of the central band, where $(\sin z/z)^2 = 1$, is therefore over 20 times as great as that at the next adjacent maximum, and most of the light is concentrated in the central band. It is easy to show that other minima occur when

$$\sin \alpha = \frac{3\lambda}{D}, \quad \frac{4\lambda}{D}, \text{ etc.,}$$

with maxima of decreasing intensity approximately halfway between them. Thus the more complete theory predicts not merely the positions of the minima, or the dark lines in the diffraction pattern, but the positions of the maxima as well, and their relative intensities.

The angle corresponding to the first minimum at either side of the center is called the *half-angular breadth* of the central band. The angular breadth of the entire central band is twice as great. The half-angular breadth is proportional to the wave length λ and inversely proportional to the slit width D. It was stated earlier that in order to obtain a divergence of the diffracted beam like that shown in Fig. 9-11, the slit width should be about 5 wave lengths. We can now justify this statement. If the slit width D equals 5 wave lengths, the half-angular breadth of the central band is

$$\sin \alpha = \frac{\lambda}{D} = \frac{\lambda}{5\lambda} = 0.20,$$

$$\alpha = 12^\circ \text{ (very nearly)},$$

which is approximately the angular divergence shown in Fig. 9-11. Let y represent the half-*linear* width of the central band, and assume that the screen is 20 cm beyond the slit. The half-angle α is approximately

$$\alpha = \sin \alpha = \frac{y}{20} = 0.20 \text{ radian},$$

and hence

$$y = 20 \times 0.20 = 4 \text{ cm}.$$

The central band is therefore 8 cm wide, although the width of the slit is only 5 wave lengths of light.

When the slit width is just one wave length,

$$\sin \alpha = 1, \quad \alpha = 90°,$$

and the central band spreads out over an angle of 180°.

9-3 The plane diffraction grating. The elementary theory of the diffraction grating was given in Sec. 8-8, where we referred to the device merely as a "grating." We can now see that the grating combines a problem in diffraction with one in interference. That is, each slit in the grating gives rise to a diffracted beam in which the intensity distribution is a function of slit width, and these diffracted beams then interfere with one another to produce the final pattern.

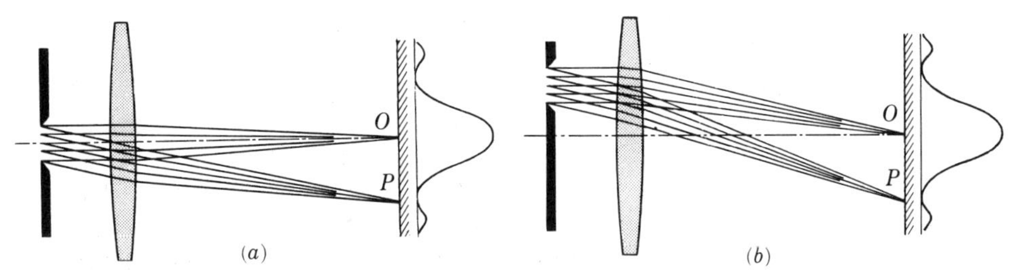

FIG. 9-17. The diffraction pattern of the slit is the same in (a) and (b).

To begin with, let us point out one property of the Fraunhofer diffraction pattern of a slit that was not mentioned in Sec. 9-2. Fig. 9-17 (a) shows a single slit in front of a lens, with a screen in the second focal plane of the lens. Parallel monochromatic light is assumed incident from the left, and the center line of the slit lies on the lens axis. Two diffracted beams are indicated, one giving rise to the central maximum in the diffraction pattern, the other to the first minimum below the center. The central maximum lies on the lens axis.

In Fig. 9-17 (b), a slit of the same width as in part (a) is located above the lens axis. Neglecting lens aberrations, the intensity distribution in the diffraction pattern of this slit is exactly the same as in part (a), and the pattern is centered at the same point on the screen, that is, on the axis of the *lens*, not on a line opposite the center of the slit. A moment's thought will show that this must be the case, since *all* rays parallel to the lens axis are imaged at its second focal point.

Now suppose we have two slits as in Fig. 9-18, each of width D and with a center-to-center spacing d. This is the arrangement used by

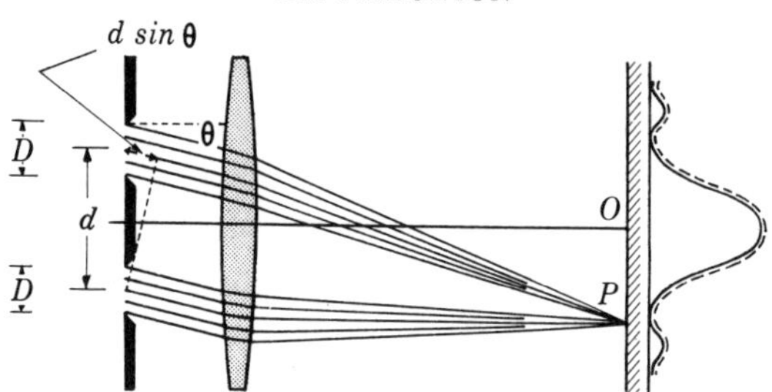

FIG. 9-18. Interference between the light beams diffracted by two slits.

Young in studying interference effects. The present discussion is a more complete theory of Young's experiment, and it will give us not merely the *positions* of the maxima and minima in the interference pattern, but the intensity distribution as well. From the discussion in the preceding paragraph, it follows that either slit *alone* would give rise to diffraction patterns of equal intensity and in the same position on the screen. The intensity distribution in these patterns is shown by the heavy solid and dotted lines at the right of Fig. 9-18. Actually, of course, the graphs should coincide. The question now is, what will be the nature of the diffraction pattern due to the two slits? It would appear at first that with two slits the pattern on the screen would merely be twice as bright as with one slit. What is actually observed is shown in the photograph in Fig. 9-2 (b). In Fig. 9-2 (a) one sees the diffraction pattern of a single slit. In Fig. 9-2 (b), where the screen receives light from two slits, a number of much sharper lines appear, with the distances between their minima less than that between the minima in the diffraction pattern.

Although the diffraction patterns of the individual slits are identical as far as their intensity and position are concerned, this is not true of the *phases* of the waves in the light from the two slits (except at the mid-point of the pattern). The mathematical analysis in Sec. 9-2 shows that the phase of the resultant electric vector in a diffracted beam, at any point on the screen, is the same as the phase of the central ray in the beam. By symmetry, the phase of the electric vector in the wave from the upper slit, at point O in Fig. 9-18, is the same as the phase of the electric vector in the wave from the lower slit. The amplitudes of the waves are equal, so the resultant at O has twice the amplitude, and four times the intensity, of the wave from a single slit.

At a point such as P, while the resultant amplitudes of the waves from the two slits are equal (although different from the amplitude at O) the phases of the waves are not the same. The central ray from the upper slit travels a greater distance than that from the lower slit, by an amount

$d \sin \theta$ (see Fig. 9-18). Since there is a phase change of 2π radians in each wave length, the phase difference at P between the two waves is

$$\phi = 2\pi \frac{d \sin \theta}{\lambda}. \tag{9-6}$$

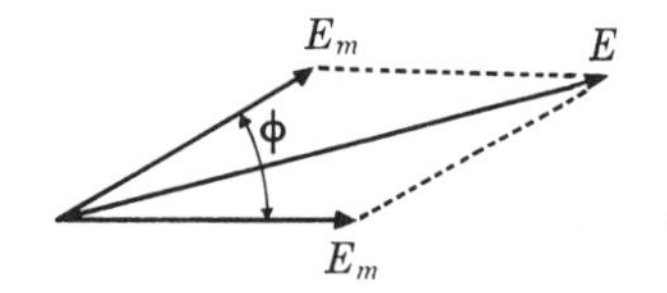

Fig. 9-19.

Let E_m represent the amplitude at P of the wave from either slit. (See Eq. (9-3).) The problem of computing the amplitude of the resultant of the two waves is the same as that of finding the resultant of two currents or voltages of equal amplitude, differing in phase by an angle ϕ. From Fig. 9-19, which may be considered a rotating vector diagram, the resultant E is seen to be

$$E = 2E_m \cos \frac{\phi}{2}.$$

The intensity at any point is proportional to E^2, or to

$$E_m{}^2 \times 4 \times \cos^2 \frac{\phi}{2}. \tag{9-7}$$

Thus the resultant amplitude varies from point to point for two reasons, first because E_m varies from point to point according to Eq. (9-3), and second because ϕ varies as given by Eq. (9-6). Notice that at any specified point P the angles α and θ are equal. It seems preferable to use different symbols for them, the former when we are considering the light *diffracted* by a single slit, the latter when considering the *interference* between the light from two (or more) slits.

Part (a) of Fig. 9-20 is a graph of the intensity distribution in the light diffracted by a single slit of width D. It is the same as the right side of Fig. 9-16, except that the horizontal scale has been expanded to cover only slightly more than the half-angular width of the central band. The ordinate of the curve is proportional to $E_m{}^2$, and for the purposes of this discussion let us assume it is *equal* to $E_m{}^2$.

The solid line in Fig. 9-20 (b) is a graph of $\cos^2 \frac{\phi}{2}$. The abscissa can be considered either as the phase difference ϕ, or as $\sin \theta$, since by Eq. (9-6) the two are proportional.

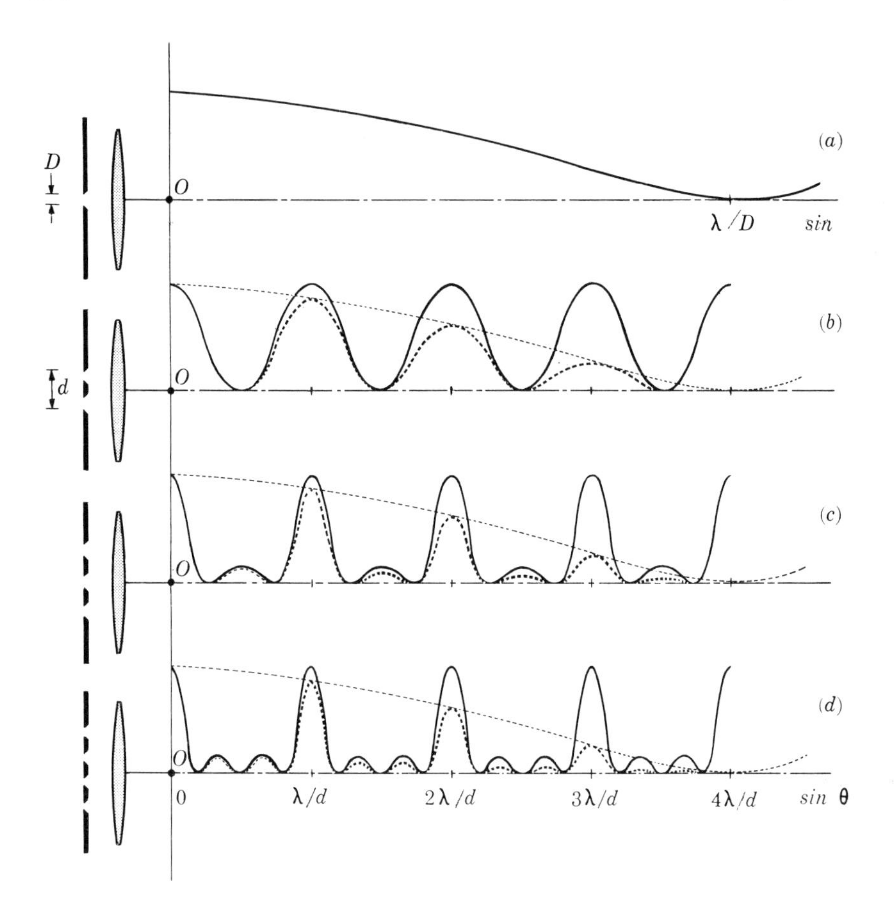

FIG. 9-20. Diffraction and interference patterns of 1, 2, 3, and 4 slits. Compare with the photographs in Fig. 9-2.

The resultant intensity at any point (Eq. (9-7)) is obtained by multiplying the ordinate of the curve in part (a), (i.e., E_m^2), by 4 times the ordinate of the solid curve in part (b) $\left(\text{i.e., } \cos^2 \frac{\phi}{2}\right)$. This product is shown by the heavier dotted curve in part (b), except that the factor 4 has been omitted to keep the diagram of reasonable height.

The maxima in the interference pattern occur when the phase difference ϕ is

$$\phi = 0,\ 2\pi,\ 4\pi,\ \text{etc.},$$

or, from Eq. (9-6), when

$$\sin\theta = m\frac{\lambda}{d}\cdot \qquad (m = 0, 1, 2, 3, \text{etc.}) \tag{9-8}$$

These angles are in agreement with those calculated from the simple theory in Sec. 8-8.

The diagram has been drawn to scale for a distance d between the slits equal to 4 times the slit width D. The first minimum in the *diffraction* pattern (curve (a)) occurs at an angle

$$\sin\alpha = \frac{\lambda}{D}\cdot$$

Since $d = 4D$, four interference fringes are included in one half-width of the diffraction band. That is, the interference fringes form a much finer pattern, superposed on the relatively broad diffraction bands as shown in the photograph of Fig. 9-2 (b).

Now suppose we have three slits instead of two, as in Fig. 9-20 (c). We shall omit details, but the problem is essentially the same as with two slits. It is necessary to find the resultant of three vectors of equal amplitude, differing in phase by an angle ϕ. The resultant intensity is

$$E^2 = E_m^{\,2}\,(1 + 2\cos\phi)^2,$$

of which the heavier dotted line in Fig. 9-20 (c) is a graph. The maximum value of $(1 + 2\cos\phi)^2$ is 9, and the vertical scale of the figure should be increased by this factor. It will be seen that although the maxima occur at the same angles with three slits as with two, they are much more intense and much narrower. Also, small subsidiary maxima appear between them, as shown in the photograph in Fig. 9-2 (c).

Fig. 9-20 (d) shows the nature of the pattern when four slits are used. The vertical scale should be increased by a factor of 16. The maxima are still brighter and narrower, and the subsidiary maxima are even smaller. Fig. 9-2 (d) is a photograph of the interference pattern of four slits.

The extension to a very large number of slits, such as one finds in an actual grating, should be evident. The maxima become very sharp indeed, and the subsidiary maxima are of negligible intensity.

In practice, the parallel beam incident on the grating is usually produced by a collimating lens, in the first focal plane of which is a narrow illuminated slit. Each of the maxima is then a sharp image of the slit, of the same color as that of the light illuminating the slit, assumed thus far to be monochromatic. If the slit is illuminated by light consisting of a mixture of wave lengths, every wave length gives rise to a set of slit images deviated by the appropriate angles. If the slit is illuminated with white light, a continuous group of images is formed side by side or, in other words, the white light is dispersed into continuous spectra. In contrast with the single spectrum produced by a prism, a grating forms a number of spectra on either side of the normal. Those which correspond to $m = 1$ in Eq. (9-8) are called *first order*, those which correspond to $m = 2$ are called *second order*, and so on. Since for $m = 0$ the deviation is zero, all colors combine to produce a white image of the slit in the direction of the incident beam.

In order that an appreciable deviation of the light may be produced, it is necessary that the grating spacing shall be of the same order of magnitude as the wave length of light. Gratings for use in or near the visible spectrum are ruled with from 10,000 to 30,000 lines per inch, and it is evident that the construction of such a grating is no simple task. *Transmission* gratings, of the type which we have been discussing, may be made by ruling fine lines on glass with a diamond point. The roughened spaces produced by the ruling point form the opaque regions of the grating. *Reflection* gratings are also in common use, made in the same way by ruling on a polished surface with a diamond point. The light *reflected* from the narrow rulings, or the spaces between them, interferes to produce maxima and minima in exactly the same way as the light transmitted by a transmission grating.

Because of the many difficulties attendant upon their construction, there are, as a matter of fact, only a relatively small number of original ruled gratings in existence. For many purposes, *replica* gratings, whose manufacture has been brought to a high degree of perfection, are perfectly satisfactory. A thin layer of collodion solution is poured over the surface of a ruled grating and allowed to harden. The collodion film, when stripped from the grating, retains an impression of the rulings of the original grating, and is mounted between glass plates if a transmission grating is desired, or against a silvered surface to form a reflection grating. Evidently any number of replicas may be made from a single original.

The diffraction grating is widely used in spectroscopy instead of a prism as a means of dispersing a light beam into spectra. If the grating spacing is known, then from a measurement of the angle of deviation of any wave length, the value of this wave length may be computed. In the case of a prism this is not so; the angles of deviation are not related in any simple way to the wave lengths but depend on the characteristics of the material of which the prism is constructed. Since the index of refraction of optical glass varies more rapidly at the violet than at the red end of the spectrum, the spectrum formed by a prism is always spread out much more in pro-

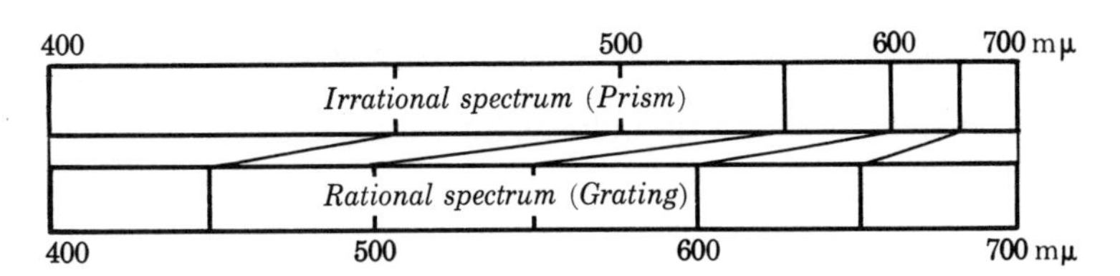

FIG. 9-21. A prism forms an irrational spectrum; a grating, a rational spectrum.

portion at the violet end than it is at the red. The spectrum formed by a prism is said to be *irrational*, while that formed by a grating is called *rational*. The effect is illustrated in Fig. 9-21, in which an irrational prism spectrum is compared with a rational grating spectrum having the same dispersion. Notice also that while a prism deviates red light the least and violet the most, the reverse is true of a grating, since in the latter case the deviation increases with increasing wave length.

Examples. (1) The limits of the visible spectrum are approximately 400 mμ to 700 mμ. Find the angular breadth of the first order visible spectrum produced by a plane grating having 15,000 lines per inch, when light is incident normally on the grating.

The grating spacing, d, in centimeters, is

$$d = 2.54/15{,}000 = 1.69 \times 10^{-4} \text{ cm.}$$

The angular deviation of the violet is

$$\sin\theta = \frac{4 \times 10^{-5}}{1.69 \times 10^{-4}} = 0.237,$$

$$\theta = 13^\circ\, 40'.$$

The angular deviation of the red is

$$\sin\theta = \frac{7 \times 10^{-5}}{1.69 \times 10^{-4}} = 0.415,$$

$$\theta = 24^\circ\, 30'.$$

Hence the first order visible spectrum includes an angle of

$$10^\circ\, 50'.$$

(2) Show that the violet of the third order visible spectrum overlaps the red of the second order.

The angular deviation of the third order violet is

$$\sin\theta = 3 \times 4 \times 10^{-5}/d$$

and of the second order red it is

$$\sin\theta = 2 \times 7 \times 10^{-5}/d.$$

Since the first angle is smaller than the second, whatever the grating spacing, the third order will always overlap the second.

(3) What is the maximum slit width D, if the complete second order spectrum is to be formed?

For practical purposes, all of the light diffracted by a single slit can be considered to lie in the central diffraction band. The half-angular breadth of this band must then be at least as great as the maximum angle of deviation of the second order spectrum, and actually it must be somewhat greater, since the intensity falls to zero at the edge of the diffraction band. The maximum angle θ in the second order spectrum is given by

$$\sin \theta = 2 \times \frac{7 \times 10^{-5}}{1.69 \times 10^{-4}} = 0.830,$$

$$\theta = 56°.$$

The half-angle α of the central diffraction band is

$$\sin \alpha = \frac{\lambda}{D} \cdot$$

Hence, setting $\alpha = \theta$, we obtain

$$.830 = \frac{7 \times 10^{-5}}{D},$$

$$D = 0.85 \times 10^{-4} \text{ cm}.$$

This is just half the distance between slits.

9-4 The concave grating. The plane grating requires the use of two lenses, the first to render parallel the light incident on the grating, and the second to bring the diffracted rays to a focus. These lenses add to the complexity of a spectrograph and, furthermore, if investigations are to be made in the ultraviolet, the lenses may have to be made of some material other than glass, since ordinary optical glass is not transparent much outside the visible spectrum. Both lenses may be dispensed with in the concave reflection grating, first developed by J. H. Rowland.

A concave grating is ruled on a polished concave spherical surface, the rulings being the intersections with the surface of equidistant planes parallel to the principal axis of the surface. The surface acts at the same time as a grating and as a concave mirror.

Let GG, Fig. 9-22, be the intersection with the plane of the figure of the concave spherical surface of the grating. The grating rulings are perpendicular to the plane of the paper. Let C be the center of curvature of the grating surface, R its radius of curvature, and $CSVI$ a circle of diameter equal to R, tangent to the grating at its vertex V. If a light source or slit S is placed at any point on the circumference of $CSVI$, the spectra

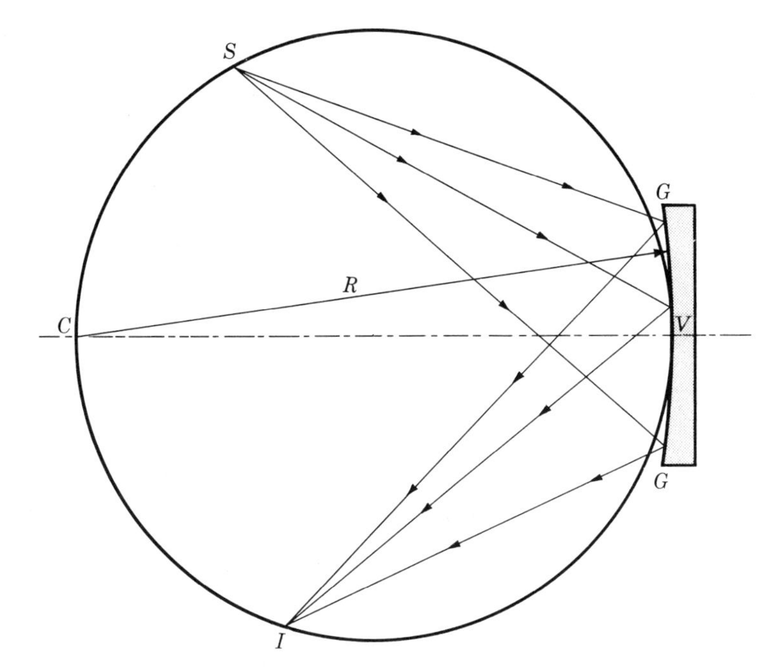

FIG. 9-22. The concave grating.

produced by the grating will be imaged at other points such as I along the circumference of this same circle.

In the large gratings in the spectroscopy laboratory at M. I. T., the light source is mounted at C, and plates for photographing the spectra may be placed along the circumference of the circle $CSVI$.

9-5 Diffraction of x-rays by a crystal. Although x-rays were discovered by Roentgen in 1895, it was not until 1913 that x-ray wave lengths were measured with any degree of precision. Experiments had indicated that these wave lengths might be of the order of 10^{-8} cm, which is about the same as the interatomic spacing in a solid. It occurred to Laue in 1913 that if the atoms in a crystal were arranged in a regular way, a crystal might serve as a three-dimensional diffraction grating for x-rays. The experiment was performed by Friederich and Knipping and it succeeded, thus verifying in a single stroke both the hypothesis that x-rays *were* waves (or at any rate wavelike in some of their properties) and that the atoms in a crystal were arranged in a regular manner. Since that time, the phenomenon of x-ray diffraction by a crystal has proved an invaluable tool of the physicist, both as a method of measuring x-ray wave lengths and of studying the structure of crystals.

Fig. 9-23 is a diagram of a simple type of crystal, that of sodium chloride (NaCl). The black circles represent the sodium, and the open circles the chlorine atoms. Fig. 9-24 is a diagram of a section through the crystal. Planes such as those parallel to *aa, bb, cc,* etc., can be constructed through the crystal in such a way that they pass through relatively large numbers of atoms. These sets of equidistant planes correspond to the lines in a plane grating.

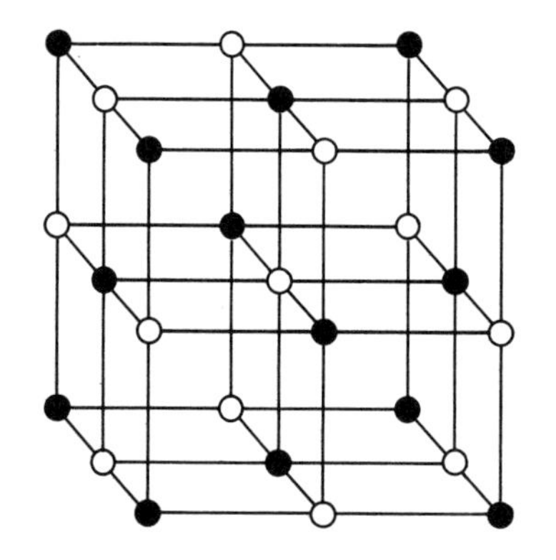

FIG. 9-23. Model of arrangement of atoms in a crystal of NaCl. Black circles, Na; open circles, Cl.

A simple way of interpreting the phenomenon of x-ray diffraction by a crystal was suggested by W. L. Bragg. When a beam of x-rays is incident on a crystal, a small part of the beam is, in effect, reflected from each of the crystal planes such as those in Fig. 9-24, with the angle of reflection equal to the angle of incidence. This reflection is brought about by the process of scattering Fundamentally, the phenomenon is the same as that responsible for the scattering of light by a gas. The only difference is that the molecules of a gas are arranged at random, while those in a crystal have a fixed, regular spacing. As the incident beam passes through the crystal, the electrons in the atoms of the crystal

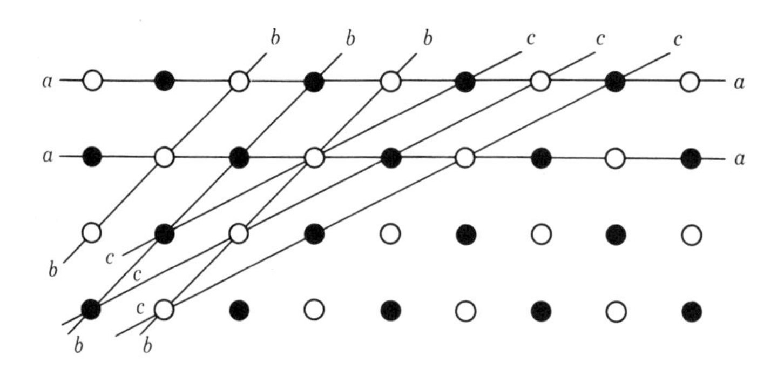

FIG. 9-24. Crystal planes such as *aa, bb, cc,* serve as a three-dimensional diffraction grating for x-rays.

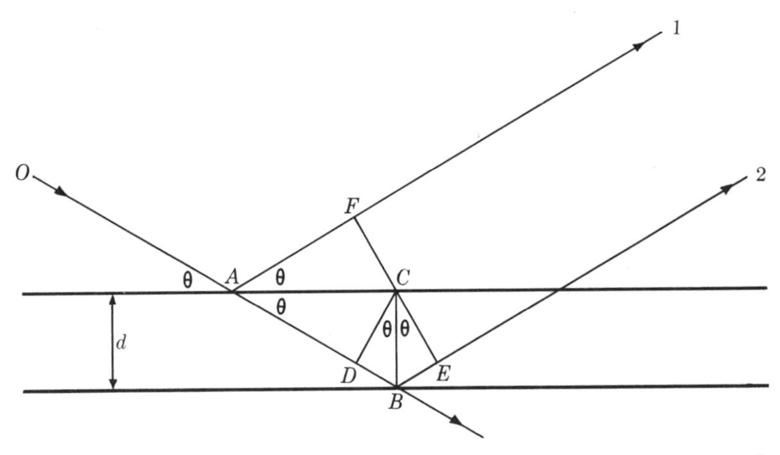

FIG. 9-25. Strong reflection of x-rays from a crystal results when $\sin\theta = m\dfrac{\lambda}{2d}$.

are set into vibration and become sources of secondary scattered waves. The phase relations of the scattered waves are such that they combine to give a relatively intense wave train in that particular direction in which the incident wave would be reflected if each crystal plane were a mirror. Only a very small part of the energy of the incident wave is, however, reflected at each plane.

Let the horizontal lines in Fig. 9-25 represent any two crystal planes. The line *oab* is one of a bundle of parallel rays incident on the crystal. The angle θ between the ray and the crystal planes is called the *glancing angle.* A small part of the incident energy is reflected at a, another small part at b, and so on. If the glancing angle has the correct value, the wave trains reflected from all the planes parallel to those shown will be in phase and will combine to give a strong reflection at the angle θ.

To find this angle, construct *bc* from b at right angles to the crystal planes. Through c draw *cd* at right angles to the incident ray, and *fce* at right angles to the reflected rays 1 and 2. Remembering that each ray represents a train of waves at right angles to the ray, it will be seen that if the waves corresponding to rays 1 and 2 are in phase at points f and e, we have the proper conditions for reenforcement. By construction, $af = ad$, so ray 2 has to travel a distance $db + be$, or $2db$, farther than ray 1. For reenforcement, this distance must equal some integral number of wave lengths, say $m\lambda$. If d represents the grating spacing, the distance $2db = 2d \sin\theta$. Hence for strong reflection,

$$\boxed{\sin\theta = m\frac{\lambda}{2d}}. \tag{9-9}$$

This relation is known as *Bragg's law*. The angle θ can be determined by observation of the reflected beam, and hence λ can be computed if d is known or vice versa.

Fig. 9-26 is a photograph made by directing a narrow beam of x-rays at a thin section of a crystal of quartz and allowing the diffracted beams to strike a photographic plate. Each spot corresponds to the reflection from a particular set of crystal planes

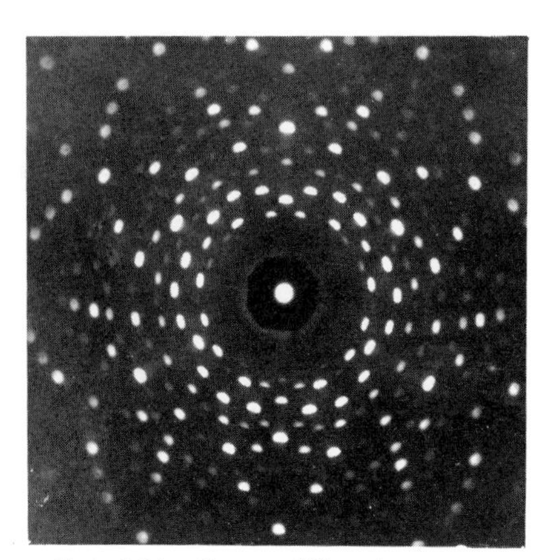

FIG. 9-26. Laue diffraction pattern formed by directing a beam of x-rays at a thin section of quartz crystal.
(Courtesy of Dr. B. E. Warren)

9-6 Fraunhofer diffraction by a circular aperture. Diffraction by a circular aperture is more common than is diffraction by a slit, since most lenses, as well as the pupil of the eye, are of circular cross section. The method of analysis is the same as that for a slit, described in Sec. 9-2, with the exception that the wave front transmitted by the aperture is subdivided into narrow annular zones instead of strips. The mathematics, however, is considerably more complicated and will not be given. The problem was first solved by Airy in 1834. The diffraction pattern consists of a bright central disk, bordered by alternate dark and bright rings. The intensity is greatest at the center of the disk, and decreases to a first minimum of zero at a half-angle α given by

$$\sin \alpha = \frac{1.22\lambda}{D} = \frac{0.61\lambda}{R}, \tag{9-10}$$

where D is the diameter of the aperture and R is its radius. Notice that except for the factor 1.22 this equation has precisely the same form as that for the half-angular breadth of the diffraction pattern of a slit. Approximately 85% of the energy transmitted by the aperture is found within the central disk.

When a lens is placed just beyond the aperture, it functions in precisely the same way as in Fig. 9-12 (d), which will serve equally well as a diagram for this case. The pattern formed in the second focal plane of the lens is geometrically similar to that which would appear on a distant screen, but is reduced in scale. Figs. 9-4 (a) and (b) are photographs of the Fraunhofer diffraction patterns of a number of point sources, formed by placing a lens just beyond a small circular opening with a photographic film in the second focal plane of the lens. We see that a geometrical point object, even in the absence of all lens aberrations, is imaged not as a geometrical point but as a disk surrounded by dark and bright rings. The half-angle of the central disk, from Eq. (9-10), is inversely proportional to the radius of the aperture in front of the lens, so that the *smaller* this aperture, the *larger* the diffraction pattern.

9-7 Fresnel zones. When the distance from an aperture to a screen is very large, or when the aperture is very small, rays from all points of the aperture to any point on the screen can be considered parallel and the problem is classified as one in Fraunhofer diffraction. When the distance to the screen is so small, or the aperture so large, that the rays cannot be considered parallel, we speak of the problem as one in Fresnel diffraction. An exact mathematical analysis of this case is very complicated, but by means of an approximation suggested by Fresnel a semiquantitative treatment can be worked out which explains the main features of the diffraction pattern and which uses only elementary mathematics. Essentially, the Fresnel approximation consists of dividing the wave surface into small but finite zones, and working with these instead of the infinitesimal zones of the more exact theory.

In Fig. 9-27 (a), light of wave length λ originates at a distant point source at the left of the diagram and falls on a screen. The wave fronts, one of which is shown at AA', may be considered planes. Point P is any point on the screen, and R is the perpendicular distance from P to the wave front. With P as center, and radii

$$R + \frac{\lambda}{2}, \quad R + 2\frac{\lambda}{2}, \quad R + 3\frac{\lambda}{2}, \text{ etc.,}$$

construct spheres which intersect the wave front AA' as circles. These circles intersect the plane of the diagram at a and a', b and b', c and c', and so on. This construction divides the wave surface into a number of regions called *half-period elements* or *Fresnel zones*, so situated that the distance to the point P increases by one-half a wave length as one passes from the inner to the outer boundary of each zone. A front view of the Fresnel zones is shown in Fig. 9-27 (b). The inner zone is a circle and the outer zones are concentric annular rings.

The radii of the zones may be computed as follows. In the right triangle POa, we have

$$r_1^2 + R^2 = \left(R + \frac{\lambda}{2}\right)^2 = R^2 + R\lambda + \left(\frac{\lambda}{2}\right)^2,$$

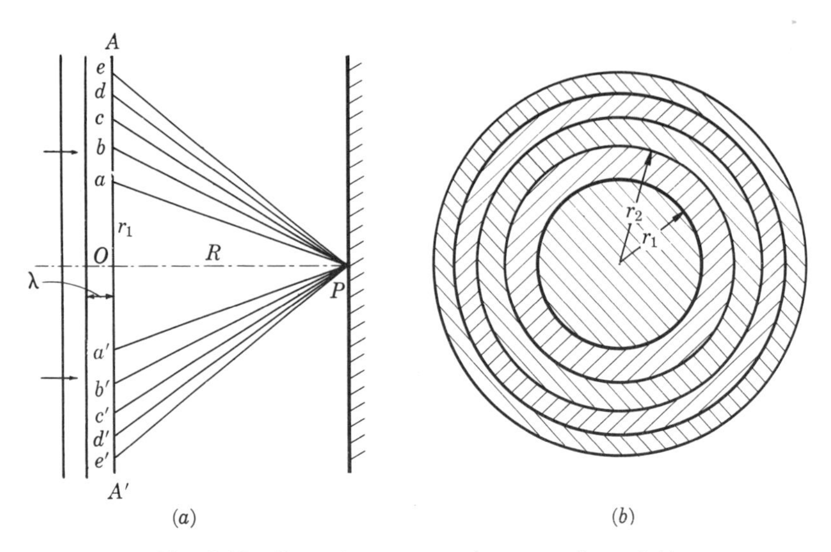

FIG. 9-27. Fresnel zones on a plane wave front AA'.

$$r_1^2 = R\lambda + \left(\frac{\lambda}{2}\right)^2.$$

If R is sufficiently large in comparison with λ we may neglect the term $\left(\frac{\lambda}{2}\right)^2$, and

$$r_1^2 = R\lambda, \quad r_1 = \sqrt{R\lambda}. \tag{9-11}$$

As an example, if $R = 50$ cm and $\lambda = 500$ m$\mu = 5 \times 10^{-5}$ cm, then

$$\begin{aligned} r_1 &= \sqrt{50 \times 5 \times 10^{-5}} \\ &= 0.05 \text{ cm} \\ &= 0.5 \text{ mm}. \end{aligned}$$

The radii of successive outer zones, say r_2, r_3, etc., are found by the same method to be

$$r_2 = \sqrt{2R\lambda} = \sqrt{2}\, r_1,$$

$$r_3 = \sqrt{3R\lambda} = \sqrt{3}\, r_1, \text{ etc.}$$

The radii are therefore in the ratio $\sqrt{1}$, $\sqrt{2}$, $\sqrt{3}$, and so on.

Note carefully that the radii of the Fresnel zones on a wave surface are not of any fixed size, but their magnitude depends both on the wave length of the light and on the distance from the wave surface to the point P. The longer the wave length, or the more distant the point, the larger are the zones.

Let us next find the phase relation between the wave trains reaching P from the various zones. Consider an instant at which the phase at P of the wave from the center of the first zone is zero. At this instant, the phase at P of the wave from the outer edge of the first zone will be π, since the edge is farther away by one-half a wave length and the waves start from AA' in phase. Accordingly, the "mean" phase of the entire wave train from the first zone, at this instant, may be taken as $\pi/2$. At this same instant, the phases at P of the waves from the inner and outer edges of the second zone are π and 2π respectively, so that the mean phase of the wave train from this zone is $3\pi/2$. The mean phases of the wave trains from successive outer zones are $5\pi/2$, $7\pi/2$, and so on. Evidently, then, the *phase difference* between the wave trains from successive zones is π, or in other words, the wave trains are alternately *out of phase* and *in phase* with that from the central zone.

The amplitude of the wave from each zone is proportional to the area of the zone. We have shown that the radii of the zones are r_1, $\sqrt{2}\, r_1$, $\sqrt{3}\, r_1$, etc. The area of the central zone is πr_1^2. The area of the second zone is $2\pi r_1^2 - \pi r_1^2 = \pi r_1^2$. That of the third is $3\pi r_1^2 - 2\pi r_1^2 = \pi r_1^2$, etc. Hence the areas of all zones are equal, and the amplitudes of all of the wave trains are equal at their points of origin.

Call $E_1, E_2, E_3, \ldots E_n$, the amplitudes at the point P of the electric vectors in the wave trains reaching P from the first, second, third . . . and nth zones. The distance each wave train must travel before reaching P is smallest for the central zone and increases slightly for successive outer zones. Furthermore, the angle which the direction of P makes with the forward direction of the wave also increases gradually for successive zones. For both of these reasons, the amplitude at P of the wave train from any one zone is slightly smaller than that of the wave from the zone just inside it. That is, E_2 is slightly smaller than E_1, E_3 is slightly smaller than E_2, and so on.

The resultant amplitude E at P, by the principle of superposition, is

$$E = E_1 - E_2 + E_3 - \cdots, \tag{9-12}$$

where the alternate plus and minus signs occur because the waves are alternately out of and in phase.

Since the amplitudes decrease slowly and regularly, the amplitude of the wave from any zone may be placed equal to the mean value of the amplitudes of the waves from the zones immediately preceding and following it. Thus one may write

$$E_2 = \frac{E_1 + E_3}{2}, \quad E_4 = \frac{E_3 + E_5}{2}, \text{ etc.} \tag{9-13}$$

Eq. (9-12) may be put in the following form

$$E = \frac{E_1}{2} + \left(\frac{E_1}{2} - E_2 + \frac{E_3}{2}\right) + \left(\frac{E_3}{2} - E_4 + \frac{E_5}{2}\right) + \cdots.$$

From Eq. (9-13), each term in parentheses is zero. Then if E_n is the amplitude of the wave from the last zone and E_{n-1} that of the wave from the next to the last zone, the series reduces to

$$E = \frac{E_1}{2} + \frac{E_n}{2}, \tag{9-14}$$

if n is odd, and to

$$E = \frac{E_1}{2} + \frac{E_{n-1}}{2} - E_n \tag{9-15}$$

if n is even.

If the entire wave front is unobstructed, the number of terms in the series is infinite. Since the terms decrease regularly both E_{n-1} and E_n are zero, and

$$E = \frac{E_1}{2}.$$

The amplitude at P due to an unobstructed wave of infinite extent is therefore *only one-half of that which would be produced by the first Fresnel zone alone.*

9-8 Fresnel diffraction by a circular aperture. Fig. 9-28 represents an opaque plate in which there is a circular aperture, with a screen S relatively close to the aperture so that we have Fresnel diffraction. Parallel monochromatic light is incident on the aperture from the left. Consider first the quantity of light reaching point P, at the center of the geometrical shadow, as the diameter of the aperture is varied.

Suppose the diameter is such as to expose only the first, central Fresnel zone. Let E_1 represent the amplitude at P of the electric vector in the wave train from this element. If the entire wave front were exposed, the amplitude at P, as we have shown, would be only $E_1/2$. Hence when all

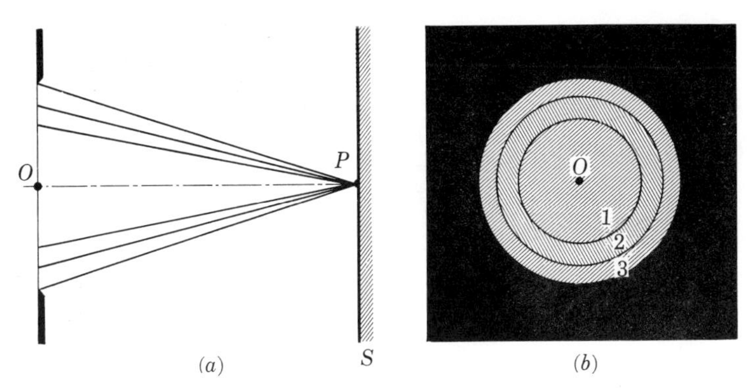

FIG. 9-28. Fresnel diffraction by a circular aperture.

of the wave front except the first Fresnel zone is obstructed by an opaque plate, the amplitude at P is twice as great, and the amount of light four times as great, as if the wave were unobstructed.

Now let the diameter of the aperture be increased until it includes the first two Fresnel zones. The amplitude at P is then $E_1 - E_2$, which, since E_1 and E_2 are nearly equal, is practically zero. *Increasing* the size of the aperture therefore *decreases* the light reaching point P, and it is possible for a Fresnel diffraction pattern to have a dark center, in contrast to a Fraunhofer diffraction pattern where the center is always bright.

If the aperture is still further enlarged so as to expose the first three Fresnel zones, as in Fig. 9-28, the amplitude is $\frac{E_1}{2} + \frac{E_2}{2}$, which again is large. It will be seen that, in general, if the opening includes an odd number of zones the center of the geometrical shadow will be bright, while if the number of zones is even the center will be dark.

The photographs in Fig. 9-5 were made by drilling a number of holes, ranging in diameter from 1 mm to 4 mm, in a brass plate, illuminating the plate from the left with parallel monochromatic light, and placing a photographic film at a distance of about a meter to the right of the plate, corresponding to the screen S in Fig. 9-28. Each hole in the plate then registers its own diffraction pattern on the film. The four holes at the top of the figure are so small as to expose less than the first Fresnel zone. In other words, these patterns are still in the region of Fraunhofer diffraction. Notice that they all have bright centers. The second hole from the left, in the second row, is large enough to expose the first two Fresnel zones and the center of the pattern is dark. The last hole in this row exposes three zones and the center is again bright.

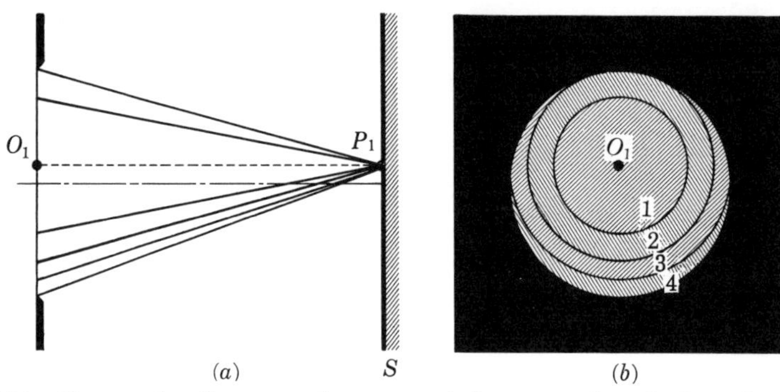

FIG. 9-29. For a point P_1 not at the center of the geometrical shadow, the Fresnel zones are exposed nonsymmetrically.

Do not confuse the bright and dark circles in these diffraction patterns with the imaginary Fresnel zones on the wave surface. The photographs in Fig. 9-5 represent the actual variations in the amount of light reaching different points on the film, or on the screen S in Fig. 9-28. The circles in part (b) of Fig. 9-28 are purely imaginary lines on the wave surface transmitted by the aperture, as they would appear if they were real and if one placed his eye at P and looked to the left toward the aperture.

For light of a given wave length, the radii of the Fresnel zones increase with the square root of the distance from the wave front to the point at which the illumination is to be computed. Hence if an aperture is maintained of fixed size it will expose a greater or smaller number of zones as the screen is moved toward or away from it. The center of the shadow of the opening, as the distance of the screen from the opening is altered, will be alternately bright and dark as the opening alternately includes an odd or an even number of zones.

Consider next the light reaching a point such as P_1 in Fig. 9-29 (a), which lies a short distance to one side of the center of the geometrical shadow. The Fresnel zones that must be used to find the amplitude at P_1 have their centers on the perpendicular from P_1 to the wave front, and the aperture is not located symmetrically with respect to the zones. For concreteness, assume that the radius of the aperture is equal to the outer radius of the third element, so that the *center* of the pattern is bright. As seen from P_1, the aperture and the elements appear as in Fig. 9-29 (b). The first and second zones are still exposed, together with about half of the third and fourth zones. Zones 1 and 2 will therefore cancel one another, and there will be approximate cancellation of zones 3 and 4. The amplitude is therefore less than at the center and the central bright spot

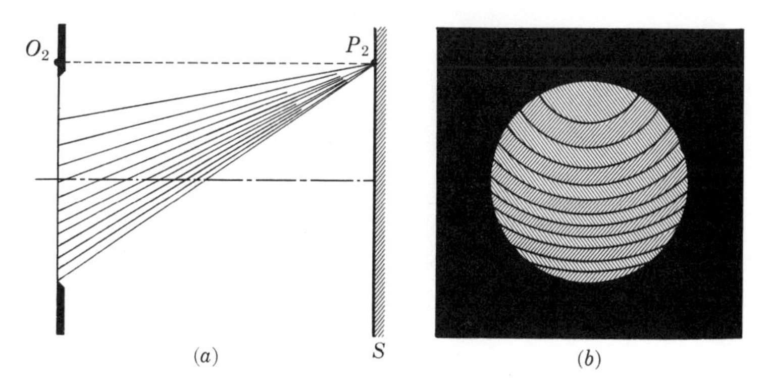

FIG. 9-30. Fresnel zones exposed by a circular aperture for a point within the geometrical shadow.

is surrounded by a darker ring. (See the pattern at the right of the second row in Fig. 9-5.) Still other bright and dark rings may be formed, depending on the size of the aperture.

In addition to the bright and dark rings in the central portion of the pattern, we also find some light reaching the screen within the region of geometrical shadow. Assume again for concreteness that the radius of the aperture equals the outer radius of the third Fresnel zone. Point P_2, in Fig. 9-30, lies just above the edge of the geometrical shadow and the exposed zones on the wave surface appear as in Fig. 9-30 (b). About one-third of the first zone is exposed, a somewhat larger fraction of the second zone, and smaller and smaller fractions of succeeding zones. The sum of the amplitudes of all the wave trains reduces to approximately one-half of the first term, since the last terms of the series are very small. The amplitude is therefore about 1/6 of E_1, and while this is small, it is not zero, so that some light does reach points within the region of geometrical shadow.

9-9 Diffraction by a circular obstacle. Let the circular aperture in Fig. 9-28 be replaced by a circular obstacle as in Fig. 9-31. Assume that the three inner zones are covered. The first zone sending light to point P is then zone 4, and to find the amplitude at P we must perform the same summation as in Eq. (9-12), beginning with E_4. This sum is evidently $E_4/2$, and the amount of light reaching P is proportional to $(E_4/2)^2$. The amplitude at P due to the entire wave front is $E_1/2$ and the amount of light is proportional to $(E_1/2)^2$. Since E_1 and E_4 are not very different, the *center* of the shadow is nearly as bright as if the obstacle were not there.

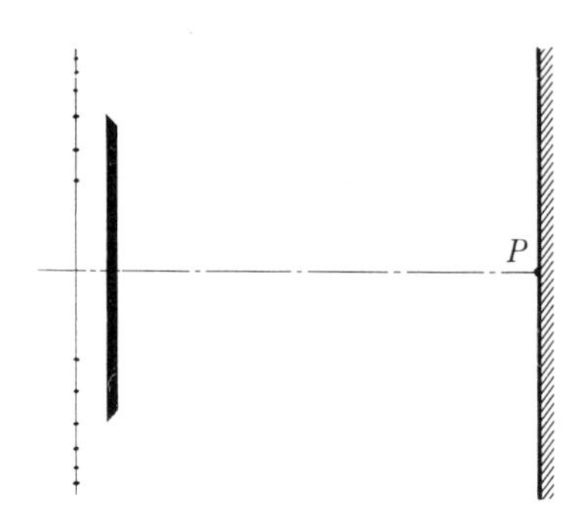

FIG. 9-31. A circular obstacle cuts off the light from the inner Fresnel zones.

From a point just to one side of the center of the shadow, those zones which are covered are the same as those exposed in Fig. 9-29 (b). Zones 1 and 2 are completely covered. A small part of zone 3 is exposed, a somewhat larger part of zone 4, and all of the remaining zones. The first term in the series is therefore small, and of course the last term is zero. The resultant amplitude is therefore small and the central bright spot is surrounded by a darker ring. Other bright and dark rings may also be seen if the obstacle is small. Fig. 9-6 is a photograph of the shadow of a small ball bearing, mounted on the point of a needle.

It is interesting to note that Poisson, who first deduced by means of the argument above that a bright spot should be formed at the center of the shadow of a circular obstacle, concluded that this seemingly absurd prediction definitely disproved the wave theory of light. The experiment was tried by Arago, however, and the bright spot was found.

9-10 Diffraction by a straight edge. Let M in Fig. 9-32 (a) represent a straight edge perpendicular to the plane of the paper. Point P is the trace of its geometrical shadow on a screen. A train of plane monochromatic waves is incident from the left. The wave surface is to be subdivided in a somewhat different way from that used for a circular opening. With P as center, and radii

$$R + \frac{\lambda}{2}, \quad R + 2\frac{\lambda}{2}, \quad R + 3\frac{\lambda}{2}, \text{ etc.,}$$

strike circles intersecting the wave front at a and a', b and b', etc., and divide the wave front into half-period zones by lines through these points parallel to the straight edge. A front view of the zones is shown in Fig. 9-32 (b). Let E_1 be the amplitude at P of the wave train from the first zone above or below the center line, E_2 the amplitude of the wave from each of the second zones, and so on.

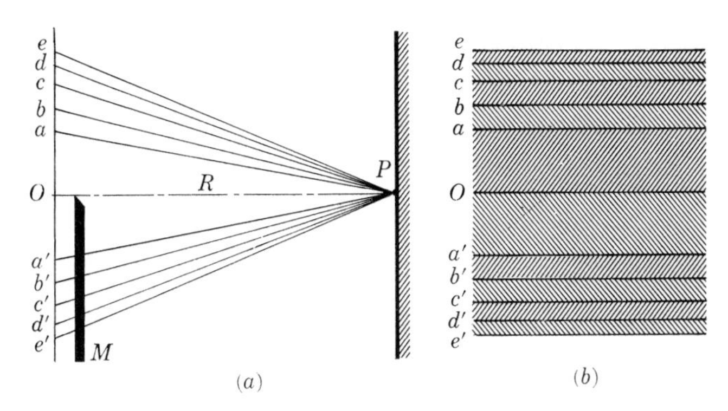

FIG. 9-32. Fresnel zone construction when the obstacle has a straight edge.

Suppose that the entire wave front is unobstructed. The resultant amplitude at P due to the upper half of the wave, say E, is

$$E = E_1 - E_2 + E_3 - \cdots$$

which reduces to

$$E = \frac{E_1}{2}.$$

In the same way, the amplitude due to the lower half of the wave, say E', is

$$E' = \frac{E_1}{2}.$$

Hence the amplitude at P due to the entire unobstructed wave front is

$$E + E' = E_1.$$

Next let the lower half of the wave be covered so that P lies at the edge of the geometrical shadow. The amplitude at P is now that due to the upper half of the wave alone, or $E_1/2$, and the intensity is 1/4 that of the unobstructed wave.

Consider next points on the screen just within and just outside the geometrical shadow. Instead of drawing a number of diagrams, it is somewhat simpler to keep the point P fixed, and imagine the straight edge to be moved slightly up or down. Then the edge of the geometrical shadow moves by the same amount and the fixed point P may be caused to lie below or above the edge of the shadow, as we wish.

First let the straight edge be moved *up* to the point a, so that P lies slightly *below* the edge of the shadow. The first zone above the center is now covered, in addition to all of the wave front below the center. The amplitude at P is

$$E = E_2 - E_3 + E_4 - \cdots = \frac{E_2}{2},$$

and the amount of light is proportional to $\left(\frac{E_2}{2}\right)^2$. As will be seen from the diagram, the first zone is considerably wider than the others, so that E_2 is considerably smaller than E_1. It is not zero, however, and while the light reaching the screen falls off rapidly within the geometrical shadow, it does not become zero suddenly at the edge.

Let the straight edge now be moved down so that P lies above the edge of the geometrical shadow. Suppose the first zone below the center is exposed. The upper half of the wave front still contributes an amount $E_1/2$ to the amplitude, and the element below the center contributes an amount E_1. The resultant amplitude (since these are in phase with one another) is $3E_1/2$, and the intensity is proportional to $9E_1^2/4$, or $2.25E_1^2$. The light incident at a point just outside the geometrical shadow is therefore 2.25 times as great as that which would be produced by the unobstructed wave.

When the straight edge moves farther down (or P moves up) to expose the first two zones below the center, the amplitude is

$$E = \frac{E_1}{2} + E_1 - E_2 = \frac{3}{2}E_1 - E_2.$$

This is less than the previous value, so that the bright band just outside the shadow is followed by a darker one. Proceeding in this way, we find outside the geometrical shadow a series of maxima and minima which become closer and closer together until eventually the screen is uniformly illuminated. A photograph of the shadow of a straight edge cast by the light from a distant pinhole is reproduced in Fig. 9-7. Notice that the first bright band just outside the shadow is considerably brighter than the region of uniform illumination.

It should be borne in mind that in order to observe these diffraction fringes in a shadow, the light waves producing them must proceed from a point source. If the light source is broad, so that many sets of wave trains traveling in different directions strike the straight edge, then while each wave train by itself produces the effects described, the numerous sets of fringes so overlap one another that no distinct pattern is obtained.

The diffraction effects produced by a fine wire in a beam of plane waves can be treated in the same way. The resultant intensity on a screen behind the wire is found by combining the amplitudes of those zones which the wire does not obstruct. Some photographs are reproduced in Fig. 9-9.

Problems—Chapter 9

(1) Find the half-angular breadth of the central bright band in the Fraunhofer diffraction pattern of a slit 14×10^{-5} cm wide, when the slit is illuminated by a parallel beam of monochromatic light of wave length (a) 400 mμ, (b) 700 mμ.

(2) The Fraunhofer diffraction pattern of a single slit, reproduced twice size in Fig. 9-3, was formed on a photographic film in the focal plane of a lens of focal length 60 cm. The wave length of the light used was 546.1 mμ. Measure the distance between the second minimum on the right and the second minimum on the left, and compute the width of the slit.

(3) A slit 0.25 mm wide is placed in front of a positive lens and illuminated by plane waves of wave length 500 mμ. In the Fraunhofer diffraction pattern formed in the focal plane of the lens, the distance from the third minimum on the left to the third minimum on the right is found to be 3 mm. Find the focal length of the lens.

(4) (a) Find the three smallest values of z for which $\sin^2 z/z^2$ is a maximum. Do the maxima in Fig. 9-16 lie halfway between the minima? (b) Find the relative intensities of the central maximum and the next two adjacent maxima, in the Fraunhofer diffraction pattern of a single slit.

(5) The interference pattern of two identical narrow slits separated by a distance $d = 0.1$ mm is observed on a screen at a distance of 1 m from the plane of the slits. The slits are illuminated by monochromatic light of wave length 590 mμ traveling perpendicular to the plane of the slits. Five bright bands are observed on each side of the central maximum, but beyond these bands the intensity is very weak. (a) Calculate the approximate width, D, of each of the slits. (b) Calculate the distance between each of the 5 bright bands.

(6) Plane monochromatic waves of wave length 600 mμ are incident normally on a plane transmission grating ruled with 5000 lines/cm. Find the angles of deviation in the first, second, and third orders.

(7) A plane transmission grating is ruled with 4000 lines/cm. Compute the angular separation in degrees, in the second order spectrum, between the α and δ lines of atomic hydrogen, whose wave lengths are respectively 656 mμ and 410 mμ.

(8) Light which is a mixture of the two wave lengths 500 mμ and 520 mμ is incident normally on a plane diffraction grating of the transmission type. The grating spacing is 10^{-4} cm and a lens of focal length 200 cm is used to observe the spectrum on a screen. Find the separation (in cm) of the two lines in the first-order spectrum.

(9) A parallel beam of light comprising the wave length range between 350 mμ and 750 mμ is incident normally on a plane transmission grating. Just beyond the grating is a lens of focal length 150 cm. The width of the first-order spectrum, in the focal plane of the lens, is 6 cm. What is the grating spacing?

(10) Compute the radius of the central disk of the diffraction pattern of the image of a star formed by (a) a camera lens 1 inch in diameter and of focal length 3 inches, (b) a telescope lens 6 inches in diameter and of focal length 5 ft.

(11) The Fraunhofer diffraction pattern of a circular aperture, enlarged to four times actual size, is shown at the left in Fig. 9-4 (a). The pattern was formed in the focal plane of a lens having a focal length of 1 m, and the wave length of the light used was 546.1 mμ. Compute the diameter of the aperture.

(12) A lens, 10 cm in diameter and 100 cm in focal length, images a distant sphere 2 cm in diameter on a photographic plate. At what object distance would the geometrical image of the sphere be of the same size as the central disk of the diffraction pattern formed by light of wave length 410 mμ from a point source at that distance?

(13) Monochromatic light of wave length 600 mμ originates at a distant point source and passes through a circular opening. The Fresnel diffraction pattern is observed on a screen 1 m beyond the opening. Find the diameter of the circular opening if it exposes (a) the central Fresnel zone only, (b) the first four Fresnel zones.

(14) Monochromatic light of wave length 563.3 mμ originates at a distant point source and passes through a circular opening 2.60 mm in diameter. The Fresnel diffraction pattern is observed on a screen 1 m beyond the opening. (a) Will the center of the diffraction pattern appear bright or dark? (b) Through what minimum distance from its first position must the screen be moved in order to reverse the condition found in part (a)?

(15) Monochromatic light of wave length 400 mμ from a distant point source falls on an opaque plate in which there is a small circular opening. As a screen is moved toward the plate from a large distance away, the Fresnel diffraction pattern on the screen first has a dark center when the distance from plate to screen is 160 cm. Find the diameter of the central disk in the Fraunhofer diffraction pattern, if a lens of focal length 160 cm is placed just to the right of the circular opening.

CHAPTER 10

LIMIT OF RESOLUTION

10-1 The Rayleigh limit of resolution. It will be recalled that the expressions for the magnification of a telescope or a microscope derived in Chap. 5 involved (except for certain numerical factors) only the focal lengths of the lenses making up the optical system of the instrument. It appears at first sight as though any desired magnification might be attained by a proper choice of these focal lengths. Unless the instrument is properly designed, however, while the image becomes larger (or subtends a larger angle) it does not gain in detail, even though all lens aberrations have been corrected. This limit to the useful magnification is set by the fact that light is a wave motion and the laws of geometrical optics do not hold strictly for a wave front of limited extent. Physically, the image of a point source is not the intersection of *rays* from the source, but the diffraction pattern of those *waves* from the source that pass through the lens system.

We have shown in Sec. 9-6 that the light from a point source, diffracted by a circular opening, is focused by a lens not as a geometrical point but as a disk of finite radius surrounded by dark and bright rings. An optical system is said to be able to *resolve* two point sources if the corresponding diffraction patterns are sufficiently small or sufficiently separated to be distinguished.

From a study of the appearance of the diffraction patterns of closely spaced point sources, Lord Rayleigh concluded that two equally bright point sources could just be resolved by an optical system if the central maximum of the diffraction pattern of one source coincided with the first minimum of the other. This is equivalent to the condition that the distance between the centers of the patterns shall equal the radius of the central disk. The Rayleigh limit of resolution is illustrated in Figs. 10-1 and 9-4. Fig. 9-4 (a) is a photograph of four point sources, made with the camera lens "stopped down" to an extremely small aperture. Fig. 10-1 is a graph of the intensity distribution in the diffraction patterns. (The curves have very nearly the same shape as those for a slit.) The diffraction patterns at the left are separated by a distance much greater than the radius of the central disk, and are clearly resolvable. The distance between the centers of the patterns at the right is very nearly equal to the radius of the central disk and evidently, if this distance were any less, the patterns would overlap to such an extent that they could not be distinguished as two.

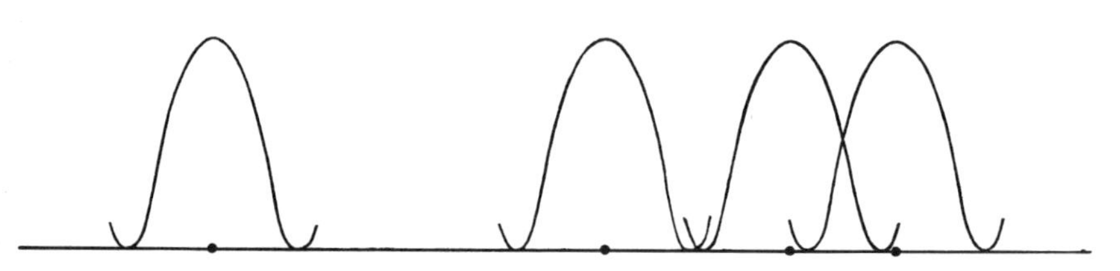

FIG. 10-1. Graph of the intensity distribution in the diffraction patterns of Fig. 9-4(a).

Fig. 10-2 shows a positive lens forming images of the two point sources P_1 and P_2. In front of the lens is a circular aperture of radius R. The images of P_1 and P_2 are diffraction patterns like those in Fig. 9-4 (a) and (b), with centers at P_1' and P_2'. Since 85% of the light is concentrated in the central disk, we can disregard the outer rings and consider the central disks as the images of the geometrical points P_1 and P_2. If the images are just resolved, the distance z' between their centers is equal to the radius of either one.

For generality, Fig. 10-2 has been drawn with a medium of index n at the left of the lens and a medium of a different index n' at the right. The angle α' is the half-angle subtended by the diffraction disk centered at P_1'. In the expression for this half-angle given in Eq. (9-10), the quantity λ referred to the wave length in the medium in which the image was formed. If λ_0 is the wave length in vacuum, the wave length at the right of the lens is $\lambda' = \dfrac{\lambda_0}{n'}$. Hence Eq. (9-10) becomes

$$\sin \alpha' = \frac{0.61\lambda_0}{n'R}. \tag{10-1}$$

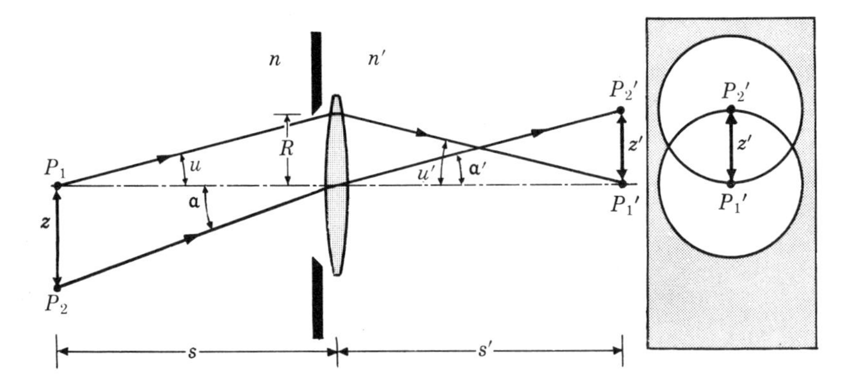

FIG. 10-2. Diffraction patterns of two point sources. (Transverse dimensions are greatly exaggerated.)

Consider the ray from P_2 to P_2', passing through the center of the lens. Since the lens surfaces are parallel to one another at the center of the lens, it follows from Snell's law that

$$n \sin \alpha = n' \sin \alpha'.$$

Hence from the two preceding equations, the angle α subtended at the lens by the just resolvable object points P_1 and P_2 is given by

$$\sin \alpha = \frac{0.61\lambda_0}{nR}. \tag{10-2}$$

The linear separation z' between the centers of the diffraction disks, which is equal to the radius of either disk, is

$$z' = s' \tan \alpha',$$

where s' is the image distance. In most cases of practical interest the half-angle u' subtended at P_1' by the aperture is sufficiently small so that we can set

$$\sin u' = \tan u' = \frac{R}{s'}.$$

The angle α' is also small, and

$$\sin \alpha' = \tan \alpha'.$$

When the three preceding equations are combined with Eq. (10-1) we get

$$z' = \frac{0.61\lambda_0}{n' \sin u'}. \tag{10-3}$$

Finally, we make use of Abbe's sine condition (page 70) which states that the magnification m produced by the lens is

$$m = \frac{z'}{z} = \frac{n \sin u}{n' \sin u'}.$$

Combining this with Eq. (10-3) gives for the linear distance z between two just resolvable object points,

$$z = \frac{0.61\lambda_0}{n \sin u}. \tag{10-4}$$

To summarize, the *linear* distance z between two just resolvable equally bright point objects, according to the Rayleigh criterion, is

$$z = \frac{0.61\lambda_0}{n \sin u}. \tag{10-4}$$

The *angle* α subtended by the objects is given by

$$\sin \alpha = \frac{0.61\lambda_0}{nR}. \tag{10-2}$$

The linear separation z' of the centers of the images is

$$z' = \frac{0.61\lambda_0}{n' \sin u'}, \tag{10-3}$$

and the angle α' subtended by the centers of the images is

$$\sin \alpha' = \frac{0.61\lambda_0}{n'R}. \tag{10-1}$$

The minimum distance between two resolvable points, from Eq. (10-4), is inversely proportional to $n \sin u$, the product of the index of the medium in which the object is situated and the sine of the half-angle of the cone of rays admitted by the instrument. This product is called the *numerical aperture* of the instrument and is abbreviated NA.

$$NA = n \sin u. \tag{10-5}$$

The greater the numerical aperture, the smaller the distance between two just resolvable points.

10-2 Limit of resolution of the eye. Fig. 10-3 is a schematic diagram of the eye, not drawn to a uniform scale, and with transverse dimensions and angles greatly exaggerated. Points P_1 and P_2, at the ends of an arrow of length z, represent two just resolvable object points at the minimum distance of distinct vision, 250 mm. The centers of the diffraction disks in the images of these points on the retina are at P_1' and P_2', and the distance z' between their centers is equal to the radius of the central disk. Actually, as we shall show, the distance z is about $\frac{1}{10}$ mm and z' is about $\frac{1}{100}$ mm.

Although the diameter of the pupil of the eye varies with the level of illumination, as shown in Fig. 6-4, it is customary in calculations of its limit of resolution to assume a pupillary diameter of 2 mm, or a radius R of 1 mm. We assume also that all the refraction of light entering the eye takes place at the cornea. The diameter of the eyeball is about 1 inch or 25 mm, and the index n' of the vitreous humor, in which the image is formed, is about 1.33. Let us assume a wave length λ_0 of 550 mμ, to which the eye is most sensitive. Objects viewed with the unaided eye are in air, so $n = 1.00$. At the minimum reading distance of 25 cm or

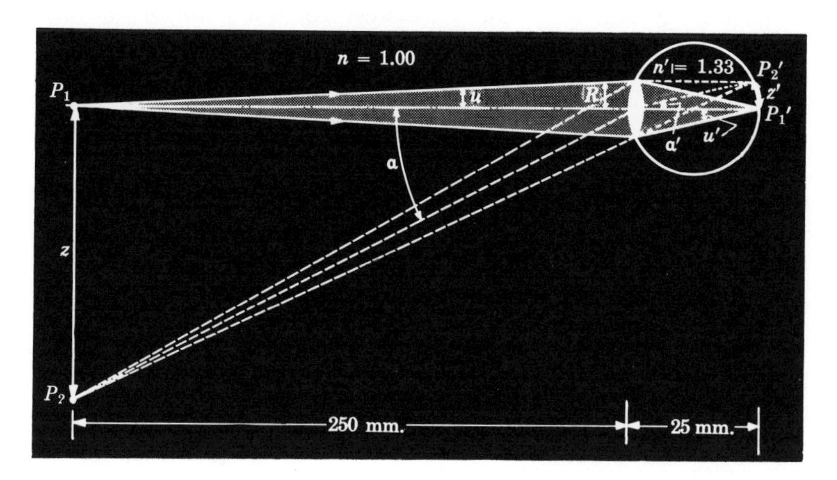

FIG. 10-3. Limit of resolution of the eye. (Transverse dimensions are greatly exaggerated.)

250 mm, the sine of the half-angle u of the cone of light admitted by the eye is very nearly

$$\sin u = \frac{1}{250} = 0.004,$$

and since $n = 1$, the maximum numerical aperture of the eye, for a pupillary radius of 1 mm, is

$$NA = 0.004.$$

Hence according to the Rayleigh criterion, the linear separation of two just resolvable point objects at a distance of 25 cm is

$$\begin{aligned} z &= \frac{0.61\lambda_0}{NA} \\ &= \frac{0.61 \times 550 \times 10^{-7}}{0.004} \\ &= 6.6 \times 10^{-3} \text{ cm} \\ &\approx \tfrac{1}{10} \text{ mm}. \end{aligned}$$

This separation, about three ten-thousandths of an inch, is in good agreement with the actual limit of resolution of a normal eye.

From Eq. (10-3), the distance z' between the centers of the diffraction disks on the retina, for two just resolvable point objects, is

$$z' = \frac{0.61\lambda_0}{n' \sin u'}$$

$$= \frac{0.61 \times 550 \times 10^{-7}}{1.33 \times 1/25}$$

$$= 6.3 \times 10^{-4} \text{ cm}$$

$$\approx \tfrac{1}{100} \text{ mm.}$$

It is interesting to note that the distance between cones in the fovea is also just about $\frac{1}{100}$ mm. The structure of the retina is thus very nicely adapted to the limit of resolution of the eye. If the cones were farther apart full advantage could not be taken of the available detail in the retinal image, while nothing would be gained if the distance between cones was much smaller than the radius of the diffraction disk of a point object.

The angle α between two point objects which can just be resolved by the eye is

$$\alpha = \sin \alpha = \frac{0.61\lambda_0}{nR}$$

$$= \frac{0.61 \times 550 \times 10^{-7}}{1 \times 10^{-1}}$$

$$= 3.4 \times 10^{-4} \text{ radian}$$

$$\approx 1 \text{ minute.}$$

Thus two point objects, to be resolved, must subtend at the eye an angle of at least 1 minute of arc. This corresponds to two points 1 inch apart at a distance of 100 yards, and again is seen to be of the correct order of magnitude.

The calculations above assumed a pupillary diameter of 2 mm. With smaller diameters, the numerical aperture of the eye decreases and the diffraction disks on the retina become larger. Hence the ability to resolve detail diminishes. It also diminishes somewhat with larger diameters because of increasing aberrations.

10-3 Limit of resolution of a microscope. The primary function of a microscope (or telescope) is not to "magnify" an object, but to make it possible to observe finer detail in the object than with the unaided eye. Two points which are so close together that their diffraction disks overlap could be resolved if there were (a) some way of making their diffraction disks smaller, and (b) at the same time distinguishing these smaller disks. The optical and morphological structure of the eye make both (a) and (b) impossible. The radius z' of the retinal diffraction disk of a point object, from Eq. (10-3), is proportional to the wave length λ_0 of the light used, and inversely proportional to the index n' of the vitreous humor and to the half-angle u' of the cone of light converging toward the retinal image. There is no way to utilize waves shorter than those to which the eye is sensitive, and no way to increase the half-angle of the convergent light cone. (Dilation of the pupil does not help, because of increasing geometric aberrations.) Furthermore, even if it were possible to make the diffraction disks smaller, no more detail could be seen, since even in an unaided eye the disks are already as small as the distance between cones in the fovea. The only procedure available, then, is to make the image larger. But merely to enlarge the retinal image is not enough, since, if the diffraction disks are enlarged in the same proportion, the image does not gain in detail. Hence a microscope or telescope must be so designed that enlarging the retinal image does not at the same time enlarge the diffraction disks.

It will be recalled that the *exit pupil* of a microscope or telescope is the image of the objective formed by the ocular. All of the light entering the objective, and refracted by the ocular, passes through the exit pupil. Unless the pupil of the eye is placed at the exit pupil of the instrument, the field of view is restricted and unless the exit pupil is as large as the pupil of the eye, the brightness of the image is reduced. We can now see that the exit pupil also determines the size of the retinal diffraction disks, since if it is any smaller than the pupil of the eye the full aperture of the pupil is not utilized and the half-angle of the convergent light cone within the eye is reduced. Therefore if the retinal diffraction disks are not to be enlarged when an instrument is used, the exit pupil of the instrument must be at least as large as the pupil of the eye. Nothing is gained by making it larger, except that proper positioning of the eye is not as critical.

Fig. 10-4 is a series of diagrams of the optical system of a compound microscope, drawn greatly out of proportion to bring out the features of interest. Objective and ocular are both shown as simple lenses. When the radius of the objective lens and the focal length of the ocular are so related that the diameter of the exit pupil is just equal to that of the pupil of the eye, the overall magnification M is said to be *normal.* This case is illustrated in Fig. 10-4 (a). P_1 and P_2 are two point objects, and P_1' and P_2' are the centers of their diffraction disks on the retina. Since the pupil of the eye is filled with light, the half-angle u' is the same as in

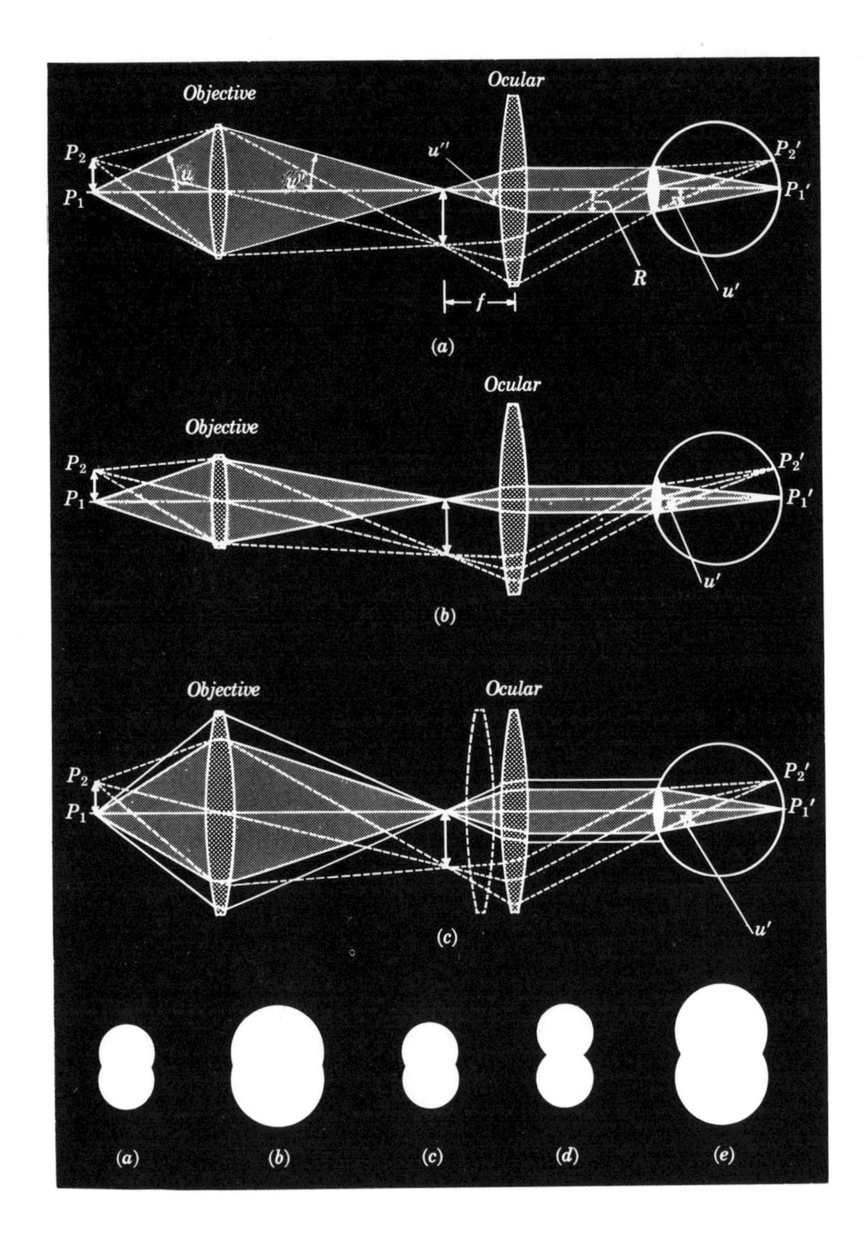

FIG. 10-4 Limit of resolution of a microscope. (a) Normal magnification; diameter of exit pupil same as pupil of eye. (b) Diameter of exit pupil smaller than pupil of eye. (c) Diameter of exit pupil greater than pupil of eye.

Fig. 10-3 and the radii of the diffraction disks are the same as in that figure, that is, about $\frac{1}{100}$ mm. Assuming that these disks are just resolvable by the Rayleigh criterion, the distance between their centers is also $\frac{1}{100}$ mm and they appear as in part (a) at the bottom of the figure.

We next consider the result of replacing the objective in part (a) with one of the same focal length, but having a smaller radius, as in Fig. 10-4 (b). The radius of the exit pupil (the image of the objective formed by the ocular) is now smaller than the radius of the pupil of the eye. The half-angle u' of the light cone converging toward the retina is reduced, and the radii of the diffraction disks are increased. The *magnification* of the instrument, which depends only on *focal lengths*, is the same in parts (a) and (b), so the distance between the centers of the diffraction disks is the same in both figures. The appearance of the disks is shown at the bottom of the diagram in part (b), and evidently they are *not* resolvable by the Rayleigh criterion. For a given magnification, then, the ability of the microscope to resolve detail is impaired if the radius of the objective lens is so small that the exit pupil does not fill the pupil of the eye. The image brightness is also reduced, since obviously less light enters the eye in part (b) than in part (a), while the image size is the same.

In part (c) of Fig. 10-4, the radius of the objective is greater than in part (a), its focal length and that of the ocular remaining the same. The radius of the exit pupil is now greater than that of the pupil of the eye, but this is of no advantage, since the eye cannot admit the wider beam. The outer part of the cone of light admitted by the objective is therefore not utilized and no benefit is derived from using a larger objective lens. The half-angle u', the radii of the diffraction disks, and their center-to-center spacing, are the same as in (a) and the disks are shown at (c) in the lower part of the figure.

Suppose, however, that with the larger objective of part (c) we had also used an ocular of shorter focal length, as indicated by the dotted lines in part (c) The overall magnification of the microscope would then be *increased* and the radius of the exit pupil would be *decreased*. If the focal length of the ocular is chosen so that the radius of the exit pupil is decreased to just equal that of the pupil of the eye, the diffraction disks still have the same radius as in part (a), but because of the greater magnification their centers are farther apart than in (a), as shown by the circles in part (d) at the bottom of the figure. The disks are now more than just barely resolvable, or, to put it another way, the object points P_1 and P_2 could be closer together before it became impossible to resolve them. In other words, full advantage can not be taken of the radius of the objective unless the magnification is great enough to reduce the radius of the exit pupil to that of the pupil of the eye.

As a final example, suppose one were to use with the objective of part (a), the ocular of shorter focal length shown by dotted lines in part (c). (The reader can construct his own diagram.) The radius of the exit pupil would be less than in part (a), the radii of the diffraction disks would be greater, and because of the greater magnification the center-to-center distance of the diffraction disks would be increased in the same proportion, as in part (e) at the bottom of the diagram. Two object points that were just resolvable with the ocular in part (a) would still be just resolvable with the higher power ocular. The image would be larger, but no more detail could be observed. Also, since the same amount of light would be distributed over a larger image, the brightness would be less.

From the standpoint of its limit of resolution, it is instructive to express the overall magnification M of a compound microscope as follows. This quantity was shown in Chap. 6 to equal the product of the linear magnification m of the objective, and the angular magnification γ of the ocular.

$$M = m\gamma. \tag{10-6}$$

Let us represent by u'' the slope angle, after refraction by the objective, of a ray entering the objective with a slope angle u. (See Fig. 10-4 (a).) Let n represent the index of the medium at the left of the objective. In any actual microscope, the medium at the right of the objective is air, of index 1.00. Then Abbe's sine condition, applied to the objective, gives

$$m = \frac{n \sin u}{\sin u''}.$$

The angle u'' is always small (this is not necessarily true for the angle u), and it will be seen from a study of Fig. 10-4 (a) that to a good approximation

$$\sin u'' = \frac{R}{f},$$

where R is the radius of the exit pupil and f is the focal length of the ocular.

The angular magnification produced by the ocular is

$$\gamma = \frac{25}{f} \text{ (f in centimeters).}$$

When the three preceding equations are combined with Eq. (10-6) we get

$$M = \frac{n \sin u}{R/25} \text{ (R in centimeters).} \tag{10-7}$$

When used at normal magnification, the full apertures of both objective and eye are utilized. Then $n \sin u$ is the (maximum) numerical aperture of the objective, and $R/25$ is the (maximum) numerical aperture of the eye. Hence

Normal magnification of a compound microscope =

$$\frac{\text{Maximum } NA \text{ of objective}}{\text{Maximum } NA \text{ of eye}}. \tag{10-8}$$

But the ratio of the numerical apertures represents the relative abilities of the objective and the unaided eye to resolve detail, and hence at normal magnification the size of the retinal image is increased in exactly this ratio. Essentially, the numerical aperture of the eye is increased to that of the objective of the microscope. We have shown that the maximum numerical aperture of the unaided eye, with a pupillary radius of 1 mm, is $\frac{1}{250}$. It follows from Eq. (10-8) that *the normal magnification of a compound microscope equals 250 times the numerical aperture of the objective lens.*

The numerical aperture of the objective of a compound microscope can be made much larger than that of the eye. Reference to Fig. 6-13 shows that an object viewed by a microscope is placed just in front of the first focal point of the objective. By designing an objective whose focal point is very close to the first lens, the half-angle u of the cone of light admitted by the objective can be made very nearly 90°, but obviously it cannot exceed 90° and $\sin u$ cannot exceed unity. The practical limit is about 0.95 The numerical aperture can be increased by filling the space between the object and the first lens with a fluid of index greater than 1.00 (the so-called "oil immersion" objective). In this way the numerical aperture may be increased to about 1.60. All high power microscope objectives are of this type, and one is illustrated in Fig. 6-15.

A numerical aperture of 1.60 is 400 times as great as that of the unaided eye ($NA = 0.004$, for a pupillary radius of 1 mm), and hence the smallest distance between two point objects that can just be resolved by an objective of $NA = 1.60$ is $\frac{1}{400}$th as great as for the unaided eye. From Eq. (10-4), this distance is

$$z = \frac{0.61 \times 550 \times 10^{-7}}{1.60}$$

$$= 210 \times 10^{-7} \text{ cm} = 210 \text{ m}\mu.$$

This is of the order of one-half a wave length of light!

With an objective of $NA = 1.60$, the normal magnification is $1.6 \times 250 = 400$. Higher magnifications do not increase detail and entail a loss of light, but are frequently used if plenty of light is available because a larger image can be studied more easily even if no more detail is evident.

To summarize:

(a) *For a given magnification M*, an objective should be used of NA at least

$$NA\,(\text{obj}) = M \times NA\,(\text{eye}) = 0.004\,M.$$

When the NA of the objective has exactly this value, the magnification is normal. (Fig. 10-4 (a).)

If the NA of the objective is less than that computed above, resolution is impaired and the image brightness is decreased. (Fig. 10-4 (b).)

If the NA is greater, there is no gain in resolution. (Fig. 10-4 (c).)

(b) *With an objective of given NA*, an ocular should be used which will give a magnification M of

$$M = \frac{NA\,(\text{obj})}{NA\,(\text{eye})} = 250 \times NA\,(\text{obj}).$$

The magnification is then normal. (Fig. 10-4 (a).)

If the magnification is less than normal, full advantage is not taken of the NA of the objective. (Fig. 10-4 (c).)

If the magnification is greater than normal, there is no gain in resolution and a decrease in brightness.

10-4 Limit of resolution of a telescope. A detailed discussion of the limit of resolution of a telescope will not be given, since exactly the same principles are involved as for a compound microscope. When the magnification is normal, the radius of the exit pupil equals the radius of the pupil of the eye, as in Fig. 10-5. If, with the same ocular as in Fig. 10-5, one uses an objective of smaller radius, the size of the exit pupil is decreased as in Fig. 10-4 (b) and the limit of resolution is impaired. If the radius of

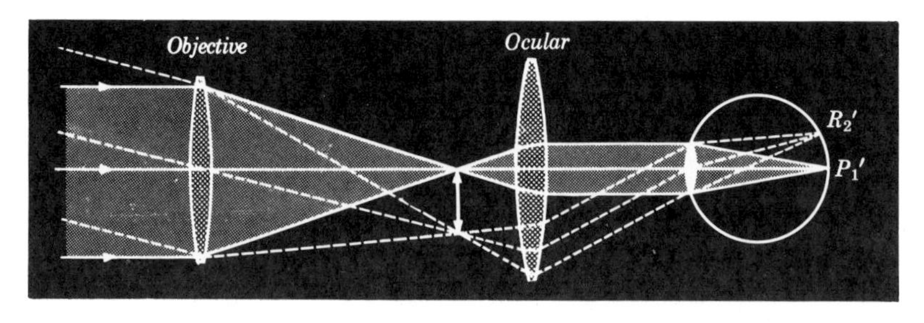

FIG. 10-5. Limit of resolution of a telescope.

the objective is larger than in Fig. 10-5, the size of the exit pupil increases as in Fig. 10-4 (c), but the eye can not admit the wider beam. To take full advantage of the potential limit of resolution of the larger objective, the magnification of the telescope should be increased by the use of an ocular of higher power, thus reducing the size of the exit pupil to that of the pupil of the eye.

Because with a telescope it is the minimum *angle* rather than the minimum *distance* between two just resolvable object points which is of interest, the normal magnification of a telescope is expressed in terms of the radii of the objective lens and of the pupil of the eye, rather than their numerical apertures. The minimum angle between two points which can just be resolved by a lens, as shown by Eq. (10-2), is directly proportional to the radius of the lens. It was shown in Sec. 6-8 that the magnification of a telescope equals the ratio of the radius of the objective lens to the radius of the exit pupil. When the magnification is normal, the latter is just equal to the radius of the pupil of the eye, so

$$\text{Normal magnification of a telescope} = \frac{\text{Radius of objective}}{\text{Radius of pupil of eye}}. \qquad (10\text{-}9)$$

Hence at normal magnification the size of the retinal image is increased in the same ratio that the ability of the objective to resolve (angular) detail exceeds that of the unaided eye. Essentially, the radius of the pupil of the eye is increased to that of the objective. For example, if the radius of the objective lens of a telescope is 5 mm, or 5 times that of the pupil of the eye, the normal magnification is $5\times$ and the smallest angle that can be resolved is ⅕th as great as with the unaided eye.

The notation 6×30 applied to a telescope, means that the angular magnification is $6\times$ and the diameter of the objective is 30 mm. The *normal* magnification, for an objective 30 mm in diameter, or 15 mm in radius, is $15\times$. When a magnification of only $6\times$ is used, full advantage is not being taken of the large objective lens or, to put it another way, an objective only 12 mm in diameter is large enough for a telescope of angular magnification $6\times$. The reason for the use of such a large objective is, first, that the larger exit pupil makes the proper positioning of the eye less critical and second, it avoids a reduction in image brightness under low levels of illumination, when, from Fig. 6-4, the pupillary diameter may increase to 5 mm or more. With a magnification of $6\times$ and an objective only 12 mm in diameter, a 5-mm pupil would not be filled with light and the brightness of the field of view would be less than with the unaided eye, while the pupil is just filled if the objective diameter is 30 mm. A 7×50 instrument can fill a pupil 7 mm in diameter.

10-5 The electron microscope. Mathematically, the reason for the use of a medium of larger index n with an oil immersion objective is to increase the numerical aperture, $n \sin u$. Physically, the function of the medium can be better appreciated by writing the expression for the linear limit of resolution in the form

$$z = \frac{0.61\,(\lambda_0/n)}{\sin u}.$$

The quantity (λ_0/n) in the numerator is simply the wave length *in the medium in which the object is situated*, and this equation states that the smaller this wave length, the smaller is the separation of two points which can just be resolved. It is as if the light waves were the "tools" with which our optical system is provided and the smaller the tools, the finer the work which the system can do. Immersing the object in an oil of high index, then, extends the limit of resolution by furnishing the objective with shorter waves. This limit may be extended to a value several hundred times smaller than that attainable with an optical instrument by using electrons, rather than light waves, to form an image of the object being examined. To understand the principles of the electron microscope, we must first see how the concept of a wave length is associated with an electron and, second, how a beam of electrons may be focused by an electron lens.

A complete discussion of the wave nature of electrons lies in the field of wave mechanics and is beyond the scope of this book. The subject is discussed briefly in Sec. 11-6, which may be referred to at this point. The principles of wave mechanics lead to the conclusion, amply verified by the proper type of experiment, that in certain circumstances an electron can be considered to have a wave length that is related to its momentum, mv, by the equation

$$mv = \frac{h}{\lambda},$$

$$\lambda = \frac{h}{mv}, \tag{10-10}$$

where λ is the wave length of the electron (called the "De Broglie wave length") and h is Planck's constant. (See Sec. 11-6.)

If the velocity v was acquired by acceleration of the electron through a potential difference V,

$$\tfrac{1}{2}\,mv^2 = eV,$$

$$v = \sqrt{\frac{2eV}{m}},$$

$$mv = m\sqrt{\frac{2eV}{m}} = \sqrt{2meV}.$$

Hence

$$\lambda = \frac{h}{\sqrt{2meV}}.$$

The numerical values of the constants on the right side of the equation are:

$$h = 6.62 \times 10^{-34} \text{ joule-sec},$$

$$m = 9.11 \times 10^{-31} \text{ kgm},$$

$$e = 1.602 \times 10^{-19} \text{ coul}.$$

Therefore

$$\lambda = \frac{12.24}{\sqrt{V}} \times 10^{-10} \text{ m} = \frac{12.24}{\sqrt{V}} \times 10^{-8} \text{ cm}.$$

For example, if $V = 100$ volts, $\lambda = 1.22 \times 10^{-8}$ cm $= 0.122$ mμ, and if $V = 10{,}000$ volts, $\lambda = 0.122 \times 10^{-8}$ cm $= 0.0122$ mμ. Hence, with even moderate accelerating voltages, the wave lengths of electrons are of the order of magnitude of those of x-rays, or about 1000 times smaller than the wave lengths of visible light.

Consider next the design of an electron lens. A beam of electrons can be focused either by a magnetic or an electric field of the proper configuration, and both types are used in electron microscopes. Fig. 10-6 illus-

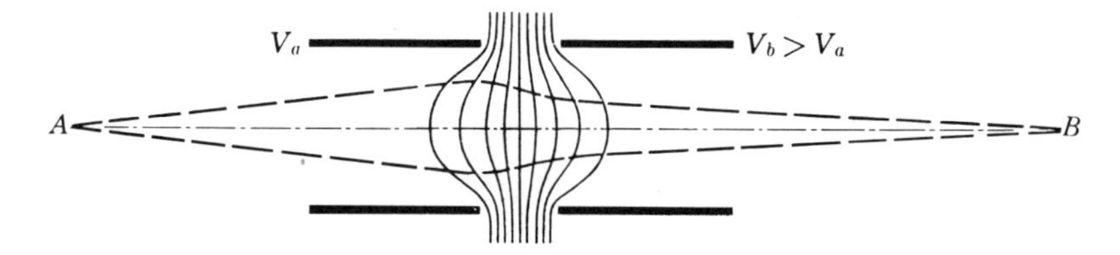

FIG. 10-6. An electrostatic electron lens. The cylinders are at different potentials, V_a and V_b. A beam of electrons diverging from point A is focused at point B.

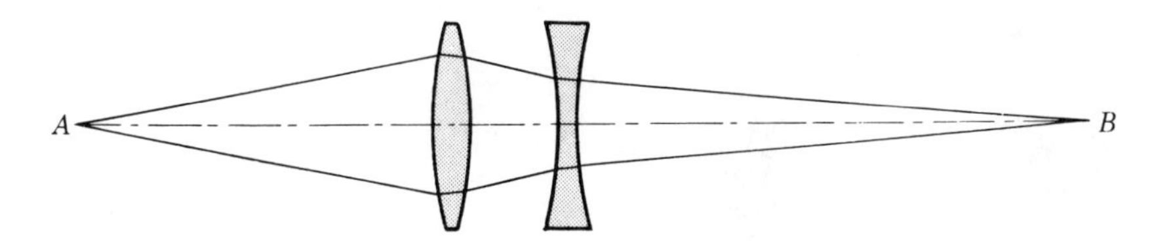

FIG. 10-7. Optical analogue of the electron lens in Fig. 10-6.

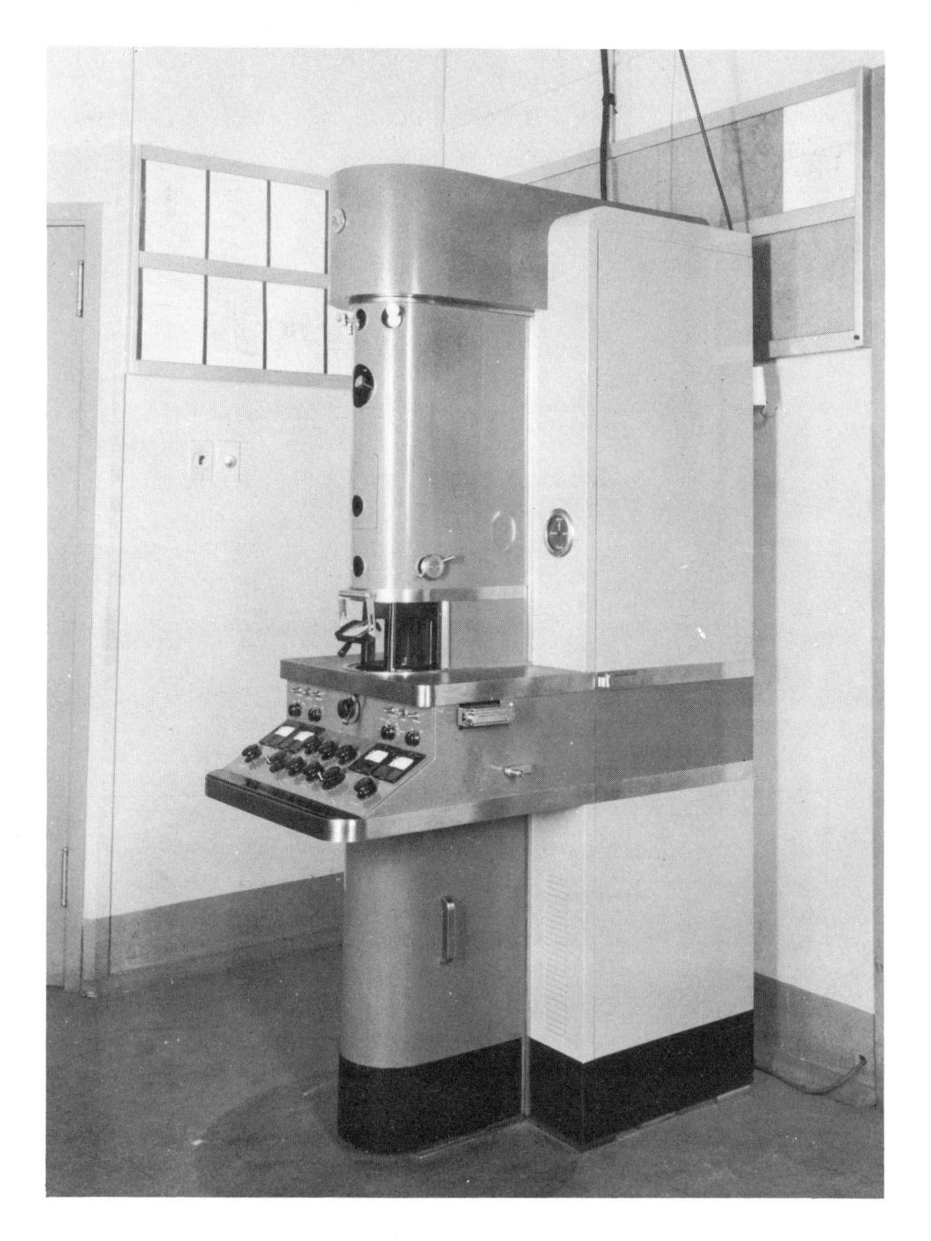

FIG. 10-8. A modern electron microscope.

(Courtesy of Radio Corporation of America)

Fig. 10-9. A desk-type electron microscope.
(Courtesy of Radio Corporation of America)

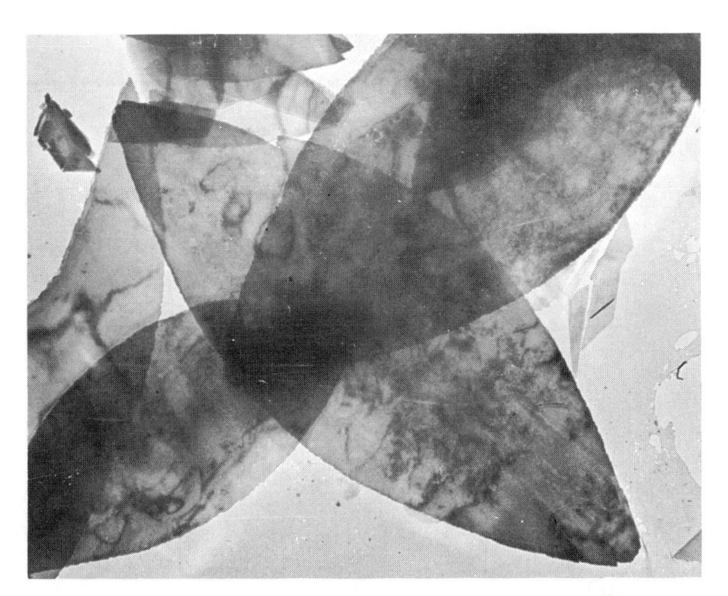

Fig. 10-10. Electron micrograph of aluminum oxide, magnified 53,500 times.
(Courtesy of Radio Corporation of America)

trates an electrostatic lens. Two hollow cylinders are maintained at different potentials. A few of the equipotentials are indicated, and the trajectories of a beam of electrons traveling from left to right are shown by the dotted lines. The optical analogue of this electrostatic lens is shown in Fig. 10-7. It will be evident without going into further detail that by the proper design of such lenses, the elements of an optical microscope, such as its condensing lens, objective, and ocular, can all be duplicated electronically.

The source of electrons in an electron microscope is a heated filament. Electrons emitted by the filament are accelerated by an electron gun and strike the object to be examined. This must necessarily be a thin section so that some of the electrons can pass through it. The thicker portions of the section absorb more of the electron stream than do the thinner portions, just as would a lantern slide in a projection lantern. Needless to say, the entire apparatus must be evacuated. Two types of electron microscope are illustrated in Figs. 10-8 and 10-9.

The final image may be formed on a photographic plate or on a fluorescent screen which can be examined visually or photographed with a still further gain in magnification. Commercial electron microscopes give satisfactory definition at an overall magnification, electronic followed by photographic, as great as 50,000×. Fig. 10-10 is an electron micrograph of aluminum oxide, magnified 53,500 times.

It should be pointed out that the ability of the electron microscope to form an image does not depend on the wave properties of the electrons; their trajectories can be computed by treating them as charged particles, deflected by the electric or magnetic fields through which they move. It is only when considerations of the limit of resolution arise that the electron wave lengths come into the picture. The situation is analogous to that in the optical microscope. The paths of light rays through an optical microscope can be computed by the principles of geometrical optics, but the resolution of the microscope is determined by the wave length of the light used.

10-6 Limit of resolution of a grating. The limit of resolution of an *image-forming* instrument such as a microscope or telescope is defined as the linear or angular separation of two point objects which can just be resolved. The function of an *analyzing* instrument such as a spectrograph, on the other hand, is to resolve two images of the same slit, formed by waves of slightly different wave lengths, and its limit of resolution can be defined as the minimum difference in wave length for which the images can be resolved. The Rayleigh criterion is applied to both types of instrument.

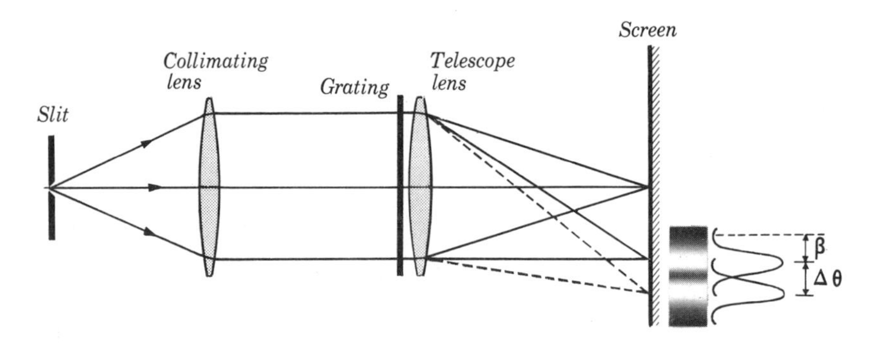

FIG. 10-11. Limit of resolution of a grating.

In Fig. 10-11, a beam of light to be analyzed diverges from a slit and is rendered parallel by a collimating lens, beyond which are a grating and a telescope lens. The diagram shows the slit images formed in the first order spectrum by light of two slightly different wave lengths, λ and $\lambda + \Delta\lambda$.

If the two images are to be resolved, the distance between their centers must be at least as great as the half-angular breadth of either. We first compute their center-to-center spacing. The angular deviation θ of the center of the diffraction pattern formed in the m'th order, by light of wave length λ, is given by

$$\sin\theta = \frac{m\lambda}{d},$$

where d is the grating spacing.

The angle $\Delta\theta$ between the centers of the two patterns is found by differentiation.

$$\cos\theta\,\Delta\theta = \frac{m}{d}\,\Delta\lambda.$$

The graphs in Fig. 9-20 show how the half-angular breadth of the maxima in the diffraction pattern formed by a grating becomes smaller as the number of lines in the grating is increased. We shall omit the details of calculating this half-angle, but the result is

$$\sin\beta = \frac{\tan\theta}{mN},$$

where β is the half-angle, θ is the angle of deviation, m is the order of the spectrum, and N is the total number of lines in the grating.

If the patterns are just resolved, we can set

$$\Delta\theta = \beta = \sin\beta,$$

and hence
$$\cos\theta\,\frac{\tan\theta}{mN} = \frac{m}{d}\,\Delta\lambda\,.$$

But $\cos\theta\,\tan\theta = \sin\theta = m\lambda/d$, and the equation above reduces to

$$\Delta\lambda = \frac{\lambda}{mN}\cdot \tag{10-11}$$

That is, the smallest wave length difference $\Delta\lambda$ that can be resolved by a grating spectrograph is proportional to the wave length λ, and inversely proportional to the order of the spectrum and to the total number of lines N in the grating. Notice that it is independent of the grating spacing d. Of course, the diameters of the collimating and telescope lenses must be great enough so that the entire ruled area of the grating is filled with light, otherwise not all of the lines in the grating are utilized.

10-7 Limit of resolution of a prism. The elements of a prism spectrograph are illustrated in Fig. 10-12. Light diverging from a slit passes through a collimating lens, a prism, and a second lens which images the slit in its second focal plane. The full lines in the emergent beam represent light of wave length λ, and the dotted lines light of wave length $\lambda + \Delta\lambda$. If the wave length difference is small, both beams follow essentially the same path through the prism and let us assume that both pass through at minimum deviation. The angles of deviation of the two beams are δ and $\delta + \Delta\delta$. If the slit images are just resolvable, the angle $\Delta\delta$ between the centers of the images must be at least as great as the half-angular breadth α of the diffraction pattern of a slit formed by a lens. From Eq. (9-1),

$$\alpha = \frac{\lambda}{D},$$

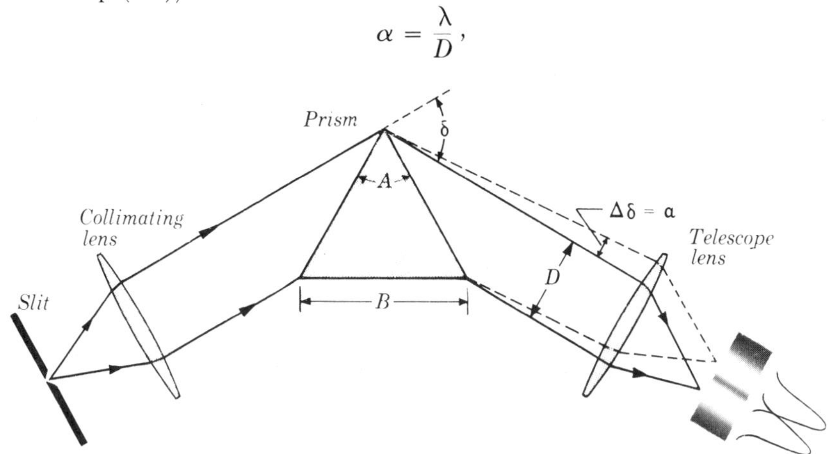

FIG. 10-12. Limit of a resolution of a prism.

where D is the diameter of the second lens, assuming its full aperture is utilized.

The difference between the angles of deviation is found as follows. The angle of minimum deviation δ by a prism is, from Eq. (2-11),

$$\sin\frac{A+\delta}{2} = n\sin\frac{A}{2}\cdot \tag{10-12}$$

Both the index n of the prism and the angle of deviation are functions of wave length. Differentiating Eq. (10-12) with respect to λ gives

$$\frac{1}{2}\cos\frac{A+\delta}{2}\frac{d\delta}{d\lambda} = \sin\frac{A}{2}\frac{dn}{d\lambda},$$

$$d\delta = \frac{2\sin\frac{A}{2}}{\cos\frac{A+\delta}{2}}\frac{dn}{d\lambda}\,d\lambda,$$

and approximately, for the small but finite quantities $\Delta\delta$ and $\Delta\lambda$,

$$\Delta\delta = \frac{2\sin\frac{A}{2}}{\cos\frac{A+\delta}{2}}\frac{dn}{d\lambda}\,\Delta\lambda. \tag{10-13}$$

The term $\frac{dn}{d\lambda}$ refers to the slope of the index vs wave length curve (Fig. 2-27) at the wave length λ.

From Fig. 10-13, it can be seen that

$$\sin\theta = \frac{D}{l}\cdot$$

But

$$2\theta + A + \delta = \pi,$$

$$\theta = \frac{\pi}{2} - \frac{A+\delta}{2},$$

$$\sin\theta = \cos\frac{A+\delta}{2}\cdot$$

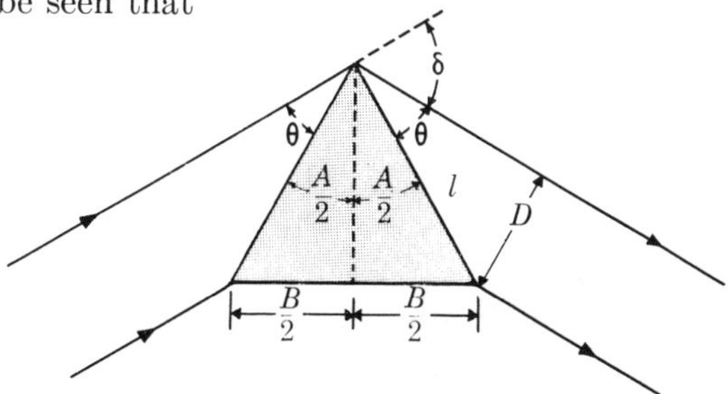

FIG. 10-13.

Hence

$$\cos\frac{A+\delta}{2} = \frac{D}{l}\cdot \tag{10-14}$$

Also, from Fig. 10-13,

$$2\sin\frac{A}{2} = \frac{B}{l}, \tag{10-15}$$

where B is the width of the prism base.

Combining Eqs. (10-14) and (10-15) with Eq. (10-13) gives

$$\Delta\lambda = \frac{\lambda}{B\dfrac{dn}{d\lambda}}. \tag{10-16}$$

The smallest wave length difference $\Delta\lambda$ that can be resolved is proportional to the wave length and inversely proportional to the prism base B and the slope of the index-wave length curve. The diameters of both lenses must, of course, be great enough to utilize the full height of the prism. The reason that the limit of resolution depends only on the width of the prism base and not on its apex angle A may be seen by constructing a diagram like Fig. 10-12, replacing the prism with another of the same base but with a greater height and hence a smaller apex angle. The deviation will be less but the diameters of the lenses will be greater, and the smaller angle $\Delta\delta$ between the slit images is just compensated by the smaller half-angle α in the diffraction pattern.

Example. (a) The wave lengths of the sodium D lines are 589.593 mμ and 588.996 mμ. What is the minimum number of lines a grating must have in order to resolve these lines in the first order spectrum?

From Eq. (10-11),

$$N = \frac{\lambda}{m\Delta\lambda}.$$

From the given data,

$$\Delta\lambda = 589.593 - 588.996 = 0.597 \text{ m}\mu.$$

Hence

$$N = \frac{589}{.597} = 980 \text{ lines}.$$

(b) How large a prism is needed to resolve these lines, if the rate of change of index with wave length, at a wave length of 589 mμ, is 5.30×10^{-5} per millimicron?

From Eq. (10-16),

$$B = \frac{\lambda}{\Delta\lambda\dfrac{dn}{d\lambda}}$$

$$= \frac{589}{0.597 \times 5.30 \times 10^{-5}}$$

$$= 185 \times 10^{5} \text{ m}\mu$$

$$= 1.85 \text{ cm}.$$

Problems—Chapter 10

(1) The headlights of a distant automobile may be considered as point sources. The distance between the headlights is 1.5 m and the automobile is 6000 m (about 4 mi) away. (a) What is the distance between the centers of the images of the sources on the retina? (b) What is the radius of the central diffraction disk of each image? (c) What is the maximum distance at which the headlights could be resolved? Assume a wave length of 550 $m\mu$ and a pupillary radius of 1 mm.

(2) A microscope is to be used to resolve two equally bright point objects separated by 555 $m\mu$. (a) What must be the numerical aperture of the objective, if light of wave length 546.1 $m\mu$ is used? (b) How great an overall magnification must the microscope have in order to take full advantage of the numerical aperture of the objective? (c) If the ocular of the microscope has a focal length of 2.5 cm, what should be the lateral magnification of the objective?

(3) The focal length of the objective of a certain microscope is 4 mm and its numerical aperture is 0.80. The objective forms an image 160 mm beyond its second focal point. (a) What is the angular magnification of the ocular that should be used with this objective to take full advantage of its numerical aperture? (b) What is the smallest separation of two point objects that can just be resolved?

(4) (a) What must be the numerical aperture of a microscope objective capable of resolving two points objects separated by 4.2×10^{-4} mm? Assume the wave length of the light used to be 550 $m\mu$. (b) If the lateral magnification produced by the objective is 20× and if the pupil of the observer's eye is 2 mm in diameter, find the focal length of the lowest power ocular which, used with this objective, will allow him to resolve the two objects.

(5) An object on the stage of a microscope is examined by light of wave length 410 $m\mu$. The numerical aperture of the objective is 0.5 and normal magnification is used. Find the diameter of the object, if its geometrical image is the same size as the central disk in the diffraction pattern that a point object would produce.

(6) A microscope is to be used to examine a cross section of the laminae of a Lippmann color photograph, as shown in Fig. 8-9. If the wave length of the light which exposed the picture was 400 $m\mu$ and if the index of refraction of the photographic emulsion was 1.5, what is the minimum required numerical aperture for the microscope? The photograph is examined by light of wave length 550 $m\mu$.

(7) The diameter of the objective lens of a telescope is 2 cm and its focal length is 25 cm. Three oculars are available for use with the telescope, of angular magnifications 5×, 10×, and 20×. The telescope is directed toward two point sources 3 cm apart, at a distance of 1000 m from the telescope. (a) Fill in the spaces in the table below. Assume a pupillary radius of 1 mm and a wave length of 500 $m\mu$. Treat all lenses as thin. (b) What is the normal magnification of the telescope? (c) Can the two point sources be resolved in each case?

Magnification of ocular	5×	10×	20×
Focal length of ocular			
Angular magnification of telescope			
Radius of exit pupil			
Radius of diffraction disks			
Distance between centers of diffraction disks on retina			
Relative brightness of retinal image		1	

(8) A telescope having an angular magnification of 10× has an objective lens 20 mm in diameter. At what maximum distance can the headlamps of an approaching automobile be separately distinguished with the aid of this telescope? Assume a wave length of 550 mμ and a separation of 1.5 m for the headlamps.

(9) Two pinholes 1.5 mm apart are placed in front of a bright light source and viewed through a telescope with its objective stopped down to a diameter of 4 mm. What is the maximum distance from the telescope at which the pinholes can be resolved?

(10) Two equally bright stars subtend an angle of 1 sec. Assuming an effective wave length of 550 mμ, (a) what is the smallest diameter of a telescope objective lens that will permit these stars to be separately distinguished? (b) What is the normal magnification of the telescope? (c) If the objective lens has a focal length of 180 cm, what focal length ocular should be used?

(11) The diameter of the objective lens of a telescope is 30 mm and the angular magnification of the instrument is 5×. If the diameter of the pupil of the observer's eye is 5 mm, what fraction of the area of the objective is actually utilized?

(12) The concave mirror of a reflecting telescope is 50 cm in diameter and has a focal length of 250 cm. (a) The moon, as seen from the earth, subtends an angle of 30 min. Find the diameter of the image of the moon formed by the mirror. Neglect aberrations. (b) Find the diameter of the image of a star formed by this mirror.

(13) Two lenses, A and B, have equal focal lengths but A has twice the diameter of B. (a) What is the ratio of the energy collected by lens A to that collected by lens B? (b) What is the corresponding ratio of the areas of the diffraction patterns? (c) What is the ratio of the energy per unit area in the two patterns?

(14) A piece of spectroscopic equipment uses a diffraction grating 2 inches in width. Two gratings are available; one (A) has 20,000 lines per inch, and the

other (B) has 10,000 lines per inch. (a) Compare the deviation in the first-order spectrum of A with that in the second-order spectrum of B. (b) Compare the smallest wave length differences that can be resolved under the conditions above.

(15) How many lines must a grating have in order to resolve in the second order two wave lengths which differ by 0.01 mμ, if the wave lengths are approximately (a) 400 mμ, (b) 700 mμ?

(16) A diffraction grating is ruled with a spacing d such that 2 orders (only) are possible for a certain wave length λ. Suppose this grating to be examined under a microscope, using monochromatic light of this same wave length λ. (a) What is the least numerical aperture necessary so as just to be able to resolve the rulings? (b) What is the normal magnification of such a microscope?

(17) (a) What is the smallest wave length difference that can be resolved at a wave length of 500 mμ by a prism spectrometer using an equiangular prism 5 cm on a side, constructed of the silicate flint glass whose dispersion curve is given in Fig. 2-27? (b) What is the width of a grating ruled with 2000 lines/cm, that has the same limit of resolution in the second order as does the prism?

CHAPTER 11

LINE SPECTRA

11-1 Line spectra. The preceding chapters have been concerned chiefly with the *propagation* of light, that is, with the phenomena of reflection, refraction, dispersion, interference, diffraction, and polarization. This chapter and the next will deal with the *emission* of light by light sources, a part of the subject that is closely correlated with the problem of atomic structure and which, because of the failure of "classical" electromagnetic theory to account for the experimental facts, has led to the development of the quantum theory and wave mechanics.

We have seen how a prism or grating spectrograph functions to disperse a beam of light into a spectrum. If the light source is an incandescent solid or liquid the spectrum is *continuous*, that is, light of all wave lengths is present. Spectra of this type will be discussed in the next chapter. If, however, the source is a gas through which an electrical discharge is passing, or a flame into which a volatile salt has been introduced, the spectrum is of an entirely different character. Instead of a continuous band of color only a few colors appear, in the form of isolated parallel lines. (Each "line" is an image of the spectrograph slit, deviated through an angle dependent on the frequency of the light forming the image.) A spectrum of this sort is termed a *line spectrum*. The wave lengths of the lines are characteristic of the element emitting the light. That is, hydrogen always gives a set of lines in the same position, sodium another set, iron still another, and so on. The line structure of the spectrum extends both into the ultraviolet and infrared regions, where naturally photographic or other means are required for its detection.

The positions of some of the more prominent lines of a number of elements are illustrated in the photographs in Fig. 11-1.

It might be expected that the frequencies of the light emitted by a particular element would be arranged in some regular way. For instance, a radiating atom might be analogous to a vibrating string, emitting a fundamental frequency and its overtones. At first sight there does not seem to be any semblance of order or regularity in the lines in Fig. 11-1, and for many years unsuccessful attempts were made to correlate the observed frequencies with those of a fundamental and its overtones. Finally, in 1885, Johann Jakob Balmer (1825-1898) found a simple formula which gave the frequencies of a group of lines emitted by atomic hydrogen. Since the spectrum of this element is relatively simple, and fairly typical of a number of others, we shall consider it in more detail in the next section.

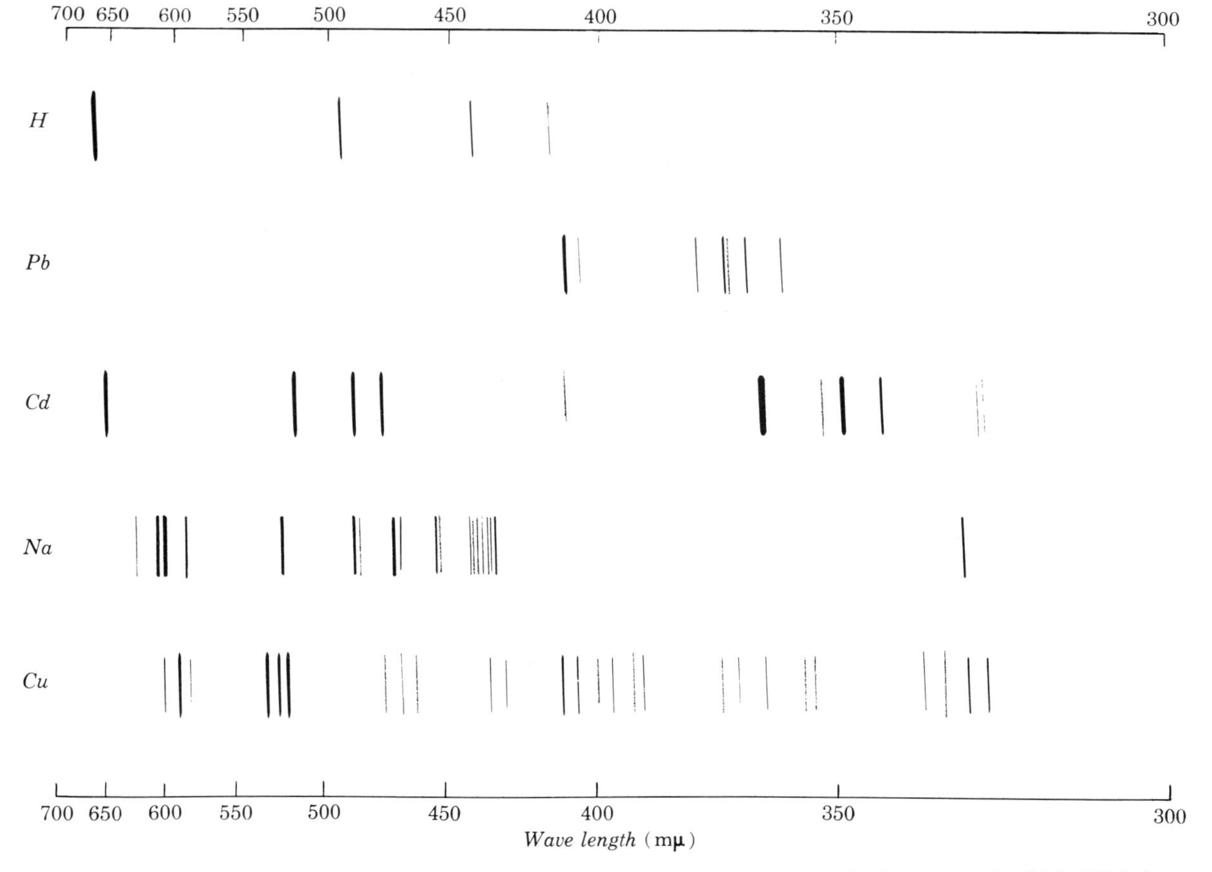

FIG. 11-1. Line spectra. The intense line in the cadmium spectrum at 652.3 mμ is the one with which Michelson compared the standard meter.
(Courtesy of Dr. R. A. McNally)

11-2 Spectral series. Under the proper conditions of excitation, atomic hydrogen may be caused to emit the sequence of lines illustrated in Fig. 11-2. This sequence is called a *series*. There is evidently a certain order in this spectrum, the lines becoming crowded more and more closely together as the limit of the series is approached. The line of longest wave length or lowest frequency, in the red, is known as H_α, the next, in the blue-green, as H_β, the third as H_γ and so on. Balmer found that the wave lengths of these lines were given very accurately by the simple formula,

$$1/\lambda = R\ (1/2^2 - 1/n^2), \tag{11-1}$$

where λ is the wave length, R is a constant called the Rydberg constant and n may have the integral values 3, 4, 5, etc. If λ is in meters,

$$R = 1.097 \times 10^7\ \text{m}^{-1}.$$

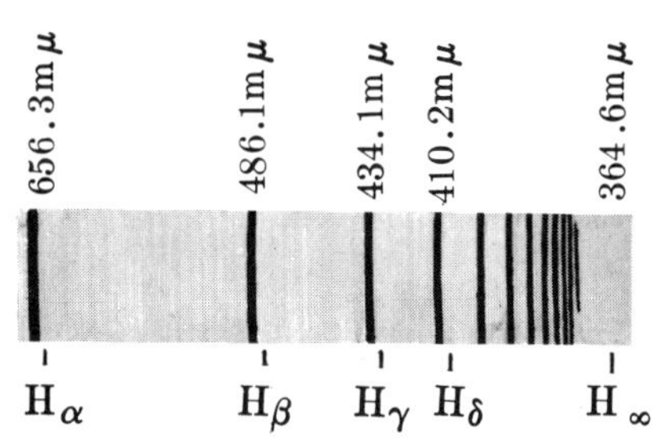

FIG. 11-2. The Balmer series of atomic hydrogen.

(Reproduced by permission from Atomic Spectra and Atomic Structure by Gerhard Herzberg, Copyright 1937 by Prentice-Hall, Inc.)

Letting $n = 3$ in Eq. (11-1), one obtains the wave length of the H_α line.

$$1/\lambda = 1.097 \times 10^7\ (1/4 - 1/9)$$

$$= 1.522 \times 10^6\ \text{m}^{-1},$$

whence

$$\lambda = 656.3\ \text{m}\mu.$$

If $n = 4$, one obtains the wave length of the H_β line, etc. For $n = \infty$, one obtains the limit of the series, at $\lambda = 364.6$ mμ. This is the shortest wave length in the series.

Still other series spectra for hydrogen have since been discovered. These are known, after their discoverers, as the Lyman, Paschen, and Brackett series. The formulas for these are

$$\text{Lyman series:}\quad 1/\lambda = R\ (1/1^2 - 1/n^2) \qquad n = 2, 3, \cdots$$

$$\text{Paschen series:}\quad 1/\lambda = R\ (1/3^2 - 1/n^2) \qquad n = 4, 5, \cdots$$

$$\text{Brackett series:}\quad 1/\lambda = R\ (1/4^2 - 1/n^2) \qquad n = 5, 6, \cdots$$

The Lyman series is in the ultraviolet, and the Paschen and Brackett series are in the infrared. The Balmer series evidently fits into the scheme between the Lyman and the Paschen series.

The Balmer formula, Eq. (11-1) may also be written in terms of the frequency of the light, recalling that

$$c = f\lambda, \quad \text{or} \quad \frac{1}{\lambda} = \frac{f}{c}\cdot$$

Thus Eq. (11-1) becomes

$$f = Rc\,(1/2^2 - 1/n^2) \tag{11-2}$$

or

$$f = Rc/2^2 - Rc/n^2. \tag{11-3}$$

Each of the fractions on the right side of Eq. (11-3) is called a *term*, and the frequency of every line in the series is given by the difference between two terms.

There are only a few elements (hydrogen, singly ionized helium, doubly ionized lithium) whose spectra can be represented by a simple formula of the Balmer type. Nevertheless, it is possible to separate the more complicated spectra of other elements into series, and express the frequency of each line in the series as the difference of two terms. The first term is constant for any one series, while the various values of the second term are found by assigning successive integral values to a quantity corresponding to n in Eq. (11-3), which appears in the (somewhat more complicated) expression for this term.

11-3 The Zeeman effect. After it had been found possible to represent the frequencies of line spectra by a mathematical formula, the next step was to correlate these frequencies with the structure of the atom and the mechanism by which the atom's energy was transformed into the energy of the emitted light. We shall first describe an effect discovered in 1896 by Pieter Zeeman (1865-1943). The theory of the Zeeman effect was worked out by Hendrik Antoon Lorentz (1853-1928) on the hypothesis that the light emitted by an atom had its origin in vibrating electrons within the atom. We can describe here only the simpler aspects of the phenomenon, and their explanation by classical electromagnetic theory. A complete analysis calls for quantum mechanical treatment and is beyond the scope of this book. Briefly, the Zeeman effect has to do with the fact that the frequency of the light emitted by atoms in a gas discharge is altered when the gas is in a magnetic field.

Let us assume that light of frequency f is emitted by electrons vibrating in linear harmonic motion with that frequency. The line OP in Fig. 11-3 (a) represents such an oscillation of amplitude A in some arbitrary direction. This oscillation can be resolved into two components, one of amplitude $A \cos\theta$ along the X-axis and another of amplitude $A \sin\theta$, lying in the Y-Z plane and making an angle ϕ with the Y-axis. The latter linear vibration can in turn be resolved into two circular motions, one clockwise and one counterclockwise, in a circle of radius $\frac{1}{2}A \sin\theta$ as indicated in Fig. 11-3 (b), where the Y-Z plane is in the plane of the diagram.

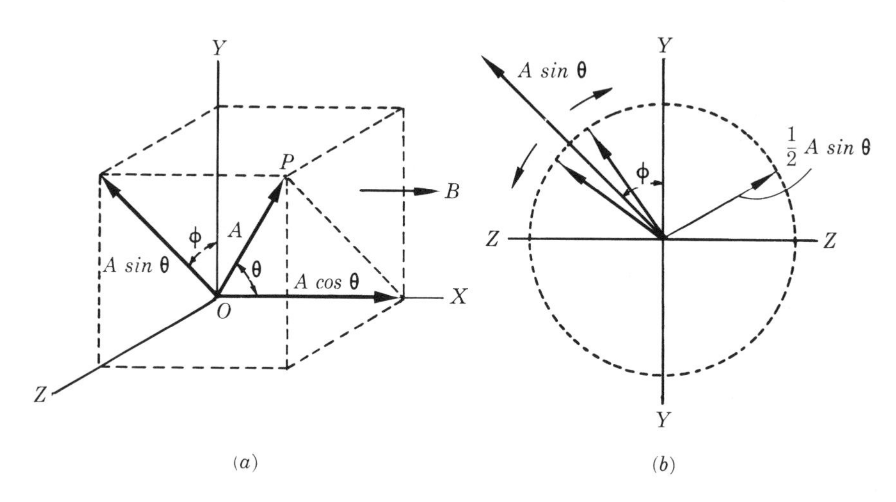

FIG. 11-3. (a) A linear vibration of amplitude A can be resolved into component linear vibrations of amplitudes A cos θ and A sin θ. (b) The linear vibration of amplitude A sin θ is resolved into two circular vibrations.

Now suppose a magnetic field of flux density B is set up, parallel to the X-axis and hence perpendicular to the Y-Z plane. The component vibration represented by $A \cos\theta$ is unaffected by the field since its direction is parallel to the field. The circular components, however, are in planes at right angles to the field. An electron rotating clockwise around the circle in Fig. 11-3 (b) experiences a force radially inward, equal to eBv. Let us represent the radius $\frac{1}{2}A \sin\theta$ by r. Then

$$v = \omega r = 2\pi f r,$$

and the force may be written as $2\pi eBfr$. An electron revolving in a counterclockwise direction experiences a radially outward force of the same magnitude.

Before the introduction of the magnetic field, the electron was retained in its circular orbit by a centripetal force of magnitude

$$F = m\omega^2 r = 4\pi^2 f^2 m r.$$

The change in the centripetal force brought about by the magnetic field is in all practical cases small in comparison with the original force. We may therefore approximate finite changes by differentials and compute the change in frequency arising from a change in the force F. This is

$$dF = 8\pi^2 f m r \, df,$$

and when dF is replaced by $\pm 2\pi eBfr$ we get

$$\pm 2\pi eBfr = 8\pi^2 fmr\, df,$$

$$df = \pm \frac{eB}{4\pi m}. \qquad (11\text{-}4)$$

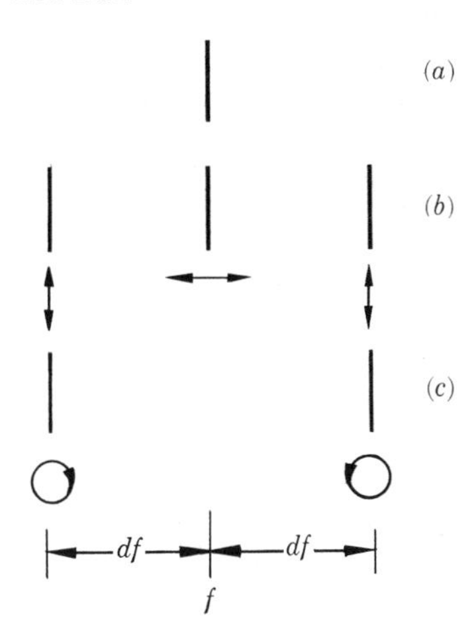

FIG. 11-4. Normal Zeeman triplet (a) No magnetic field. (b) Viewed perpendicular to magnetic field. (c) Viewed parallel to magnetic field. Arrows under lines show state of polarization.

What can now be predicted regarding the light emitted by the vibrating electron? Suppose first that the light source is viewed along the direction of the Z-axis, transverse to the magnetic field. The component $A \cos \theta$ should give rise to waves of the original frequency f, linearly polarized with the E-vector parallel to the direction of vibration, that is, in the X-Z plane. The two circular vibrations, viewed from this direction, appear like linear vibrations and should emit waves of frequencies $f + df$ and $f - df$, linearly polarized with the E-vector in the Y-Z plane. *In certain cases* this is exactly what is observed when the light is dispersed by a spectroscope, as illustrated in Fig. 11-4 (b). The shift in frequency, and the type of polarization, are exactly as predicted by the theory.

When the light source is viewed along the X-axis, parallel to the magnetic field, the component $A \cos \theta$ is viewed "end-on." Hence no light should be emitted in this direction by this component. The other two components should emit *circularly* polarized light of frequencies $f + df$ and $f - df$. Again, in certain cases, this is what is observed as shown in Fig. 11-4 (c).

Instead of computing the shift in frequency from Eq. (11-4) and comparing with observation, the frequency shift may be determined from spectroscopic measurements and Eq. (11-4) used to compute the electronic charge to mass ratio, e/m. Zeeman himself, on the basis of not very precise measurements, estimated this ratio to be of the order of magnitude of 10^{11} coul/kgm. The calculation was of the greatest importance at the time, since it was just at this period (1896) that the existence of electrons was first being suspected and their properties determined. With the advance of spectroscopic techniques in recent years the precision of the method has increased to a degree comparable with that of deflection experiments. The best value of e/m, from the Zeeman effect, is 1.7570×10^{11}

coul/kgm, while the best value including the results of all methods of measurement is 1.7592×10^{11} coul/kgm.

The normal Zeeman triplet, illustrated in Fig. 11-4, is observed only in a few cases. Ordinarily, the influence of a magnetic field is much more complex, and as mentioned earlier can only be interpreted by quantum mechanics.

11-4 The photoelectric effect. All of the evidence afforded by interference and diffraction effects points to the inescapable conclusion that light is some sort of wave motion. The agreement between the measured velocity of light and the computed velocity of electromagnetic waves indicates in turn that the waves are electromagnetic. The Zeeman effect, described in the preceding section, points to the origin of these waves in the vibrations of electrons. In the closing years of the last century the electromagnetic theory of light was generally accepted, and it was believed that after a few minor discrepancies had been cleared up our understanding of the nature of light would be complete. One phenomenon which did not fit satisfactorily into the picture was the photoelectric effect, but the eventual explanation of this effect by Einstein, instead of reconciling it with the classical electromagnetic theory, led to a radical revision in the accepted concepts of the nature of light and the processes of light emission and absorption.

The photoelectric effect was first observed by Heinrich Hertz in 1887. He noticed that a spark would jump more readily between two spheres when their surfaces were illuminated by the light from another spark. Hallwachs investigated the effect more fully in the following year and it usually goes by his name.

According to the modern theory of metals, a metallic conductor consists of a lattice of positive ions, permeated by a swarm of "free" electrons in random motion. An electron approaching the surface of the metal can not normally pass through the surface and escape because of a potential barrier, or potential difference, between the interior of the metal and the surrounding space. However, when light of sufficiently short wave length is incident on the surface of a metal, some of the free electrons acquire sufficient energy to penetrate the barrier and may be drawn to a positively charged collector nearby.

It is found that with a given material as emitter, the wave length of the incident light must be shorter than a critical value, different for different materials, in order that electrons may be emitted. The frequency corresponding to this critical wave length is called the *threshold frequency* of the material. The threshold frequency for most metals is in the ultraviolet, but for potassium and caesium oxides, which are now commonly used as photoemissive surfaces, it lies in the visible spectrum.

A remarkable feature of photoelectric emission is the relation between the number of escaping electrons and their maximum velocity, on one hand, and the intensity and wave length of the incident light on the other. It is found that electrons are emitted from an illuminated surface with a range of velocities from zero up to a certain maximum value. The maximum velocity of emission is found to be *independent of the intensity* of the incident light, but it does depend on the wave length of the light, being greater, the shorter the wave length. It is true that the photoelectric current increases as the light intensity is increased, but only because more electrons are emitted. With light of a given wave length, no matter how feeble it may be, the maximum velocity of the electrons from a given material is always the same, provided, of course, that the frequency is above the threshold frequency.

This behavior is quite unexplainable by classical theory. It is understandable that an electromagnetic wave could, in virtue of the electric field associated with it, exert a force on an electron and eject the latter from a metal. But increasing the intensity of such a wave means increasing the electric intensity and hence increasing the force on an electron and the velocity of emission. There is no way of accounting for the fact that with light of a given wave length, whether its amplitude is large or small, the fastest electrons all have exactly the same velocity. Equally mysterious is the fact that a weak light beam of short wave length causes the emission of electrons with greater energy than those resulting from illumination by an intense beam of longer wave length.

Still another anomaly arises when one investigates the mechanism by which an electron acquires energy from the incident light. All experimental evidence indicates that photoelectric emission begins instantaneously when light strikes the emitting surface. (The time lag, if it exists at all, has been shown to be less than 10^{-9} sec.) One can measure the rate at which energy is incident per unit area, and it turns out that in order for an electron to acquire sufficient energy to escape with its observed velocity, in a time interval of less than 10^{-9} sec, it must be capable of "collecting" in some way the energy incident on an area that is millions of times as great as the cross section of an atom.

The explanation of the photoelectric effect was given by Einstein in 1902, although his theory was so radical that it was not generally accepted until 1906 when its essential features were confirmed by experiments performed by Millikan. Extending a suggestion made two years earlier by Planck, in connection with the radiant energy emitted by solids and liquids and discussed in the next chapter, Einstein postulated that the energy in a beam of light, instead of being spread out over a wave surface as in the classical theory, was concentrated in small "packets" which are now called *light quanta* or *photons*. The concepts of wave length and frequency are still associated with photons, and the energy E of a photon is proportional to its frequency f.

$$\boxed{E = hf.} \tag{11-5}$$

The proportionality constant h is called *Planck's constant*. Its numerical value is 6.624×10^{-27} erg-second, or 6.624×10^{-34} joule-second.

When a photon collides with an electron at or just within the surface of a metal it may transfer its energy to the electron. This transfer is an "all-or-none" process, the electron acquiring either all of the photon's energy or none. The photon then simply drops out of existence. The energy acquired by the electron may enable it to penetrate the potential barrier if it is moving in the right direction. In penetrating the potential barrier the electron loses an amount of energy ϕ which is characteristic of the surface and is called the *work function* of the surface. Electrons that start at some distance below the surface may lose more than this amount, but the maximum energy with which an electron can emerge is the energy gained from a photon, minus the work function. Hence the maximum kinetic energy of photoelectrons ejected by light of frequency f is

$$\boxed{\tfrac{1}{2}\, m v^2_{\text{max}} = hf - \phi.} \tag{11-6}$$

This is Einstein's photoelectric equation, and (except for a small correction dependent on the temperature of the metal, but which does not alter the essential features of the problem) it was in exact agreement with Millikan's experimental results.

All the difficulties encountered in explaining photoelectric emission by the classical theory are easily understandable on the basis of the newer picture. The shorter the wave length of the incident light, or the higher its frequency, the greater the energy of the photons comprising it and the greater the kinetic energy of the electrons ejected. Increasing the intensity of light of a given wave length merely means that more photons strike the metal per unit time. Hence more electrons are ejected, but with no increase in their maximum energies since the energies of the photons are the same. It is unnecessary to wait for energy to accumulate when a light beam strikes a surface because a single photon, colliding with an electron, can immediately give the latter enough energy to escape. The threshold frequency of a given material is merely that frequency at which the energy of a photon is equal to the work function of the material, since an electron must acquire at least this much energy in order to penetrate the potential barrier.

FIG. 11-5. A modern phototube. (Courtesy of Radio Corp. of America)

But if the photon theory is correct, what about the phenomena of interference and diffraction, which seem inconsistent with anything but a wave theory? In all likelihood the last chapter on the subject has not yet been written, but it has proven possible to reconcile the two viewpoints by divesting the electromagnetic waves of some of their reality and considering that they serve merely as guides to direct the paths of the photons. That is, large numbers of photons follow paths that take them to regions such as the maxima of interference patterns where electromagnetic theory predicts a large amplitude of E or H, while smaller numbers travel toward regions where the field intensities are less. Whatever form the complete theory may eventually take, we must accept the experimental facts that in problems where the propagation of light is concerned, it behaves *as if* it were an electromagnetic wave, while in the interaction of light with matter it behaves *as if* it were an assemblage of corpuscles or photons.

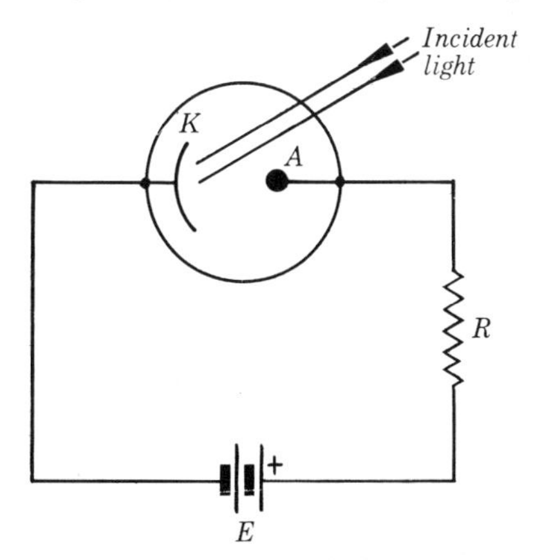

FIG. 11-6. Diagram of phototube circuit. K is the photosensitive cathode, A the anode, and E a source of emf. The photoelectric current results in an IR drop across the resistor R, proportional to the current.

Fig. 11-5 is a photograph of a modern type of phototube, and Fig. 11-6 is a diagram of the tube and its associated electrical circuit. When light strikes the cathode K photoelectrons are emitted and are drawn to the anode A, normally maintained at a potential of the order of a hundred volts above the cathode. Cathode and anode are enclosed in an evacuated container.

11-5 The Bohr atom. The preceding section has described how Einstein invoked the concept of light quanta, or photons, to account for the experimental facts of photoelectric emission. We shall now see how the Danish physicist Niels Bohr, in 1913, first applied the same ideas to the emission of light by atoms. Although the Zeeman effect seemed to indicate that light waves had their origin in vibrating or revolving electrons, the classical theory failed to account both for the existence of stable atoms and the fact that line spectra could be arranged in series having frequencies given by Balmer's formula or one of its modifications.

Experiments on the scattering of alpha-particles by thin metallic foils were performed by Rutherford and his co-workers about 1906. These led to the hypothesis that atoms consisted of a relatively massive, positively charged nucleus, surrounded by a swarm of electrons. To account for the fact that the electrons in an atom remained at relatively large distances from the nucleus, in spite of the electrostatic force of attraction of the nucleus for them, Rutherford postulated that the electrons revolved about the nucleus, the force of attraction providing the requisite centripetal force to retain them in their orbits. This assumption, however, has an unfortunate consequence. A body moving in a circle is continuously accelerated toward the center of the circle, and according to classical electromagnetic theory an accelerated electron radiates energy. The kinetic energy of the electrons would therefore gradually decrease, their orbits would become smaller and smaller and eventually they would spiral in to the nucleus and come to rest. Furthermore, according to classical theory, the electrons would emit a continuous spectrum (a mixture of all frequencies) in contradiction to the line spectrum which is observed. Such a continuous spectrum is actually emitted by the rapidly revolving electrons in a betatron.

Faced with the dilemma that electromagnetic theory predicted an unstable atom emitting radiant energy of all frequencies, while observation showed stable atoms emitting only a few frequencies, Bohr concluded that in spite of the success of electromagnetic theory in explaining large scale phenomena, it could not be applied to processes on an atomic scale. Bohr's *first postulate*, therefore, was that *an electron in an atom can revolve in certain specified orbits without the emission of radiant energy*, contrary to the predictions of the classical electromagnetic theory. The first postulate therefore "explained" the stability of the atom.

A completely stable atom, however, is as unsatisfactory as an unstable one, since atoms *do* emit radiant energy. Bohr's *second postulate* incorporated into atomic theory the quantum concepts that had been developed by Planck in connection with blackbody radiation (see Chap. 12) and applied by Einstein to the photoelectric effect. The second postulate was that *an electron may suddenly "jump" from one of its specified nonradiating orbits to another of lower energy. When it does so, a single photon is emitted whose energy equals the energy difference between the initial and final states, and whose frequency f is given by the relation*

$$\boxed{hf = E_1 - E_2,} \tag{11-7}$$

where h is Planck's constant and E_1 and E_2 are the initial and final energies.

It remained to specify the radii of the nonradiating orbits. Bohr found that the frequencies of the spectral lines of atomic hydrogen, as computed from Eq. (11-7), were in agreement with observation provided the electron were permitted to rotate about the nucleus *only in those orbits for which the angular momentum is some integral multiple of* $h/2\pi$. It will be recalled that the angular momentum of a particle of mass m, moving with tangential velocity v in a circle of radius r, is mvr. Hence the quantum condition above may be stated

$$\boxed{mvr = n\frac{h}{2\pi},} \tag{11-8}$$

where $n = 1, 2, 3$, etc.

The hydrogen atom consists of a single electron of charge $-e$, rotating about a single proton of charge $+e$. The electrostatic force of attraction between the charges,

$$F = \frac{1}{4\pi\epsilon_0}\frac{e^2}{r^2},$$

provides the centripetal force, and from Newton's second law,

$$\frac{1}{4\pi\epsilon_0}\frac{e^2}{r^2} = \frac{mv^2}{r}. \tag{11-9}$$

(We are using rationalized mks units.)

When Eqs. (11-8) and (11-9) are solved simultaneously for r and v, we obtain

$$r = \epsilon_0 \frac{n^2h^2}{\pi me^2}, \tag{11-10}$$

$$v = \frac{1}{\epsilon_0}\frac{e^2}{2nh}.$$

Let

$$\epsilon_0 \frac{h^2}{\pi me^2} = r_0. \tag{11-11}$$

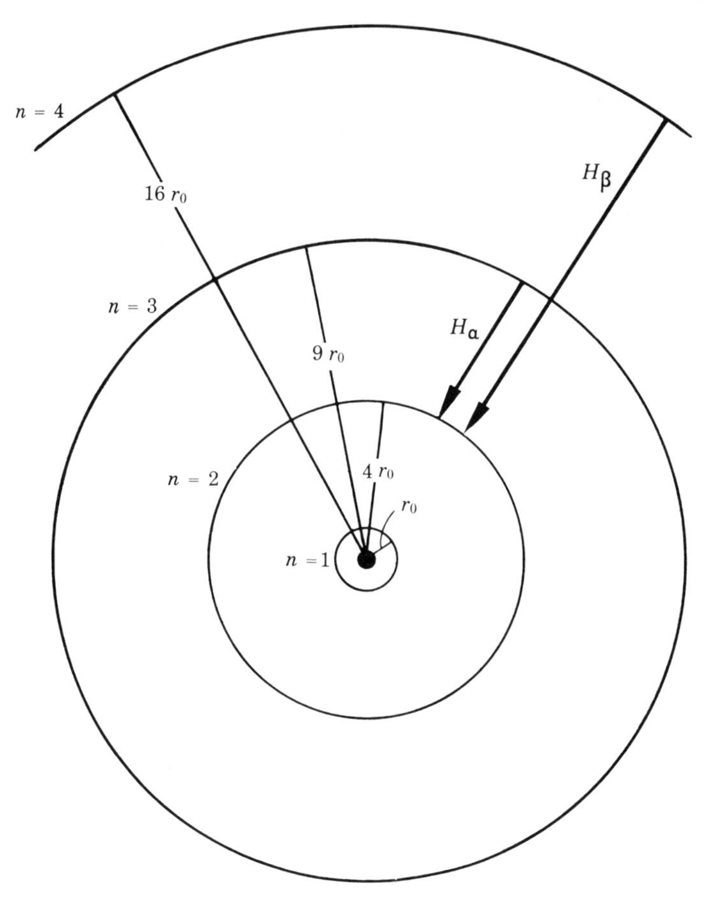

FIG. 11-7. "Permitted" orbits of an electron in the Bohr model of a hydrogen atom. The transitions or "jumps" responsible for some of the lines of the Balmer series are indicated by arrows.

Then Eq. (11-10) becomes

$$r = n^2 r_0,$$

and the permitted, nonradiating orbits, are of radii

$$r_0,\ 4r_0,\ 9r_0,\ \text{etc.}$$

The appropriate value of n is called the *quantum number* of the orbit. (See Fig. 11-7.)

The numerical values of the quantities on the left side of Eq. (11-11) are:

$$\epsilon_0 = 8.85 \times 10^{-12} \frac{\text{coul}^2}{\text{newton-m}^2}, \qquad m = 9.11 \times 10^{-31} \text{ kgm},$$

$$h = 6.62 \times 10^{-34} \text{ joule-sec}, \qquad e = 1.60 \times 10^{-19} \text{ coul}.$$

Hence r_0, the radius of the first Bohr orbit, is

$$\begin{aligned} r_0 &= \frac{8.85 \times 10^{-12} \times (6.62 \times 10^{-34})^2}{3.14 \times 9.11 \times 10^{-31} \times (1.60 \times 10^{-19})^2} \\ &= 5.3 \times 10^{-11} \text{ m} \\ &= 0.53 \times 10^{-8} \text{ cm}. \end{aligned}$$

This is in good agreement with atomic diameters as estimated by other methods, namely, about 10^{-8} cm.

The kinetic energy of the electron in any orbit is

$$KE = \tfrac{1}{2} mv^2 = \frac{1}{\epsilon_0{}^2} \frac{me^4}{8n^2h^2},$$

and the potential energy is

$$PE = -\frac{1}{4\pi\epsilon_0} \frac{e^2}{r} = -\frac{1}{\epsilon_0{}^2} \frac{me^4}{4n^2h^2}.$$

The total energy, E, is therefore

$$E = KE + PE = -\frac{1}{\epsilon_0{}^2} \frac{me^4}{8n^2h^2}. \tag{11-12}$$

The total energy has a negative sign because the reference level of potential energy is taken with the electron at an infinite distance from the nucleus. Since we are interested only in energy differences this is not of importance.

The energy of the atom is least when its electron is revolving in the orbit for which $n = 1$, for then E has its largest negative value. For $n = 2, 3, \ldots$ the absolute value of E is smaller, hence the energy is progressively larger in the outer orbits. The *normal state* of the atom is that of lowest energy, with the electron revolving in the orbit of smallest radius, r_0. As a result of collisions with rapidly moving electrons in an electrical discharge, or for other causes, the atom may temporarily acquire sufficient energy to raise the electron to some outer orbit. The atom is then said to be in an *excited state*. This state is an unstable one, and the electron soon falls or "jumps" back to a state of lower energy, emitting a photon in the process.

Let n be the quantum number of some excited state, and l the quantum number of the lower state to which the electron returns after the emission process. Then E_1, the initial energy, is

$$E_1 = -\frac{1}{\epsilon_0^2}\frac{me^4}{8n^2h^2},$$

and E_2, the final energy, is

$$E_2 = -\frac{1}{\epsilon_0^2}\frac{me^4}{8l^2h^2}.$$

The decrease in energy, $E_1 - E_2$, which we place equal to the energy hf of the emitted photon, is

$$E_1 - E_2 = hf = -\frac{1}{\epsilon_0^2}\frac{me^4}{8n^2h^2} + \frac{1}{\epsilon_0^2}\frac{me^4}{8l^2h^2},$$

or

$$f = \frac{1}{\epsilon_0^2}\frac{me^4}{8h^3}\left(\frac{1}{l^2} - \frac{1}{n^2}\right). \tag{11-13}$$

This equation is of precisely the same form as the Balmer formula (Eq. (10-2)) for the frequencies in the hydrogen spectrum if we place

$$\frac{1}{\epsilon_0^2}\frac{me^4}{8h^3} = Rc, \tag{11-14}$$

and let $l = 1$ for the Lyman series, $l = 2$ for the Balmer series, etc. The Lyman series is therefore the group of lines emitted by electrons returning from some excited state to the normal state. The Balmer series is the group emitted by electrons returning from some higher state, but which decide to stop in the second orbit instead of falling at once to that of lowest energy. That is, an electron returning from the third orbit ($n = 3$) to the second orbit ($l = 2$) emits the H_α line. One returning from the fourth orbit ($n = 4$) to the second ($l = 2$) emits the H_β line, etc. (See Fig. 11-7.)

The question naturally arises as to whether or not Eq. (11-14) is true, since every quantity in it may be determined quite independently of the Bohr theory, and apart from this theory we have no reason to expect these quantities to be related in this particular way. The quantities m and e, for instance, are found from experiments on free electrons, h may be found from the photoelectric effect, R by measurements of wave lengths, while c is the velocity of light. However, if we substitute in Eq. (11-14) the values of these quantities, obtained by such diverse means, we find that it does hold exactly, within the limits of experimental error.

The discovery of deuterium constitutes another triumph of Bohr's theory of the hydrogen atom. According to Bohr, an electron revolving about a stationary nucleus in an orbit characterized by the quantum number n has an energy given by Eq. (11-12),

$$E = -\frac{1}{\epsilon_0^2} \frac{me^4}{8n^2h^2} \cdot$$

The nucleus of a hydrogen atom, a proton, has a mass 1840 times that of an electron, and therefore the assumption that the proton remains stationary while the electron revolves around it is a fair one if great accuracy is not desired. In devising a theory to explain spectral lines, however, the utmost refinements must be introduced because measurements of the wave lengths of spectral lines are among the most precise in all experimental physics. It is natural, therefore, to take into account the fact that both the electron and proton revolve about their common center of mass. When this is done, the energy of an electron becomes

$$E = -\frac{1}{\epsilon_0^2} \frac{m}{1 + \frac{m}{M}} \frac{e^4}{8n^2h^2},$$

where M is the mass of the nucleus.

It was suspected in 1931 that ordinary hydrogen was a mixture of two isotopes, one isotope consisting of an electron revolving about a proton, the other consisting of an electron revolving about a nucleus consisting of a proton and a neutron, and having a mass almost exactly twice that of the proton.

If all of these atoms are excited in an electric discharge and the Balmer lines are measured carefully, the frequency of, say, the second Balmer line (H_β) would be slightly different for the two different hydrogens. Thus

$$f_1 = \frac{1}{\epsilon_0^2} \frac{m}{1 + \frac{m}{M_1}} \frac{e^4}{8h^3} \left(\frac{1}{2^2} - \frac{1}{4^2}\right) \left(\begin{matrix}\text{for a nucleus} \\ \text{of mass } M_1\end{matrix}\right),$$

and

$$f_2 = \frac{1}{\epsilon_0^2} \frac{m}{1 + \frac{m}{M_2}} \frac{e^4}{8h^3} \left(\frac{1}{2^2} - \frac{1}{4^2}\right) \left(\begin{matrix}\text{for a nucleus} \\ \text{of mass } M_2\end{matrix}\right),$$

where $m/M_1 = 1/1840$ and $m/M_2 = 1/3680$. The two different H_β's would be very close together because the wave length difference corresponding to the frequency difference $f_1 - f_2$ is only 1.3 Angstroms. Furthermore, the H_β line corresponding to the heavy isotope would be very much fainter than the other because in any ordinary sample of hydrogen, there is presumably only a small percentage of heavy hydrogen.

In spite of these difficulties, Urey, Murphy, and Brickwedde, in 1932, undertook the task of determining whether the heavy isotope of hydrogen existed. With the aid of a diffraction grating spectrograph they found the blue H_β line to consist of a very strong line and a faint companion. They then proceeded to prepare hydrogen with a higher percentage of the heavy isotope by allowing liquid hydrogen to evaporate and retaining the residue. With this residue in the discharge tube, the companion line was much stronger and its displacement from the other H_β line was in agreement with the value predicted by the Bohr theory. They called the heavier isotope *deuterium*.

11-6 Wave mechanics. The Bohr model of the atom was successful in explaining the observed spectra of atomic hydrogen and of a few other elements, but for atoms having a large number of orbital electrons, and for molecules, the theory was not as satisfactory. Furthermore, there seemed to be no good justification, except that it led to the right answer, for the hypothesis that only those orbits are permitted for which the angular momentum is equal to some integral multiple of $h/2\pi$. The next advance in atom-building, which followed the theory of Bohr by about ten years, was a suggestion by de Broglie that since light appeared to be dualistic in nature, behaving in some aspects like waves and in others like corpuscles, the same might be true of matter. That is, electrons and protons, which until that time had been thought to be purely corpuscular, might in some circumstances behave like waves. The rapid development of this idea in the hands of Heisenberg, Schrödinger, and many others, led to the so-called *wave mechanics* or *quantum mechanics* which has placed atomic theory on what we believe to be a secure foundation.

A single section on "wave mechanics" can not, of course give the reader any adequate comprehension of this complex and highly mathematical subject, any more than the whole field of "Newtonian mechanics" could be covered in the same amount of space. We can only point out the main lines of thought in a nonmathematical way, describe some of the experimental evidence for the wave-nature of material particles, and show how the quantum numbers that were introduced in such an artificial way by Bohr, now enter naturally into the problem of atomic structure.

The essential feature of the new wave mechanics, as we have said, is that particles of matter are also endowed with wavelike properties. An electron, then, must be considered as some sort of wave, more or less spread out through space and not simply localized at a point. The idea that the electrons in an atom move in definite Bohr orbits such as those in Fig. 11-7 has been abandoned. Instead, the new theory specifies merely that there are certain *regions* in which an electron is more or less likely to be found. The orbits themselves, however, were never an essential part of Bohr's theory, since the only quantities that determined the frequencies of the emitted photons were the *energies* corresponding to the orbits. The new theory still assigns definite energy states to an atom. In the hydrogen atom the energies are the same as those given by Bohr's theory; in more complicated atoms where the Bohr theory did not work, the wave mechanical picture is in excellent agreement with observation.

We shall illustrate how quantization arises in atomic structure by an analogy with the classical mechanical problem of a vibrating string fixed at its ends. The reader will recall that the application of Newton's second law to an element of a stretched string, slightly displaced from its equilibrium position, leads to a second order differential equation; the so-called wave equation. The differential equation is satisfied by a wave traveling along the string in either direction with a velocity determined by the tension in the string and its mass per unit length. But if the string is fixed at both ends, it is not enough to satisfy the differential equation. We can only use those solutions for which the ends of the string remain permanently at rest, or for which the ends are nodes. Nodes may occur at other points also, and the general requirement is that the length of the string shall equal some integral number of half-wave lengths. The point of interest is that the differential equation and the boundary conditions, both of which must be satisfied, lead to the appearance of *integral numbers* in the solution of the problem.

In a similar way, the principles of quantum mechanics lead to a wave equation (Schrödinger's equation) that must be satisfied by an electron in an atom, subject also to certain boundary conditions. Let us think of an electron as a wave extending in a circle around the nucleus. In order that the wave may "come out even," the circumference of this circle must include some *integral number* of wave lengths. The wave length of a particle of mass m, moving with a velocity v, is given according to wave mechanics by the equation (see page 270)

$$\lambda = \frac{h}{mv}, \tag{11-15}$$

where λ is the wave length and h is Planck's constant. Then if r is the radius and $2\pi r$ the circumference of the circle occupied by the wave, we must have

$$2\pi r = n\lambda,$$

where $n = 1, 2, 3$, etc.

Since $\lambda = h/mv$, this equation becomes

$$2\pi r = n\frac{h}{mv},$$

or

$$mvr = n\frac{h}{2\pi}.$$

But mvr is the angular momentum of the electron, and we see that the wave mechanical picture leads naturally to Bohr's postulate that the angular momentum equals some integral multiple of $h/2\pi$.

There is even more direct experimental evidence of the wavelike nature of electrons. We have described in Chap. 9 how the layers of atoms in a crystal serve as a diffraction grating for x-rays. An x-ray beam is strongly reflected when it strikes a crystal at such an angle that the waves reflected from the atomic layers combine to reenforce one another. The point of importance here is that the existence of these strong reflections is evidence of the wave nature of x-rays.

In 1927, Davisson and Germer, working in the Bell Telephone Laboratories, were studying the nature of the surface of a crystal of nickel by directing a beam of electrons at the surface and observing the electrons reflected at various angles. It might be expected that even the smoothest surface attainable would still look rough to an electron, and that the electron beam would therefore be diffusely reflected. But Davisson and Germer found that the electrons were reflected in almost the same way that x-rays would be reflected from the same crystal. The wave lengths of the electrons in the beam were computed from their known velocity, with the help of Eq. (11-15), and the angles at which strong reflection took place were found to be the same as those at which x-rays of the same wave length would be reflected.

11-7 Absorption spectra. Although the precise picture of electronic orbits about the nucleus of an atom has been abandoned in modern physics, the concept of energy levels still remains. The fundamental problem of the spectroscopist is to determine the energy levels of an atom from the measured values of the wave lengths of the spectral lines emitted when the atom proceeds from one set of energy levels to another. In the case of complicated spectra emitted by the heavier atoms this is a task requiring tremendous ingenuity. Nevertheless, almost all spectra have been analyzed and the resulting energy levels have been tabulated or plotted.

The lowest energy level of the atom is called the *normal state* and all higher levels are called *excited states.* As we have seen, a spectral line is emitted when an atom proceeds from an excited state to a lower state. The only means discussed so far for raising the atom from the normal state to an excited state has been with the aid of an electric discharge. Let us consider now another method involving the absorption of radiant energy.

A sodium atom emits the characteristic yellow light of wave lengths 589.0 mμ and 589.6 mμ (the D_1 and D_2 lines) when it undergoes the transitions from two levels called *resonance levels* to the normal state. Suppose a sodium atom in the normal state were to absorb a quantum of radiant energy of wave length 589.0 mμ or 589.6 mμ. It would then undergo a transition in the opposite direction and be raised to one of the resonance levels. After a short time, known as the *lifetime* of the excited state (which in the case of the resonance levels of the sodium atom has been found to be 1.6×10^{-8} sec) the atom returns to the normal state and emits this quantum. The emission process is called *resonance radiation* and may be easily demonstrated as follows. A strong beam of the yellow light from a sodium arc is concentrated on a glass bulb which has been highly evacuated and into which a small amount of pure metallic sodium has been distilled. If the bulb is gently warmed with a bunsen burner to increase the sodium vapor pressure, resonance radiation will take place throughout the whole bulb, which glows with the yellow light characteristic of sodium.

A sodium atom in the normal state may absorb radiant energy of wave lengths other than the yellow resonance lines. All wave lengths corre-

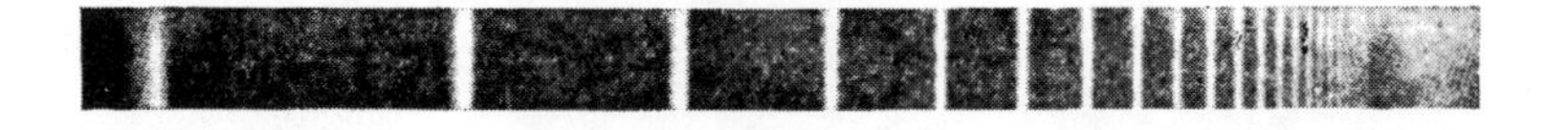

FIG. 11-8. Absorption spectrum of sodium.

sponding to spectral lines emitted when the sodium atom returns to its normal state may be absorbed. If therefore the light from a carbon arc is sent through an absorption tube containing sodium vapor, and then examined with a spectroscope, there will be a series of dark lines corresponding to the wave lengths absorbed, as shown in the reversed print of Fig. 11-8. This is known as an *absorption spectrum.*

The sun's spectrum is an absorption spectrum. The main body of the sun emits a continuous spectrum whereas the cooler vapors in the sun's atmosphere emit line spectra corresponding to all the elements present. When the intense light from the main body of the sun passes through the cooler vapors, the lines of these elements are absorbed. The light emitted by the cooler vapors is so small compared with the unabsorbed continuous spectrum, that the continuous spectrum appears to be crossed by myriads of faint dark lines. These were first observed by Fraunhofer and are therefore called *Fraunhofer lines.* They may be observed with any student spectroscope pointed toward any part of the sky.

11-8 Band spectra. Up to this point we have confined our attention to spectra emitted by individual atoms. Many gases, however, have molecules consisting of two or more atoms held more or less tightly together. Thus hydrogen, oxygen, nitrogen, carbon monoxide, etc. are diatomic gases with molecules composed of two atoms each. To obtain the spectrum of atomic hydrogen or atomic oxygen, a heavy electric discharge is needed to dissociate some of the molecules into atoms. If no dissociation takes place, the molecules themselves emit light which when analyzed with a spectroscope shows an enormous number of lines spaced so close together that they form what appear roughly to be bands. Hence the term *band spectrum.* Typical band spectra are shown in Fig. 11-9.

Each line in a band spectrum is the result of a transition between two energy levels of the molecule. The energy levels of molecules are much more numerous and much more complicated than those of atoms. They arise not only from different electron "orbits" but also from different energies of rotation and vibration. A whole set of bands corresponds to an electronic transition, a single band to a vibrational transition, and a single line in a band to a rotational transition.

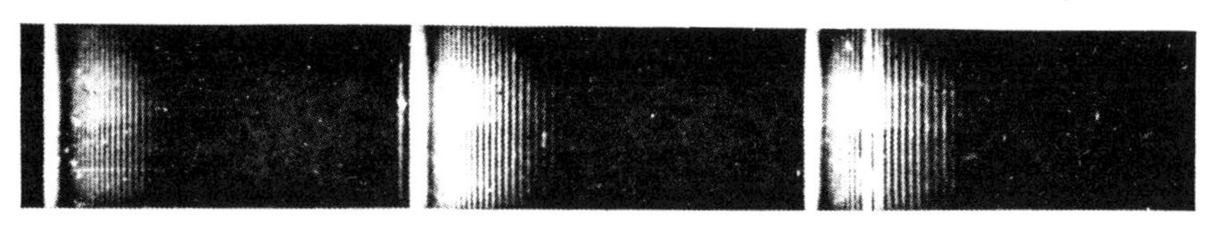

Fig. 11-9. Typical band spectra.

11-9 X-ray spectra. The elements of the periodic system may be arranged in sequence according to *atomic number,* that is, according to the number of electrons which surround the nucleus of the atom. While the notion of the explicit orbits of the Bohr theory has been abandoned, one still speaks of these electrons as being located in various "shells" about the nucleus. Each shell seems able to hold a certain maximum number of electrons. As the atomic number of the atom increases, first the inner shell fills up to its maximum number, then the second shell, then the third, and so on. (This order is not exactly followed in all cases, but is nearly so.) The innermost shell, known as the K shell, can contain at the most two electrons. The next outer shell, the L shell, can contain eight. The third, the M shell, has a capacity for 18 electrons, while the N shell may hold 32. The element sodium, for example, which contains 11 electrons, has two in the K shell, eight in the L shell, and a single electron in the M shell. Molybdenum, with 42 electrons, has two in the K shell, eight in the L shell, 18 in the M shell, and 14 in the N shell.

The outer electrons of an atom, as the M electron of sodium, are the ones responsible for the optical spectra of the elements. Relatively small amounts of energy suffice to remove these to excited states, and on their return to their normal states wave lengths in or near the visible region are emitted. The inner electrons, such as those in the K shell, require much more energy to displace them from their normal levels. As a result, we would expect a photon of much larger energy, and hence much higher frequency, to be emitted when the atom returns to its normal state after the displacement of an inner electron. This is in fact the case, and it is the displacement of the inner electrons which gives rise to the emission of x-rays.

The usual method of producing x-rays is to bombard the atoms of an element with rapidly moving electrons (cathode rays). The substance to be bombarded is made the anode in an evacuated tube, the cathode being an incandescent tungsten filament. Electrons emitted thermionically by the filament are accelerated by a high potential difference between cathode and anode. On colliding with the atoms of the anode, or *target*, some of these electrons, provided they have acquired sufficient energy, will dislodge one of the inner electrons of a target atom, say one of the K electrons. This leaves a vacant space in the K shell, which is immediately filled by an electron from either the L, M, or N shells. The readjustment of the electrons is accompanied by a decrease in the energy of the atom, and an x-ray photon is emitted with energy just equal to this decrease. Since the energy change is perfectly definite for atoms of a given element, we can predict definite frequencies for the emitted x-rays, or in other words the x-ray spectrum should be a line spectrum also. We can predict further that there should be just three lines in the series, corresponding to three possibilities that the vacant space may have been filled by an L, M, or N electron.

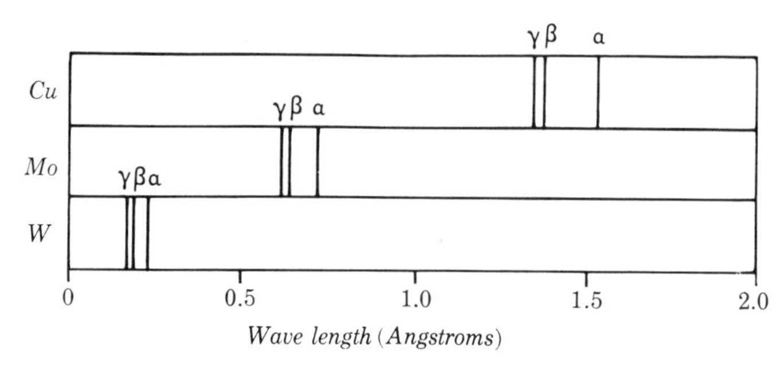

FIG. 11-10. Wave lengths of the K_α, K_β, and K_γ lines of copper, molybdenum, and tungsten.

This is precisely what is observed. Fig. 11-10 illustrates the so-called *K*-series of the elements tungsten, molybdenum, and copper. Each series consists of three lines, known as the K_α, K_β, and K_γ lines. The K_α line is produced by the transition of an *L* electron to the vacated space in the *K* shell, the K_β line by an *M* electron, and the K_γ line by an *N* electron.

In addition to the *K* series, there are other series known as the *L*, *M*, and *N* series, produced by the ejection of electrons from the *L*, *M*, and *N* shells rather than the *K* shell. As would be expected, the electrons in these outer shells, being farther away from the nucleus, are not held as firmly as those in the *K* shell. Consequently the other series may be excited by more slowly moving electrons, and the photons emitted are of lower energy and longer wave length.

In addition to the x-ray *line* spectrum there is a background of *continuous* x-radiation from the target of an x-ray tube. This is due to the sudden deceleration of those cathode rays which do not happen to eject an electron. The remarkable feature of the continuous spectrum is that while it extends indefinitely toward the long wave length end, it is cut off very sharply at the short wave length end. Again the quantum theory furnishes a satisfactory explanation of the short-wave limit of the continuous x-ray spectrum.

A bombarding electron may be brought to rest in a single process, if the electron happens to collide head-on with an atom of the target, or it may make a number of collisions before coming to rest, giving up part of its energy each time. If we assume that the energy lost at each collision is radiated as an x-ray photon, these photons may be of any energy up to a certain maximum, namely, that of an electron which gives up all of its energy in a single collision. Hence there will be a short wave limit to the spectrum. The frequency of this limit is found as usual by setting the energy of the electron equal to the energy of the x-ray photon.

$$\tfrac{1}{2}\, mv^2 = hf. \tag{11-16}$$

This is precisely the same equation as that for the photoelectric effect, except for the work function term which is negligible here since the energies of the x-ray photons are so large. In fact, the emission of x-rays may be described as an *inverse photoelectric effect*. In photoelectric emission the energy of a photon is transformed into kinetic energy of an electron; here, the kinetic energy of an electron is transformed into that of a photon.

Example. Compute the potential difference through which an electron must be accelerated in order that the short wave limit of the continuous x-ray spectrum shall be exactly 1 Angstrom.

The frequency corresponding to 1 Angstrom (10^{-10} m) is given by

$$f = \frac{c}{\lambda} = \frac{3 \times 10^8}{10^{-10}} = 3 \times 10^{18}\ \text{sec}^{-1}.$$

The energy of the photon is

$$hf = 6.62 \times 10^{-34} \times 3 \times 10^{18} = 19.9 \times 10^{-16}\ \text{joule}.$$

This must equal the kinetic energy of the electron, $\frac{1}{2}\, mv^2$, which is also equal to the product of the electronic charge and the accelerating voltage, V.

$$\tfrac{1}{2}\, mv^2 = eV = 19.9 \times 10^{-16}\ \text{joule}.$$

Since

$$e = 1.60 \times 10^{-19}\ \text{coulomb},$$

$$V = \frac{19.9 \times 10^{-16}}{1.60 \times 10^{-19}} = 12{,}400\ \text{volts}.$$

Problems Chapter 11

(1) Compute the wave lengths of the longest and shortest waves in the Lyman, Paschen, and Brackett series of atomic hydrogen.

(2) (a) Find the frequency and the wave length of the spectrum line which is emitted when the electron of a hydrogen atom returns from the fourth Bohr orbit to the third. (b) Of what series of spectrum lines is this line a member?

(3) (a) Find the difference in wave length between the lines in a normal Zeeman triplet, if the wave length of the central line is 400 mμ and the magnetic flux density is 1 weber/m^2 = 10^4 gauss. (b) How many lines would a grating have to have to resolve the triplet in the second-order spectrum?

(4) (a) Compute the energy in joules associated with a photon of wave length 600 mμ. (b) Compute the work function of a photoemissive surface, in joules, if the threshold wave length of the surface is 600 mμ. (c) Through how many volts would an electron have to be accelerated to acquire this energy? (The answer gives the work function of the surface in electron-volts.)

(5) The work function of the emitting surface of a photocell is 2 electron-volts. (An electron-volt is equal to the energy acquired by an electron accelerated through a potential difference of 1 volt.) (a) What is the longest wave length which can eject photoelectrons from this surface? (b) What is the maximum velocity with which photoelectrons are emitted when ultraviolet light of wave length 300 mμ is incident on the surface?

(6) When ultraviolet light of wave length 254 mμ is incident on the emitting surface of a certain photocell, photoelectrons leave the surface with velocities up to 10^6 m/sec. (a) What is the work function of the emitting surface? (b) What is the longest wave which can eject photoelectrons from this surface?

(7) An image of the sun is formed on the metal surface of a photoelectric cell and produces a current I. The lens forming this image is then replaced by another of the same diameter but only half the focal length. (a) How does the area of the second image compare with the first? (b) What is the photocell current in the second case?

(8) What is the minimum potential difference required to operate an x-ray tube capable of producing x-rays of wave length 10^{-3} mμ?

(9) (a) What is the short-wave limit of the continuous x-ray spectrum produced by an x-ray tube operating at 30,000 volts? (b) What is the wave length of the shortest x-rays that could be produced by a Van de Graaff generator developing 5,000,000 volts?

CHAPTER 12

THERMAL RADIATION

12-1 Thermal radiation. In the preceding chapter we have discussed the line spectra typical of materials in the gaseous state, as for example a neon sign or a sodium vapor lamp. The atoms in a gas through which an electrical discharge is passing are at relatively large distances from one another and so are capable of radiating independently. Thus the line spectrum emitted by a gas is characteristic of the atoms composing it. The present chapter will be devoted to spectra of the type emitted by an incandescent solid or liquid. In a solid or liquid the atoms are practically in contact with one another and cannot act independently. Instead of a line spectrum, containing only a relatively few frequencies, the spectra of solids and liquids are *continuous*. That is, waves of all frequencies are emitted, although with an intensity that becomes vanishingly small at both extremely long and extremely short wave lengths. The distribution of energy in the spectrum is determined chiefly by the *temperature* of the emitting surface and is not characteristic of the *material* of which it is composed.

Any source of electromagnetic waves is called a *radiator*. The energy transported by the waves is *radiant energy*. The general term for the *process* by which radiant energy is generated is *radiation*. In the particular case under discussion the thermal energy of a solid or liquid is transformed into radiant energy and the process is called *thermal radiation*.

It is a familiar fact that solids and liquids radiate visible light when their temperature is above 500-550°C. This light is, of course, radiant energy. The emission of radiant energy does not begin suddenly, however, at the temperature at which a body becomes self-luminous, but is going on at all temperatures down to absolute zero. A body is not self-luminous at low temperatures because the radiant energy emitted at low temperatures is chiefly of long wave length, outside the limits of the visible spectrum, and not until the temperature is increased sufficiently is there enough energy of wave lengths in the visible spectrum for the body to become self-luminous.

12-2 Kirchhoff's law. Fig. 12-1 represents an evacuated enclosure whose walls are maintained at a uniform temperature. Within the enclosure is a small body A. One finds by experiment that whatever the

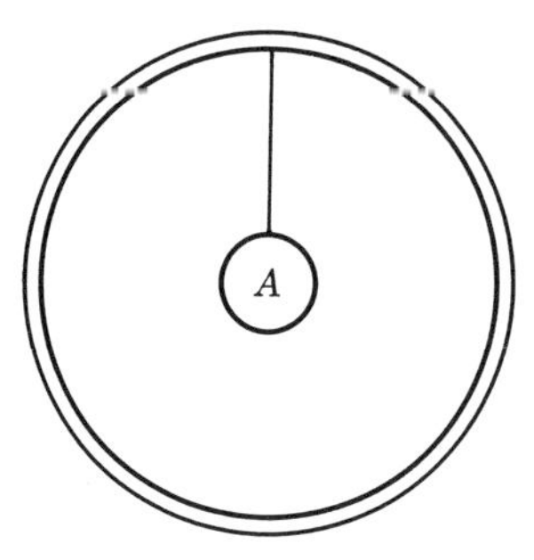

Fig. 12-1.

nature of this body, and whatever its initial temperature, it eventually comes to and remains at the same temperature as the walls of the enclosure. The mechanism by which this equality of temperature is brought about is the *emission* and *absorption* of radiant energy at the surfaces of the body and the enclosure.

It is necessary to distinguish carefully between several processes going on at the surface of the body. First, radiant energy emitted by or reflected from the inner walls of the enclosure, is incident on the surface of the body at a certain rate. Second, a part of this incident energy is reflected by the body and the remainder absorbed. (We shall assume the body to be opaque so that no energy is transmitted.) The energy absorbed is converted to thermal energy of the atoms of the body. Third, radiant energy is emitted at the surface of the body at a certain rate. This energy is supplied by the thermal energy of the atoms. After the equilibrium state has been reached, the temperature of the body remains constant and there must exist an energy balance between these processes.

The radiant energy which strikes or crosses a surface per unit of time, or the radiant energy emitted by a source per unit time, is called the *radiant flux* striking the surface or emitted by the source, and is represented by P. Radiant flux is expressed in watts in the mks system. The radiant flux incident on a surface *per unit area* is called the *irradiance* and is represented by H. The mks unit of irradiance is 1 watt/m^2. (The hybrid unit, 1 watt/cm^2, is often of more convenient size.)

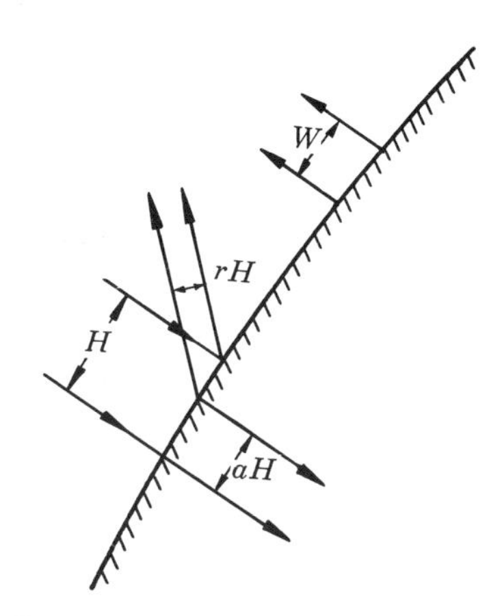

Fig. 12-2. Schematic diagram to illustrate the processes of irradiation, absorption, reflection, and emission of radiant energy at a surface. In equilibrium, $aH = W$.

Fig. 12-2 represents a portion of the surface of body A in Fig. 12-1. The magnitude of the irradiance, or the incident flux per unit area, is represented schematically by the width of the band lettered H. If the body is small compared with the size of the enclosure, the irradiance is the same at all points of the surface of the body. It will, however, depend on the temperature of the enclosure.

A certain fraction of the incident flux, dependent on the nature of the surface, is absorbed and the remainder reflected. The fraction reflected is called the *reflectance* of the surface and is represented by r. The fraction absorbed is called the *absorptance* and is represented by a. Reflectance and absorptance are pure numbers whose magnitude must lie between 0 and 1. Evidently, for an opaque surface,

$$r + a = 1.$$

The product rH, represented schematically by the width of the beam rH in Fig. 12-2, is the rate at which energy is reflected per unit area. The product aH, similarly represented, is the rate at which energy is absorbed and converted to thermal energy.

The rate at which radiant energy is *emitted* per unit area of a surface is called the *radiant emittance* and is represented by W. Radiant emittance may also be described as the radiant flux emitted per unit area. The mks unit of radiant emittance is 1 watt/m^2. The magnitude of the radiant emittance is represented in Fig. 12-2 by the width of the band lettered W. Actually, of course, radiant energy is incident at all points of the body's surface and reflection, absorption, and radiation are going on simultaneously at all points. The separation of the bands in Fig. 12-2 is schematic only.

We can now set up the conditions for a balance between the rate of absorption and the rate of emission of energy. The rate at which energy is absorbed per unit area is aH. The rate at which it is emitted per unit area is W. Hence in the equilibrium state,

$$aH = W,$$

or

$$\frac{W}{a} = H. \tag{12-1}$$

The irradiance H depends only on the temperature of the enclosure and not on the nature of the surface of the body. Hence we conclude that *the ratio of radiant emittance to absorptance is the same for all surfaces at the same temperature.* This fact is known as *Kirchhoff's law.*

An immediate consequence of Kirchhoff's law is that if the absorptance a of a given surface is large, the radiant emittance W of the surface is large also; if the absorptance is small the radiant emittance is small. This explains why the inner surfaces of the double walls of a Dewar flask ("thermos bottle") are silvered. A silvered surface has a large reflectance, hence a small absorptance and a small radiant emittance. If a hot liquid is placed in the flask, the loss of heat by radiation will be small. Or, if a cold liquid is placed in the flask, the gain of heat from outside by radiation will be small since most of the radiant energy incident on the surface is reflected.

12-3 The complete radiator or blackbody. Since from Kirchhoff's law the ratio of absorptance to radiant emittance, a/W, is the same for all surfaces at the same temperature, it follows that the surface having the maximum radiant emittance is one having the maximum absorptance. But the maximum possible value of the absorptance is $a = 1$, that is, all incident energy is absorbed and none is reflected. A surface which absorbs all of the radiant energy incident on it would appear black (provided its temperature is not so high that it is self-luminous), and it is called an *ideally black surface*, or a *blackbody*. At a given temperature, no surface can have a larger radiant emittance than a blackbody. The term "blackbody" is unfortunate, because although such an object does not reflect any energy, it does radiate, and in most cases of interest in photometry and colorimetry it radiates copiously. The alternate term, *complete radiator*, has been proposed.

No material surface absorbs all of the radiant energy incident on it. Lampblack reflects about 1%. A blackbody may be very closely approximated, however, by an enclosure with a small opening through which radiant energy may enter or leave. (Fig. 12-3.) Of the radiant energy entering the opening, a part will be absorbed by the walls and the remainder reflected. Only a small fraction of the reflected energy passes out through the opening, the remainder being again absorbed or re-reflected. After repeated reflections practically all of the incoming energy is absorbed. The *opening*, then, behaves like a blackbody in that all of the energy incident on it is absorbed.

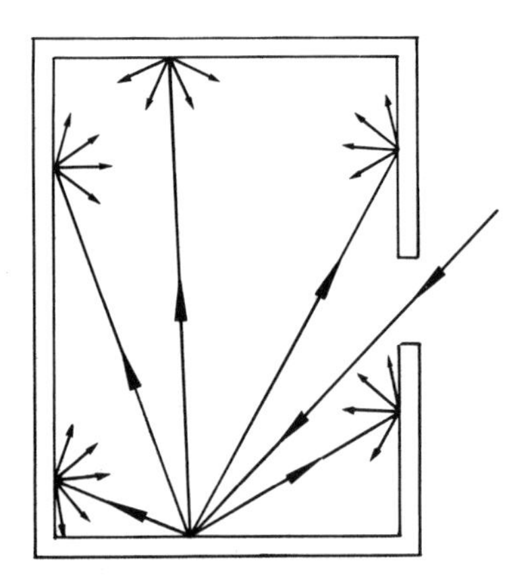

FIG. 12-3. A small opening in the walls of an enclosure is very nearly a complete absorber of radiant energy.

The interior walls of the enclosure in Fig. 12-3 are emitting radiant energy as well as absorbing it, and a part of this energy escapes through the opening. If the walls are not ideally black surfaces their radiant emittance is less than that of a blackbody, but in any case their smaller emittance is just compensated for by their larger reflectance. Hence the energy escaping from the opening is identical with that which would be emitted by a blackbody at the same temperature as the walls of the enclosure.

Let W_{bb} represent the radiant emittance of a blackbody. Then from Kirchhoff's law, since $a = 1$ for a blackbody,

$$\frac{a}{W} = \frac{1}{W_{bb}}, \quad \text{or} \quad \frac{W}{a} = W_{bb}, \tag{12-2}$$

where a and W refer to any surface whatever. That is, the ratio of the radiant emittance to the absorptance is the same for all surfaces at the same temperature, and is equal to the radiant emittance of a blackbody at the same temperature.

12-4 Planck's law.

If a beam of radiant energy from a blackbody is dispersed into a spectrum, the reading of a sensitive thermometer or thermopile placed in the spectrum is an indication of the radiant emittance of the blackbody within the particular range of wave lengths (or frequencies) which the indicating instrument intercepts. As the thermometer is moved along the spectrum one finds that its reading passes through a maximum at some one wave length (the particular wave length depends on the temperature of the blackbody) and that on either side of the maximum there is a considerable range of wave lengths throughout which the energy is distributed.

Max Planck, in 1900, developed an empirical equation that satisfactorily represented the energy distribution in the spectrum of a blackbody. After unsuccessful attempts to justify his equation by theoretical reasoning based on the laws of classical physics, Planck concluded that these laws did not apply to energy transformations on an atomic scale. Instead, he postulated that a radiating body consisted of an enormous number of elementary oscillators, vibrating at all possible frequencies. These oscillators were the source of the radiant energy emitted by the body. The energy E of any one oscillator was not permitted to take on any arbitrary value, but was proportional to some integral multiple of the frequency of the oscillator, f. That is,

$$E = nhf, \tag{12-3}$$

where n is an integer and the proportionality constant h is *Planck's constant* to which reference has been made previously. Out of this postulate, which was one of the most spectacular and important accomplishments in the history of theoretical physics, there has developed most of the modern theory of atomic processes.

Although Einstein developed his theory of the photoelectric effect from the quantum concepts first proposed by Planck, it should be noted that he made one important modification or extension of Planck's original theory. The latter's viewpoint was that while the energies of the elementary oscillators could take on only discrete values, the energy radiated by them was propagated through space according to classical electromagnetic theory. It was Einstein who first realized that quantum conditions applied to the radiant energy also, and that the latter could exist only in discrete packets or photons of energy hf. That is, while Planck quantized the oscillators of a radiating body, Einstein went one step further and quantized the radiant energy emitted by them.

The derivation of Planck's formula from his hypothesis regarding the energies of the elementary oscillators is too long and involved to be given here. We shall merely give the formula without deriving it.

Consider an infinitesimal range of frequencies between f and $f + df$, and let dW_{bb} be the radiant emittance of a blackbody within this frequency range. The radiant emittance will be proportional to the frequency range df if the latter is sufficiently small, and we may write

$$dW_{bb} = W_f df.$$

Planck's formula for the coefficient W_f is

$$W_f = \frac{2\pi h}{c^2} \frac{f^3}{e^{hf/kT} - 1}, \tag{12-4}$$

where h is Planck's constant, c is the velocity of light, T is the Kelvin temperature of the blackbody, k is the Boltzmann constant or the gas constant per molecule, and e is the base of natural logarithms.

It is interesting to note that the factor kT occurs also in connection with the energy of the molecules of a gas, where, it will be recalled, the principle of equipartition of energy assigns an energy $\frac{1}{2} kT$ to each degree of freedom. Furthermore, the Maxwell-Boltzmann equation for the velocity distribution of the molecules of a gas contains the factor

$$\frac{1}{e^{\frac{1}{2}mv^2/kT}},$$

where $\frac{1}{2} mv^2$ is the kinetic energy of a molecule. Since the product hf is the energy of a photon, both Planck's equation and the Maxwell-Boltzmann equation involve the ratio *energy*/kT. More general reasoning based on the principles of *statistical mechanics* shows that all of these problems are in reality special cases of a general distribution principle.

From the experimental point of view it is convenient to express Planck's equation in terms of wave lengths rather than frequencies. Since $c = f\lambda$, where c is the velocity of light, the ratio hf/kT becomes

$$\frac{hf}{kT} = \frac{hc}{\lambda kT} = \frac{c_2}{\lambda T},$$

where c_2 is an abbreviation for hc/k.

Also,

$$f^3 = c^3/\lambda^3$$

and

$$df = -\frac{c}{\lambda^2}\,d\lambda.$$

Then

$$dW_{bb} = W_f df = \frac{2\pi h}{c^2}\,\frac{\frac{c^3}{\lambda^3}}{e^{c_2/\lambda T} - 1}\,\frac{c}{\lambda^2}\,d\lambda,$$

$$dW_{bb} = \frac{c_1\lambda^{-5}}{e^{c_2/\lambda T} - 1}\,d\lambda, \qquad (12\text{-}5)$$

where $c_1 = 2\pi c^2 h$. The minus sign can be disregarded since we are interested only in the *magnitude* of the wave length or frequency interval. Let us define a quantity W_λ (corresponding to W_f) by the equation

$$dW_{bb} = W_\lambda d\lambda, \qquad (12\text{-}6)$$

Then

$$\boxed{W_\lambda = \frac{c_1\lambda^{-5}}{e^{c_2/\lambda T} - 1}.} \qquad (12\text{-}7)$$

If mks units are used consistently in Eq. (12-5), λ and $d\lambda$ are expressed in meters and dW_{bb} is in watts/m^2. The numerical values of c_1 and c_2 are then

$$c_1 = 3.740 \times 10^{-16}, \quad c_2 = 1.4385 \times 10^{-2}.$$

It is often more convenient to express λ and $d\lambda$ in millimicrons, retaining the watt/m^2 as the unit of dW_{bb}. The values of c_1 and c_2 are then

$$c_1 = 3.740 \times 10^{20}, \quad c_2 = 1.4385 \times 10^{7}.$$

The quantity W_λ, equal to $dW_{bb}/d\lambda$, can be described as the radiant emittance of a blackbody per unit range of wave length. It is called the *spectral emittance.* A graph of the spectral emittance of a blackbody at a temperature of 773°K (500°C) is given in Fig. 12-4 as a function of wave

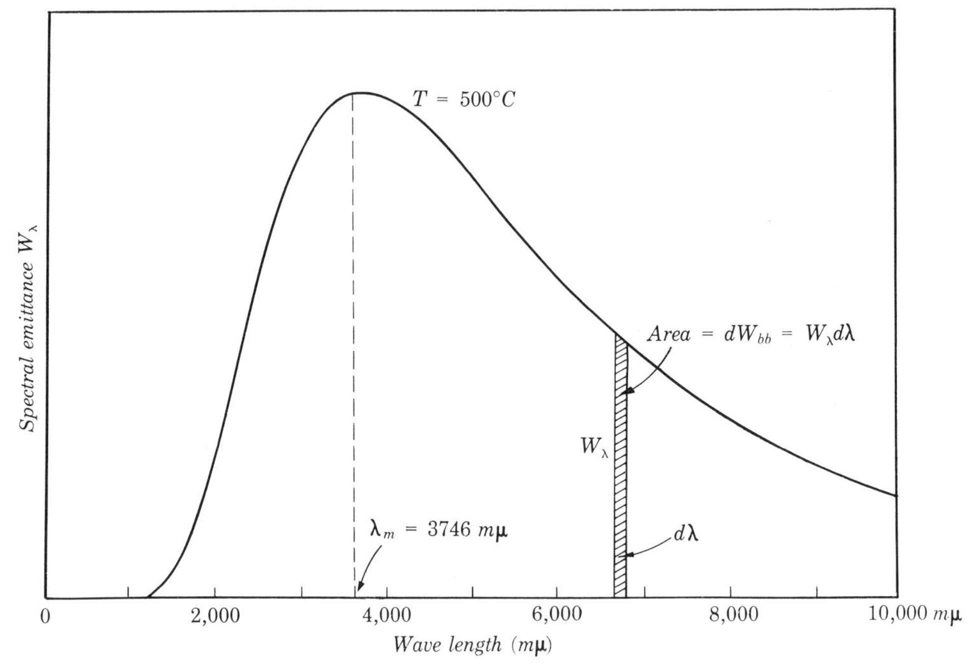

FIG. 12-4. Spectral emittance of a blackbody or complete radiator at a temperature of 500°C. The area of a small vertical strip at any wave length λ represents the radiant energy emitted per unit area and per unit time, in the wave length range between λ and $\lambda + d\lambda$.

length. The *radiant* emittance dW_{bb} at any wave length is represented by the area of a narrow vertical strip under the curve at that wave length since the height of the strip is W_λ, its width is $d\lambda$, and by definition

$$dW_{bb} = W_\lambda d\lambda.$$

The wave length λ_m at which the spectral emittance is a maximum at any temperature can be found by differentiating W_λ with respect to λ and setting the derivative equal to zero. One obtains by straightforward methods

$$\frac{c_2}{\lambda_m T} = 5\,(1 - e^{-c_2/\lambda_m T}). \tag{12-8}$$

This is a transcendental equation and cannot be solved explicitly for λ_m in terms of T. A solution is readily obtained by trial, however, and is found to be

$$\frac{c_2}{\lambda_m T} = 4.965114,$$

as may be verified by substitution in Eq. (12-8). Since $c_2 = 1.4385 \times 10^7$ when λ is in millimicrons and T in degrees Kelvin, then in these units

$$\boxed{\lambda_m T = 2.8971 \times 10^6.} \quad (12\text{-}9)$$

The wave length of maximum spectral emittance is therefore inversely proportional to the Kelvin temperature, and as the temperature is increased the maximum shifts toward shorter wave lengths. This fact is known as *Wien's displacement law.*

A number of graphs of the spectral emittance of a blackbody are given in Fig. 12-5 for various Kelvin temperatures. The curves are plotted on a logarithmic scale because of the wide range of wave lengths and emittances to be covered. The limits of the visible spectrum are indicated. The progressive shift of the maximum toward the blue accounts for the change in color of a body from red through white to blue as its temperature is increased.

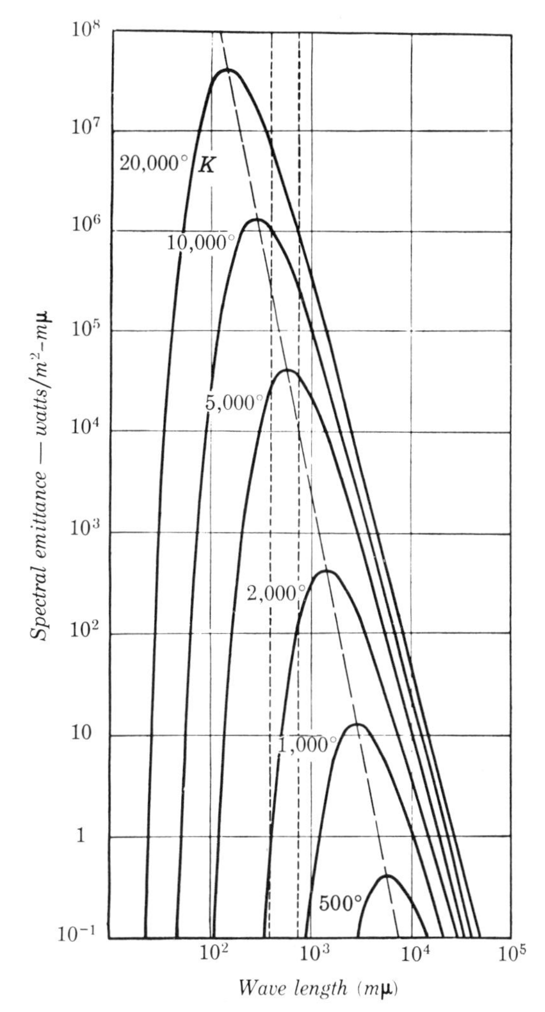

FIG. 12-5. Spectral emittance of a blackbody at various temperatures. The vertical dotted lines indicate the boundaries of the visible spectrum.

12-5 The Stefan-Boltzmann law. It is evident, either from the graphical interpretation of the area of a vertical strip in Fig. 12-4, or from the definition of dW_{bb}, that the radiant emittance W_{bb} of a blackbody can be found by integrating Eq. (12-5) from $\lambda = 0$ to $\lambda = \infty$. The details of the integration will not be given but the result is

$$W_{bb} = \int dW_{bb} = \int_0^\infty W_\lambda d\lambda = \frac{\pi^4}{15} \frac{c_1}{c_2^4} T^4,$$

or

$$\boxed{W_{bb} = \sigma T^4,} \tag{12-10}$$

where

$$\sigma = \frac{\pi^4}{15} \frac{c_1}{c_2^4}.$$

Inserting the numerical values of c_1 and c_2, we obtain

$$\sigma = 5.672 \times 10^{-8},$$

when T is in degrees Kelvin and W_{bb} in watts/m^2.

Eq. (12-10) states that the radiant emittance of a blackbody, or the radiant energy emitted per unit time and unit area, is proportional to the 4th power of the Kelvin temperature. As a matter of fact, this equation was proposed as an empirical equation by Josef Stefan in 1879, before Planck's law was known. It was later derived on theoretical grounds by Boltzmann and is often called the *Stefan-Boltzmann equation.* The constant σ is called the *Stefan-Boltzmann constant.*

We have shown in Eq. (12-2) that the radiant emittance of a surface that is not ideally black is given by

$$W = aW_{bb}.$$

Hence for such a surface

$$W = a\sigma T^4. \tag{12-11}$$

Considered from this point of view the absorptance a is equal to the ratio of the radiant emittance of a surface to the radiant emittance of a blackbody at the same temperature. As such, it is called the *emissivity* of the surface and represented by the letter e.

$$e = a = 1 - r. \tag{12-12}$$

Hence one often finds Eq. (12-11) written

$$\boxed{W = e\sigma T^4.} \tag{12-13}$$

The fact that the radiant emittance increases with the 4th power of the absolute temperature makes it very difficult to thermally insulate a body at high temperatures. Even if a polished surface is used, for which the absorptance and therefore the emissivity is small, the T^4 term becomes very large at elevated temperatures.

12-6 Heat transfer by radiation. The rate at which radiant energy is *emitted* from the surface of a body, per unit area, is given by Eq. (12-13). In general, radiant energy is also *incident* on the body's surface. A part of this energy is absorbed and the *net* rate of gain or loss of energy equals *the difference between the rates of emission and of absorption.* We shall discuss two special cases, first, that of a small body within an enclosure whose walls are at uniform temperature, and second, two surfaces at a separation small compared with their linear dimensions.

Consider first a small body such as that in Fig. 12-1, suspended within a large evacuated enclosure whose walls are at a temperature T_0. Let the area of the body be A, let a represent its absorptance, H the irradiance on its surface, and W the radiant emittance from its surface. If it is in thermal equilibrium,

$$AaH = AW.$$

We have shown that the emissivity e is equal to the absorptance a. Also, if the body is in thermal equilibrium its temperature is T_0 and the preceding equation may therefore be written

$$AeH = Ae\sigma T_0^4.$$

Hence

$$H = \sigma T_0^4,$$

and we see that the irradiance H is independent both of the nature of the surface of the body and of the walls of the enclosure, depending only on the temperature T_0.

Now suppose the body is at a temperature T, different from that of the walls. Since we have assumed the body to be small, a change in its temperature will have only a negligible influence on the radiant energy within the enclosure and the irradiance H remains the same. However, the body is no longer in equilibrium. Its radiant emittance is now

$$W = e\sigma T^4$$

and the rate of emission of radiant energy is

$$AW = Ae\sigma T^4.$$

The rate of absorption remains unchanged, so the *net* rate of emission is

$$\begin{aligned} &Ae\sigma T^4 - AeH \\ &= Ae\sigma T^4 - Ae\sigma T_0^4 \\ &= Ae\sigma\,(T^4 - T_0^4). \end{aligned} \tag{12-14}$$

If the body is not small compared with the enclosure the irradiance on its surface depends on its own temperature as well as on that of the walls, since, unless the walls are complete radiators (i.e., perfect absorbers) some of the radiant energy emitted by the body will be reflected from the walls and returned to the body. No general equation can be given that will cover all cases, but in the special case where the "body" is practically the same size as the enclosure so that the separation of their surfaces is small, the solution is relatively simple.

Fig. 12-6 represents two closely spaced surfaces 1 and 2, at temperatures T_1 and T_2, and having emissivities e_1 and e_2. The "pipelines" represent schematically the processes taking place at the surfaces and in the interspace. Let H_1 represent the irradiance on the surface at temperature T_2. A fraction a_2H_1 is absorbed, and a fraction r_2H_1 is reflected and returned to the surface at temperature T_1. The radiant emittance from surface 2 is e_2W_2, so the irradiance H_2 on surface 1 is

$$H_2 = r_2H_1 + e_2W_2.$$

Similar reasoning shows that the irradiance H_1 is

$$H_1 = r_1H_2 + e_1W_1.$$

When these equations are solved simultaneously for H_1 and H_2, we get

$$H_1 = (r_1e_2W_2 + e_1W_1)/(1 - r_1r_2),$$

$$H_2 = (r_2e_1W_1 + e_2W_2)/(1 - r_1r_2).$$

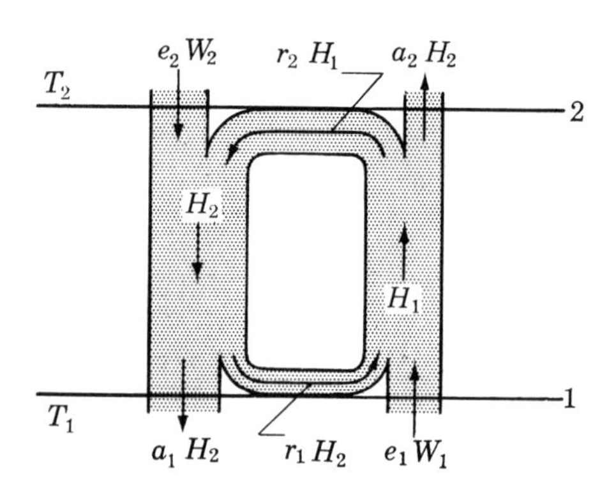

FIG. 12-6. Schematic diagram showing interchange of energy by radiation between two closely spaced surfaces.

The *net* rate of transfer of energy between the surfaces, per unit of area, is $H_2 - H_1$. From the equations above, and making use of the fact that

$$r_1 = 1 - e_1, \qquad r_2 = 1 - e_2,$$

$$W_1 = \sigma T_1^4, \qquad W_2 = \sigma T_2^4,$$

we get for the net rate of transfer per unit area,

$$H_2 - H_1 = \frac{\sigma(T_2^4 - T_1^4)}{\frac{1}{e_2} + \frac{1}{e_1} - 1}. \tag{12-15}$$

Summary

The table below may help in correlating the large number of probably unfamiliar terms that have been introduced in this chapter.

Quantity	Symbol	Unit	Definition
Radiant flux	P	watt	Radiant energy crossing or striking a surface per unit time.
Irradiance	H	watt/m^2	Radiant flux incident per unit area.
Reflectance	r		Fraction of incident flux reflected.
Absorptance	a		Fraction of incident flux absorbed.
Radiant emittance	W	watt/m^2	Radiant flux emitted per unit area.
Spectral emittance	W_λ	(watt/m^2)/mμ	Radiant emittance per unit range of wave length.
Emissivity	e		Ratio of radiant emittance of a body to that of a blackbody.

Problems—Chapter 12

(1) The irradiance on the surface of a certain opaque body is 50 watts/m². The surface absorbs 20 watts/m². (a) What is the reflectance of the surface? (b) What is the absorptance of the surface? (c) If the surface area of the body is 100 cm², what is the total radiant flux incident on it? (d) If the body is in thermal equilibrium and can interchange energy with other bodies only through emission and absorption of radiant energy, what is the radiant emittance from its surface? (e) What is the temperature of the body? (f) What would be the radiant emittance of a blackbody at the same temperature?

(2) The temperature of a blackbody is 3000°K. Compute the ratio of its spectral emittance at a wave length of 1000 mμ (in the infrared) to its spectral emittance at 500 mμ (in the visible).

(3) At what wave length is the spectral emittance of a blackbody a maximum, if its temperature is (a) 500°K, (b) 5000°K? (c) At what temperature does the maximum spectral emittance lie at a wave length of 555 mμ, where the eye is most sensitive?

(4) What is the radiant emittance of a blackbody at a temperature of (a) 300°K, (b) 600°K, (c) 1200°K?

(5) A small body is suspended within an evacuated enclosure whose walls are maintained at 300°K. Find the ratio of the power required to maintain the body at 500°K to the power required to maintain it at 400°K. Neglect heat conduction.

(6) The maximum spectral emittance of a blackbody is found to occur at a wave length of 800 mμ. If the surface area of the body is 100 cm², what power input is required to maintain it at constant temperature within a large evacuated enclosure at a temperature of 300°K? Neglect heat conduction.

(7) The emissivity of tungsten is approximately 0.35. A tungsten sphere 1 cm in radius is suspended within a large evacuated enclosure whose walls are at 300°K. What power input is required to maintain the sphere at a temperature of 3000°K, if heat conduction along the supports is neglected?

(8) (a) Compute the net rate of heat transfer by radiation, per square centimeter of area, between the silvered walls of a "thermos bottle," if the reflectance of the walls is 0.9 and the temperatures are 87°C and 27°C. (b) Compare this rate with the rate of heat flow by conduction, if the bottle were insulated with 0.5 cm of cork of thermal conductivity 0.0001 cal/sec-cm-C°.

(9) In what ratio would the heat transfer by radiation between the walls of a "thermos bottle" be increased if the reflectance of each wall were reduced from 0.9 to 0.3?

(10) The outer wall of a Dewar flask is at a temperature of 50°C. The flask is filled, first with water at 100°C and then with water at 0°C. Find the ratio of the initial rate of temperature decrease, in the first case, to the initial rate of temperature increase in the second.

(11) A steam pipe 1½ inches outside diameter and 12 ft long contains steam at 212°F. The pipe is found to lose 1100 Btu per hour in a room at 75°F. What fraction of the heat loss is due to radiation? Assume an emissivity of 0.5.

(12) A constant temperature bath is constructed in the form of a Dewar flask, with closely spaced silvered inner surfaces which may be taken as parallel plates. The reflectance of the walls is 0.9 and the area of each is 300 cm^2. If the temperatures of the inner and outer walls of the flask are 150°C and 27°C respectively, how much power must be supplied to the contents if all heat loss is due to radiation?

(13) Two large closely spaced surfaces, both of which are ideal radiators, are maintained at temperatures of 200°K and 300°K respectively. The space between them is evacuated. (a) What is the net rate of loss of heat from the warmer surface, in watts per square meter?

A thin sheet of aluminum foil of emissivity 0.1 on both surfaces is placed between the two surfaces in part (a). Assume both faces of the foil to be at the same temperature. When the steady state has been established, compute (b) the temperature of the aluminum foil, (c) the new net rate of loss of heat from the warmer surface.

CHAPTER 13

PHOTOMETRY

13-1 The luminosity of radiant flux. In the preceding chapter we have dealt with radiant flux from the purely objective or physical standpoint, apart from any visual sensation produced by the flux. The visual sensation that results when radiant flux is incident on the retina has three attributes: *hue*, *saturation*, and *brightness*. Taken together, these three make up the *sensation of color*. The term "hue" refers to the attribute that enables one to classify a color as red, green, blue, etc. A neutral gray has no hue, or better, its hue is indeterminate. "Saturation" describes the extent to which a color departs from a neutral gray and approaches a pure spectrum color. A neutral gray is completely unsaturated and a pure spectrum color is completely saturated. The attributes of hue and saturation, taken together, constitute the *chromaticness* of the sensation. A neutral gray has neither hue nor saturation and is called *achromatic*.

The term "brightness" may be defined as follows. Fig. 13-1 shows a series of rectangles printed in neutral gray, ranging from white at one end of the scale to black at the other. The white rectangle evokes the greatest sensation of brightness, the black rectangle the least, while intermediate sensations of brightness are evoked by intermediate rectangles. A chromatic sample may be compared with the achromatic gray scale and (disregarding the hue and saturation of the sample) it can be said to evoke the same brightness sensation as some member of this scale. (Try the experiment with a sample of blue or green blotting paper.) Then we may say: *Brightness is that attribute of any color sensation which permits it to be classified as equivalent to the sensation produced by some member of a series of neutral grays.*

Equal amounts of radiant flux of different wave lengths do not produce visual sensations of equal brightness. This is evident from visual exami-

FIG. 13-1. A scale of grays to illustrate the sensation of brightness. The greatest brightness sensation is evoked by the rectangle at the left, the least by that at the right.

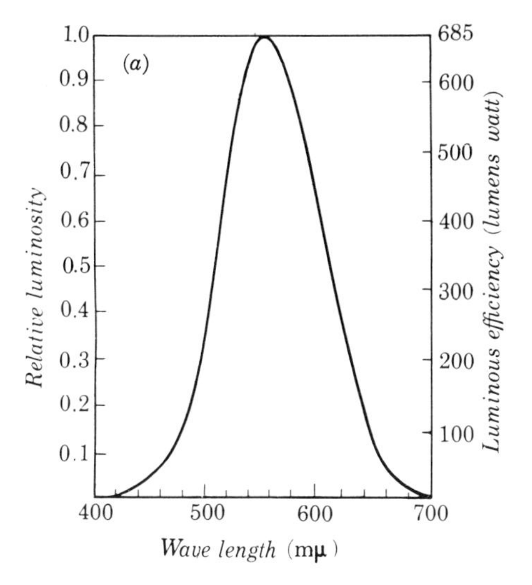

FIG. 13-2. Relative luminosity curve for the standard observer. Scale at left, relative luminosity; scale at right, luminous efficiency.

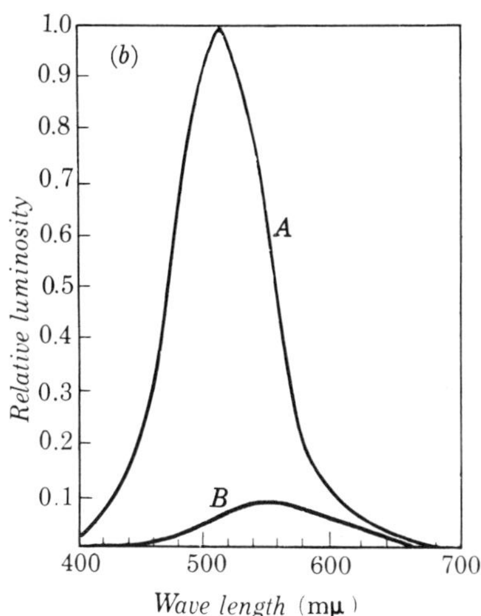

FIG. 13-3. The Purkinje effect. Relative luminosity curves under conditions of (a) subdued lighting, (b) good lighting.

nation of the spectrum of any source which emits approximately equal amounts of radiant flux per unit wave length interval throughout the visible spectrum. The solar spectrum, for example, appears brightest in the central yellow-green region, and the brightness decreases toward either end. Careful experiments have been made with a large number of observers to determine the relative effectiveness of monochromatic radiant flux in evoking the brightness sensation. The results obtained under good lighting conditions are shown in Fig. 13-2. The ordinate of the curve, on the scale at the left, indicates the relative capacities of radiant flux of the various wave lengths to evoke visual sensations of equal brightness. The scale is arbitrarily set equal to unity at the maximum of the curve.

When the lighting is very subdued, as in a photographic darkroom or under starlight, the sensitivity of the eye is greatly increased and the maximum shifts toward the left This phenomenon, known as the *Purkinje effect*, is illustrated in Fig. 13-3, where curve (a) represents the sensitivity at low levels of illumination and curve (b) is the same as that in Fig. 13-2, on a reduced vertical scale. It is believed that only the rods in the retina are stimulated when the illumination is very low.

The curves in Figs. 13-2 and 13-3 are called *relative luminosity* curves. They fall off asymptotically to zero at either side of the maximum and hence no definite limits can be set to the "visible spectrum."

For purposes of standardization and inter-laboratory comparison of photometric data, the curve in Fig. 13-2 has been adopted by the International Commission on Illumination as the *standard luminosity curve.* It is important to remember, however, that the luminosity curve is not unique and the standard curve should not be used to predict the retinal response under conditions of extremely subdued lighting.

13-2 Luminous flux. The lumen. The effectiveness of a sample of radiant flux in evoking the sensation of brightness depends not only on the quantity of radiant flux in the sample but also on how the flux is distributed among the various wave lengths the sample may contain. Fig. 13-2 shows that the brightness sensation evoked by a given quantity of radiant flux of wave length 555 mμ, for example, is greater than that evoked by the same quantity of radiant flux of wave length 500 mμ or 600 mμ.

With the aid of the standard relative luminosity curve, any sample of radiant flux can be evaluated with respect to its brightness-producing capacity. *Radiant flux, evaluated with respect to its capacity to evoke the sensation of brightness*, is called *luminous flux*. In illuminating engineering, the same definition is given of the term *light*, so that "luminous flux" and "light" may be considered synonyms for the purposes of illuminating engineering. The unit of luminous flux is the *lumen*, and the number of lumens in a sample of flux is a measure of the brightness-producing capacity of the sample.

Consider a sample of flux of wave length 555 mμ, corresponding to the peak of the standard luminosity curve. If the *radiant* flux in the sample is 1 watt, the *luminous* flux is 685 lumens.[1] (The figure of 685 may for the present be considered an arbitrary one. We shall explain in Sec. 13-11 how it is obtained.) Another way of stating the relation above is that if the luminous flux in a sample of wave length 555 mμ is 1 lumen, the radiant flux is 1/685 watt or 0.00146 watt.

Evidently, the lumen does not correspond to any fixed number of watts. Suppose the wave length of a sample of flux is 600 mμ. At this wave length the relative luminosity is only 0.6, or in other words the effectiveness of radiant flux of this wave length in evoking the brightness sensation is only six-tenths as great as the effectiveness of flux of wave length 555 mμ. It follows that if the radiant flux in a sample of wave length 600 mμ is 1 watt, the luminous flux is only 0.6×685 lumens, or 411 lumens. Conversely, if the luminous flux in a sample of this wave length is 1 lumen, the radiant flux is 1/411 watt or 0.00243 watt.

[1] This figure refers to the so-called "new" lumen.

The *luminous efficiency* of any sample of flux is defined as the ratio of the luminous flux in the sample to the radiant flux. The luminous flux, or the number of lumens, is represented by F. The radiant flux, expressed in watts in the mks system, is represented by P. Then

$$\text{Luminous efficiency} = \frac{\text{Luminous flux}}{\text{Radiant flux}} = \frac{F}{P}.$$

Evidently, the luminous efficiency of monochromatic flux of wave length 555 mμ is 685 lumens/watt. The luminous efficiency of monochromatic flux of any other wave length is 685 lumens/watt multiplied by the relative luminosity at that wave length. The scale at the right of Fig. 13-2 gives the luminous efficiency of monochromatic flux of any wave length.

Luminous efficiency, as defined above, expresses a property of a sample of radiant flux. The same term is also applied to a light source such as an incandescent or fluorescent lamp, but with a somewhat different meaning which is better described as an "overall" luminous efficiency. The latter is defined as the ratio of the luminous flux output of the source to the *total* power input. Since most light sources do not, in the first place, convert all the power supplied to them into radiant flux, and since in the second place much of the radiant flux they emit is at wave lengths where the relative luminosity is small or even zero, the overall luminous efficiency of light sources rarely exceeds 50 lumens/watt and is often much less.

If a sample of radiant flux is not monochromatic its luminous efficiency must be found by integration methods. Let dP be the radiant flux of the sample in the wave length range between λ and $\lambda + d\lambda$. Let V be the relative luminosity at the wave length λ. The luminous efficiency at this wave length is then 685 V lumens/watt and the number of lumens dF in the sample is $dF = 685\, V dP$ lumens, if dP is in watts. The radiant flux dP will be some function of the wave length and will be proportional to the wave length range $d\lambda$. Hence one can say in general

$$dP = f(\lambda)d\lambda,$$

$$dF = 685\, Vf(\lambda)d\lambda.$$

The total number of lumens in the sample, or the luminous flux F, is

$$F = \int dF = 685 \int_0^\infty Vf(\lambda)d\lambda. \tag{13-1}$$

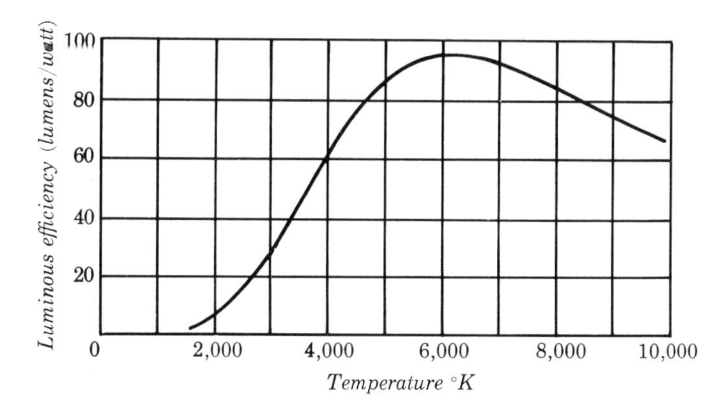

FIG. 13-4. Luminous efficiency of the radiant flux from a blackbody as a function of temperature.

The radiant flux in the sample is

$$P = \int dP = \int_0^\infty f(\lambda) d\lambda, \tag{13-2}$$

and the luminous efficiency is therefore

$$\text{Luminous efficiency} = \frac{F}{P} = 685 \frac{\int_0^\infty V f(\lambda) d\lambda}{\int_0^\infty f(\lambda) d\lambda}. \tag{13-3}$$

Since the luminosity curve cannot be expressed analytically the integration must be carried out by graphical, numerical, or mechanical methods.

The luminous efficiency of the radiant flux from a blackbody may be found from Eq. (13-3). $f(\lambda)$ is then given by Planck's equation. Fig. 13-4 shows how the luminous efficiency varies with temperature. The maximum efficiency of about 93 lumens per watt occurs at a temperature of approximately 6500°K.

13-3 Luminous intensity. Consider the point source of light S in Fig. 13-5. Let dF represent the luminous flux crossing any section of a cone of solid angle $d\omega$ steradians whose apex is at the source. The *luminous intensity* of the source, in the direction of the cone, is defined as the ratio of the flux dF to the solid angle $d\omega$, or as the *flux emitted per unit solid angle.* Luminous intensity is represented by I.

$$I = \frac{dF}{d\omega}. \tag{13-4}$$

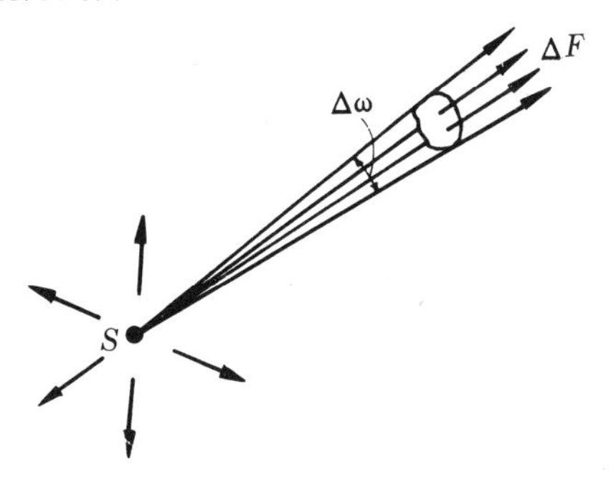

FIG. 13-5. The luminous intensity I of the point source in the direction of the small cone is defined as $I = dF/d\omega$.

(For brevity, we shall usually omit the prefix "luminous" and speak merely of "intensity.")

The unit of intensity is *one lumen per steradian.* This unit is also called *one candle.* The common expression "candlepower" should be avoided, since luminous intensity is not power.

One steradian is the solid angle subtended at the center of a sphere by an area of arbitrary shape on the surface of the sphere, equal to the square of the radius of the sphere. (See Fig. 13-6 (a).) Since the total area of a spherical surface is $4\pi R^2$, the total solid angle about a point is 4π steradians.

In general (Fig. 13-6 (b)), if ω represents the solid angle subtended at a point by an area A on the surface of a sphere of radius R, then ω (in steradians) is equal to the area A divided by the square of the radius R.

Most sources do not emit equal quantities of flux per unit solid angle in all directions, so in general the intensity of a source is different in dif-

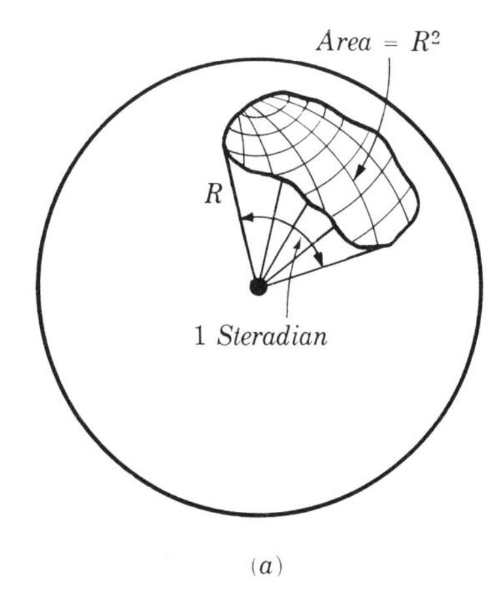

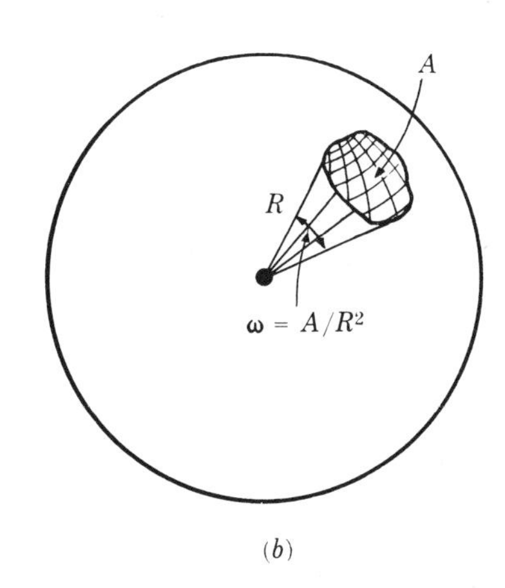

FIG. 13-6.

ferent directions. A common incandescent lamp, for example, cannot emit any flux in the direction of its base and its intensity in that direction is zero. The intensity of most incandescent lamps is a maximum in the horizontal plane through the lamp filament (assuming the lamp mounted vertically). An idealized point source that emits uniformly in all directions is called a *uniform point source.* Since the total solid angle subtended at a point is 4π steradians, a source whose intensity in all directions is I candles, or I lumens/steradian, emits a total of $4\pi I$ lumens.

Fig. 13-7 illustrates the optical system of the searchlight. A source S is located in the first focal plane of a positive lens L. Aberrations have been neglected, and the height of the source has been exaggerated in comparison with the radius of the lens. To find the number of lumens of light flux intercepted by the lens, we treat the source as a point source of intensity I_1 lumens per steradian. The solid angle subtended by the lens at the center of the source is ω_1 and the number of lumens F intercepted by the lens is therefore

$$F = I_1\omega_1.$$

The light rays from each point of the source emerge from the lens as a parallel beam, but because of the finite size of the source, these beams are not all parallel to the axis of the lens. Let ω_2 represent the solid angle of the beam diverging from the lens. Neglecting any loss of light by reflection and absorption in the lens, the number of lumens in the emergent beam is the same as the number intercepted by the lens, so the intensity I_2 in the beam, in lumens per steradian or candles, is

$$I_2 = \frac{F}{\omega_2} = I_1\frac{\omega_1}{\omega_2}.$$

It will be seen from the diagram that the solid angle ω_2 of the emergent beam is equal to the solid angle subtended by the *source,* at the center of the lens. The solid angle ω_2 of the emergent beam is smaller than the solid angle ω_1 of the beam intercepted by the lens and the intensity is proportionally greater. What the searchlight does, in effect, is to increase the intensity of the source, or the number of lumens per steradian, by "compressing" a given number of lumens into a smaller number of steradians.

The intensity of the beam, I_2, is often referred to as its "beam candlepower."

For many years, the photometric standard has been a standard of *intensity.* (A standard, it will be recalled, is a material object such as the standard kilogram or the standard meter, which embodies a unit such as

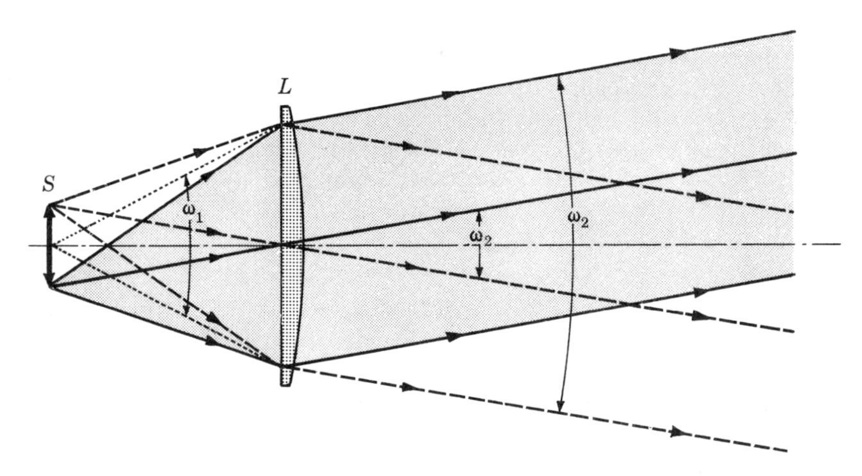

FIG. 13-7. Principle of the optical system of the searchlight.

the unit of mass or length.) The first photometric standard was the *standard candle*, an actual candle of sperm wax constructed in a specified way. The standard, of necessity, could not be permanent and the obvious difficulty of reproducing it exactly has led to its abandonment. The term "candle" as the name of the unit of intensity has been retained, however. For many years, the primary photometric standards of the United States were a number of carbon filament lamps kept in the vaults of the National Bureau of Standards. The intensities of these lamps were carefully determined in terms of the old standard candle. To prolong their life they were operated only at relatively long intervals, when another group of secondary standards was checked against them. The actual tests of the Bureau were made with still a third group of lamps whose intensity had been found by comparison with the secondary standards.

The present international photometric standard, adopted in January 1948, is the surface of a blackbody at the temperature of freezing platinum. The new standard is a standard of *luminance* rather than of intensity and will be discussed further in Sec. 13-9. The unit of intensity defined in terms of the new blackbody standard is slightly smaller than the old, or international candle, and is called a "new" candle. The "new" candle, in turn, determines the "new" lumen through the relation 1 "new" candle = 1 "new" lumen/steradian. We have used the "new" candle and the "new" lumen in this chapter, and the figure of 685 lumens/watt for the luminous efficiency of monochromatic radiant flux of wave length 555 mμ is based on the "new" lumen.

13-4 Illuminance. When luminous flux strikes a surface, we say that the surface is *illuminated.* This familiar concept is defined quantitatively as follows. At any point of the surface we construct a small area dA and let dF represent the luminous flux incident on this area. The *illuminance* E at the point is defined as the ratio of dF to dA.

$$\boxed{E = \frac{dF}{dA}\cdot} \tag{13-5}$$

That is, *illuminance* is *luminous flux incident per unit area,* and is expressed in lumens/cm², lumens/m², or lumens/ft². One lumen/m² is called a *lux,* and one lumen/ft² is called a *footcandle,* although the latter term is becoming obsolete. Note that the term *illuminance* is exactly analogous to *irradiance.* The former refers to the *luminous* flux incident per unit area, the latter to the *radiant* flux incident per unit area.

If the illuminance is the same at all points of a surface of finite area A, and if F is the total luminous flux incident on the surface, Eq. (13-5) becomes

$$E = \frac{F}{A}\cdot \tag{13-6}$$

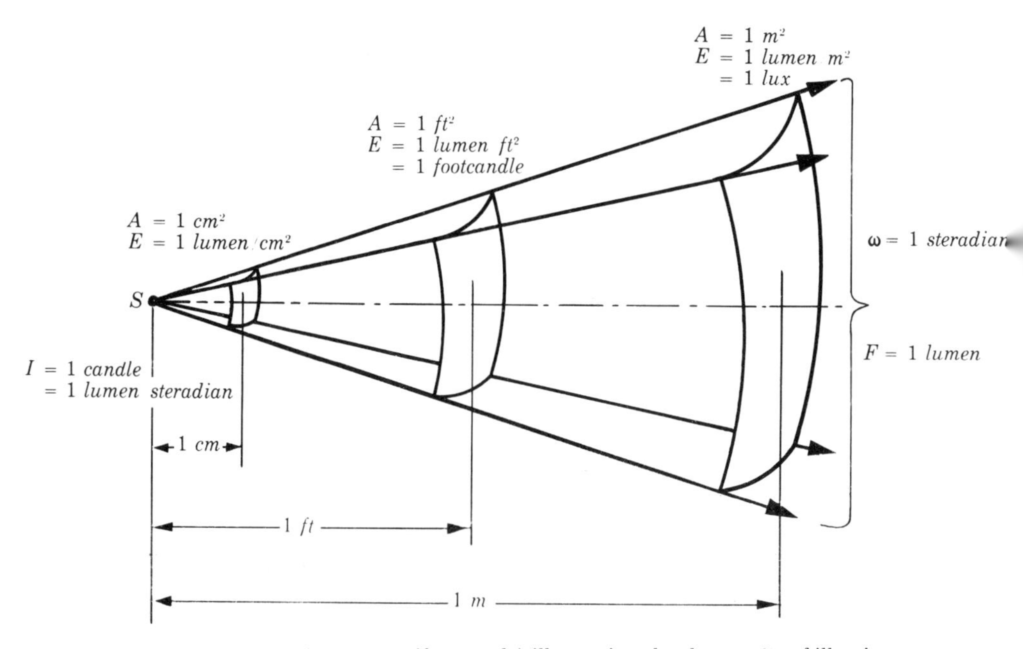

FIG. 13-8. Diagram (not to a uniform scale) illustrating the three units of illuminance.

Fig. 13-8 (not to scale) illustrates the three units of illuminance mentioned above. S is a uniform point source of intensity 1 candle or 1 lumen/steradian. The luminous flux emanating from the source within a solid angle of 1 steradian is therefore 1 lumen. If a spherical surface of radius 1 cm is constructed concentric with the source, the area cut out of this surface by the unit solid angle is 1 cm^2. Hence the illuminance on this surface is 1 lumen/cm^2. The area cut out of a spherical surface 1 ft in radius is 1 ft^2. This area also intercepts 1 lumen and the illuminance on it is 1 lumen/ft^2 or 1 footcandle. Similarly, the illuminance on a surface of 1 m^2, at a distance of 1 m from the source, is 1 lumen/m^2 or 1 lux.

A table of some typical values of illuminance is given below.

TABLE 13-1.

TYPICAL VALUES OF ILLUMINANCE

Mode of Illumination	Illuminance (Representative values)	
	Lumens/m^2 (lux)	Lumens/ft^2 (footcandles)
Sunlight plus skylight (maximum)	100,000	10,000
Sunlight plus skylight (dull day)	1,000	100
Interiors—daylight	200	20
Interiors—artificial light	100	10
Full moonlight	0.2	0.02

13-5 Illuminance produced by a point source. Let dA in Fig. 13-9 be an element of surface whose normal makes an angle θ with the distance r to a point source S. Let I be the intensity of the source in the direction of the element dA. The solid angle subtended by dA at the source is $d\omega = dA \cos \theta/r^2$, and since by definition the intensity $I = dF/d\omega$, the flux dF in the solid angle $d\omega$ is

$$dF = I d\omega = \frac{I dA \cos \theta}{r^2}. \tag{13-7}$$

All of this flux is incident on the area dA, and hence the illuminance of dA is

$$\boxed{E = \frac{dF}{dA} = \frac{I \cos \theta}{r^2}.} \tag{13-8}$$

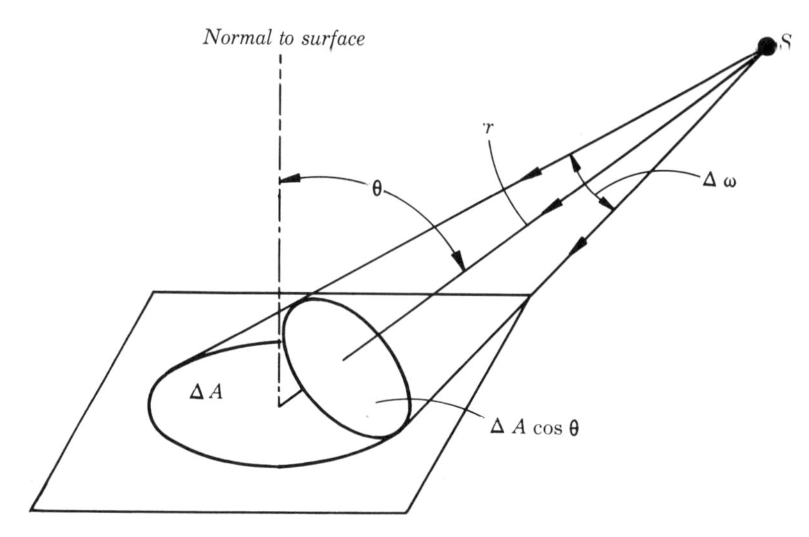

FIG. 13-9. Illuminance of an element of surface by a point source.

That is, the illuminance of an element of surface by a point source is inversely proportional to the square of the distance of the element from the source and directly proportional to the intensity of the source and the cosine of the angle between the normal to the element and the line joining the element and the source.

When a surface element is illuminated by more than one source the total illuminance is the arithmetic sum of the illuminances produced by the individual sources.

13-6 The photometer. A rough comparison of the intensities of two point sources may be made by looking at them, but when quantitative measurements must be made the sources are compared indirectly by means of the illuminance they produce on a screen. The apparatus for doing this is called a *photometer* and it is illustrated, in principle, in Fig. 13-10. One side of the screen S is illuminated only by the source P_1, the other only by the source P_2. The position of the screen may be adjusted along the line joining P_1 and P_2 until an observer judges both faces of the screen to be equally illuminated. The distances d_1 and d_2 are then measured, and from Eq. (13-8),

$$\frac{I_1}{d_1^2} = \frac{I_2}{d_2^2}, \quad \text{or} \quad \frac{I_1}{I_2} = \frac{d_1^2}{d_2^2},$$

where I_1 and I_2 are the intensities of the sources in the direction of the screen. If the intensity of either source is known, that of the other can be computed.

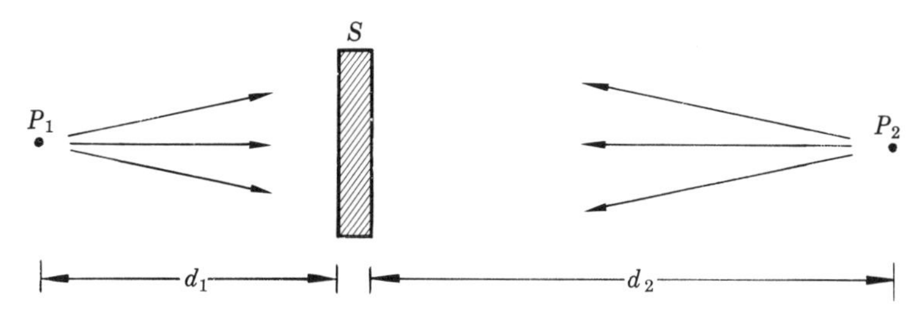

FIG. 13-10. Principle of the photometer.

The intensity of the unknown source in other directions may be found by rotating it and making measurements at various azimuths. Remembering that the intensity in any direction is the number of lumens per steradian in that direction, it will be seen that from a knowledge of the intensity in a number of directions about the source the total flux emitted by the source may be found.

In practice, the simple screen S of Fig. 13-10 is replaced by a more complicated device called a Lummer-Brodhun photometer head, which permits both sides of the screen to be viewed simultaneously. The construction of this instrument is illustrated in Fig. 13-11. The screen S is a sheet of plaster of Paris or magnesium oxide. Diffusely reflected light from either side of the screen is reflected by the right-angle prisms A and B to the photometer "cube" C, which consists of two right-angle prisms, the hypotenuse of one being ground away over a portion of its area. Light incident on the area in contact is transmitted, so that the center of the field of view receives only light reflected from the right side of the screen S. The outer portion of the field receives light from the left side of S, totally reflected by the left prism at regions where the two prisms are not in contact. When the photometer is adjusted for a balance the entire field appears uniformly bright.

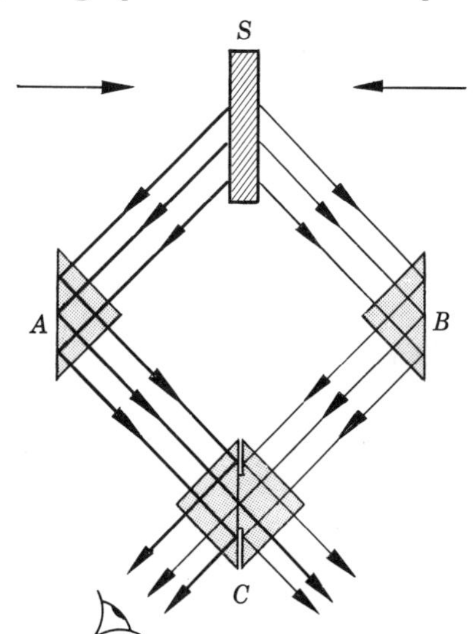

FIG. 13-11. The Lummer-Brodhun photometer.

To compensate for any differences in the reflecting characteristics of the faces of the screen S the photometer head is rotated through 180°, thus interchanging the faces of the screen, and another setting made. Averaging the two very nearly eliminates any error due to dissimilarity of the faces.

13-7 Heterochromatic photometry. If both parts of a photometric field are not of the same color it is very difficult to decide when the two appear equally bright. The process of comparing two light beams of different color is called *heterochromatic photometry.* One method makes use of the *flicker photometer*, in which the field is illuminated alternately by light from the two sources. As the frequency of the alternations is increased from a small value, it is found that at a particular frequency the color difference disappears because of persistence of vision, while a brightness flicker remains unless both sides of the photometer screen are equally illuminated. The setting of the photometer head is then made by noting the position at which the brightness flicker disappears.

13-8 Spectrophotometry. Spectrophotometry is one method of heterochromatic photometry in which the two beams to be compared are first dispersed into spectra and then compared step-by-step throughout the spectrum. Both parts of the field are then illuminated by monochromatic light at each setting. Spectrophotometric methods find their chief application in the determination of the transmittance and reflectance of colored materials. If the material is transparent its *spectral transmittance*, i.e., the percentage of incident light transmitted at each wave length, is obtained by dividing the light beam from a single source into two parts, sending one part through and the other around the sample, and then comparing the two beams wave length by wave length. The *spectral reflectance* of an opaque material is found by comparing at each wave length the quantity of light reflected from a sample with that reflected from a diffusely reflecting white surface. The subject of spectrophotometry will be discussed further in the next chapter.

13-9 Luminance. The light sources that have been considered up to this point have been sufficiently small, or sufficiently distant, so that they could be treated as point sources. In most cases of practical interest, however, one has to do with extended sources of light rather than points. These sources may be self-luminous, as is the surface of a metal heated to incandescence or the screen of a television tube, or may be diffusely reflecting or transmitting like the frosted bulb of an incandescent lamp or the ceiling of an indirectly illuminated room.

In discussing the flux emitted by an extended source it is necessary to introduce two orders of infinitesimals. Any infinitesimal element of a surface emits an infinitesimal quantity of flux, distributed over a hemisphere or within a solid angle of 2π steradians. The flux within an *infinitesimal* solid angle is an infinitesimal fraction of the flux emitted and therefore is an infinitesimal of second order. We shall represent an infinitesimal surface element by ΔA, the total flux it emits by ΔF, and the flux within an infinitesimal solid angle by $d(\Delta F)$.

Consider the pencil of rays in Fig. 13-12 (a), having its apex at a small element of area ΔA and making an angle θ with the normal to the area. Let $d\omega$ be the solid angle included by the pencil and let $d(\Delta F)_\theta$ be the flux within it. The intensity of the element in the direction θ is, by definition,[1]

$$\Delta I_\theta = \frac{d(\Delta F)_\theta}{d\omega}. \qquad (13\text{-}9)$$

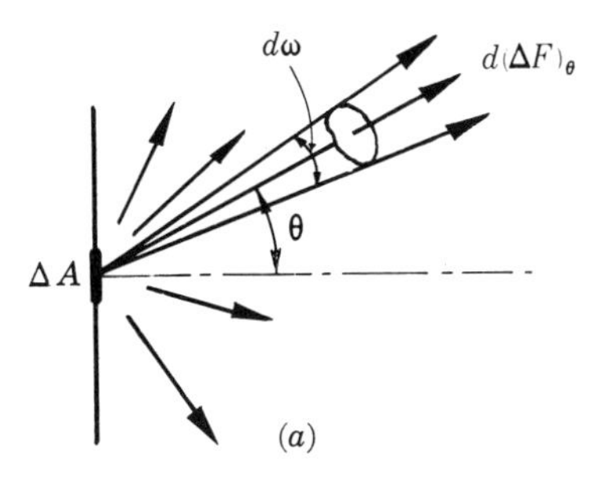

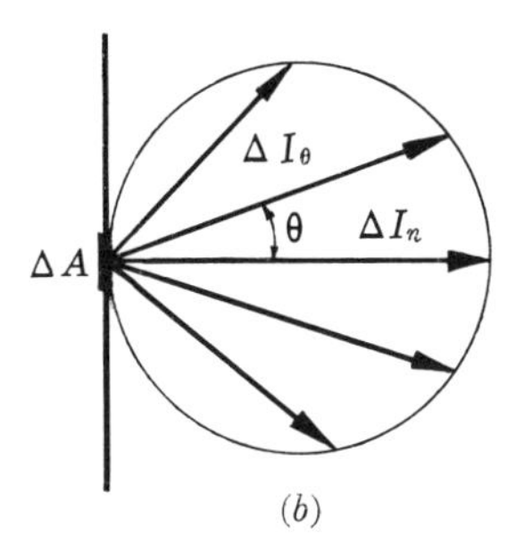

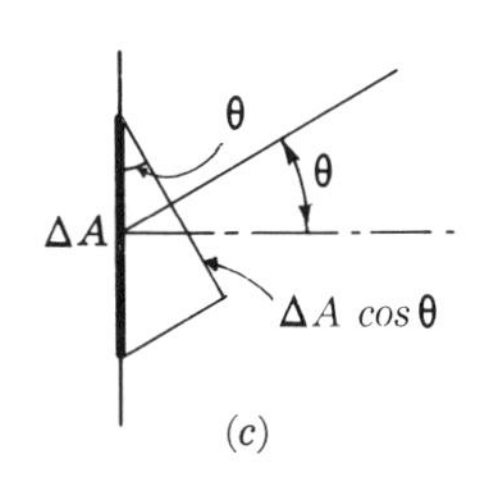

Fig. 13-12. (a) The intensity of the small surface element ΔA in the direction θ is $d(\Delta F)_\theta/d\omega$. (b) Polar graph of intensity of a surface element obeying Lambert's law.

If the element emitted like a *uniform* point source, the intensity (i.e., the flux per unit solid angle) would be the same in all directions. Actually, however, the intensity is found to be different in different directions, and for many surfaces it varies with the angle θ according to the relation

$$\Delta I = \Delta I_n \cos\theta, \qquad (13\text{-}10)$$

where ΔI_θ is the intensity in a direction making an angle θ with the normal to the surface, and ΔI_n is the intensity along the normal. This vari-

[1] In the most general case, the intensity may be a function of both θ and ϕ where ϕ is the angle between the plane defined by the pencil and the normal, and some reference plane such as that of the diagram.

ation of intensity with angle is shown graphically in Fig 13-12 (b), where the length of the vector ΔI in any direction represents the intensity in that direction. The complete graph is, of course, three-dimensional and may be obtained by imagining the diagram rotated about the normal to the surface. The intensity is a maximum along the normal, and falls to zero in the direction tangent to the surface.

Eq. (13-10) is called *Lambert's law.* A surface element (self-luminous, transmitting, or reflecting) which emits (or transmits or reflects) in this way is said to be *perfectly diffuse.* A sheet of blotting paper is a good approximation to a diffuse reflector.

The *luminance* of the surface element in any direction is defined as the ratio of the intensity of the element in that direction, ΔI_θ, to the area of the element projected on a plane perpendicular to the direction. This projected area is $\Delta A \cos\theta$, as will be seen from Fig. 13-12 (c), which is an enlarged view of a portion of Fig. 13-12 (a). The luminance in the direction θ is represented by B_θ.

$$B = \frac{\Delta I_\theta}{\Delta A \cos\theta}. \tag{13-11}$$

Luminance is *intensity per unit of projected area of source.* If the intensity is in candles the luminance of the source is in candles/cm^2, candles/m^2, or candles/ft^2.

Eq. (13-11) is a general definition of luminance, whether or not the element obeys Lambert's law. If Lambert's law *is* obeyed, then

$$\Delta I_\theta = \Delta I_n \cos\theta$$

and

$$B_\theta = \frac{\Delta I_n \cos\theta}{\Delta A \cos\theta} = \frac{\Delta I_n}{\Delta A} = \text{constant.}$$

Hence in this special case the luminance is *independent of the angle* θ, or is a constant. For such surfaces, then, the subscript θ can be dropped from B_θ and we can speak of *the* luminance of the surface element and represent it by B. Representative values of the luminance of a number of extended sources are given in Table 13-2.

The *brightness sensation* when one looks at a surface depends upon the luminance of the surface. If the surface obeys Lambert's law and its luminance is the same in all directions, it appears equally bright from whatever angle it may be viewed. Thus an incandescent sphere appears like a uniformly bright disk although the central portion is viewed normally and the edges tangentially.

TABLE 13-2.

TYPICAL VALUES OF LUMINANCE

Source	Luminance (candles/m²)
Surface of sun	2×10^9
Tungsten filament at 2700° K	10^7
White paper in sunlight	25,000
Fluorescent lamp	6,000
Candle flame	5,000
Clear sky	3,200
Surface of moon	2,900
White paper in moonlight	0.03

The concept that is now known as "luminance" was for many years designated by the term "brightness." This led to much confusion between the objective concept of "brightness" as intensity per unit of projected area, and the subjective concept of "brightness" which referred to a sensation in the consciousness of a human observer. The newer term "luminance" was adopted to avoid this confusion.

As we have stated earlier, the original photometric standard was the standard candle, an actual candle constructed according to definite specifications. This standard was later replaced by a number of carbon filament lamps as described in Sec. 13-3. The need for a more permanent standard which can be reproduced at will has led to the introduction of the *new candle*, which is defined as *a unit of such luminous intensity that the luminance of a blackbody at the temperature of freezing platinum shall be exactly 600,000 new candles/m²*. On the basis of the former standard based on groups of lamps, the luminance of a blackbody at this temperature is 589,000 candles/m², so the new candle is a slightly smaller unit than the old.

13-10 Illuminance produced by an extended source. We have shown in the preceding section that each surface element of a light source (using the term "source" in a general sense to include diffusely transmitting or reflecting surfaces) emits with an intensity that is different in different directions. We now wish to compute the illuminance produced by an extended, diffusely emitting source. In general, each surface element of the source lies at a different distance from any given point on the illuminated surface, and the direction to a given point on the illuminated surface varies from point to point of the emitting surface.

Let the emitting surface, or source, be a flat circular disk of radius a and luminance B, and let the illuminance be computed at a point P (Fig. 13-13) on the axis of the disk and on a surface at right angles to the axis.

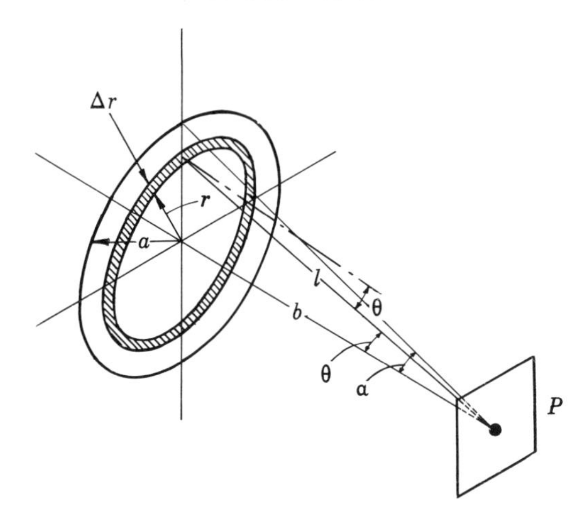

FIG. 13-13. Illuminance at an axial point P by a circular disk.

Construct an annular element of the disk of radius r, width Δr, and area $\Delta A = 2\pi r \Delta r$. All portions of this element are at the same distance l from the point P, and l makes the same angle θ with the normal at every portion of the element. The intensity of the element in the direction of P, from Eq. (13-11), is

$$\Delta I_\theta = B \Delta A \cos\theta$$
$$= B \times 2\pi r \Delta r \times \cos\theta.$$

We have written B instead of B_θ since the luminance is independent of θ if the source obeys Lambert's law. The illuminance at P, due to the annular element, is from Eq. (13-8),

$$\Delta E = \frac{\Delta I_\theta \cos\theta}{l^2}.$$

But from Fig. 13-13,

$$l = b \sec\theta,$$
$$r = b \tan\theta,$$
$$\Delta r = b \sec^2\theta \, \Delta\theta.$$

Hence

$$\Delta E = 2\pi B \sin\theta \cos\theta \, \Delta\theta.$$

The illuminance at P due to the entire disk is found by integrating this expression between the limits of $\theta = 0$ and $\theta = \alpha$, where α is the half-angle subtended at P by the disk. This leads to the result

$$E = \pi B \sin^2 \alpha. \tag{13-12}$$

If B is in candles/m^2, E is in lumens/m^2; if B is in candles/ft^2, E is in lumens/ft^2, etc.

Since $\sin^2 \alpha = \dfrac{a^2}{a^2 + b^2}$, Eq. (13-12) can be written,

$$E = \frac{B\pi a^2}{a^2 + b^2},$$

and since πa^2 equals the area of the disk, say A,

$$E = \frac{BA}{a^2 + b^2}. \tag{13-13}$$

The product of B (intensity per unit area) and A (the area of the disk) is the intensity of the disk. Hence the illuminance produced by the disk at an *axial* point is given by an equation similar to Eq. (13-8) for a point source. At distances sufficiently large for a^2 to be negligible compared with b^2, Eq. (13-13) reduces to the familiar inverse square law for a point source. For example, if a is one-tenth of b, then a^2 is only one percent of b^2 and an error of but one percent is introduced if a^2 is neglected in the denominator of Eq. (13-13). In other words, a circular disk can be considered a "point" source at any distance greater than ten times its radius, within an accuracy of 1%.

Although Eq. (13-13) was deduced for a circular disk, it may be used for an area of any shape approximating a circle.

13-11 Flux emitted by an extended source. The total luminous flux emitted (or transmitted or reflected) by a diffuse surface must be found by integration methods, since the intensity of the surface is not the same in all directions. Let ΔA in Fig. 13-14 represent a small element of an emitting surface. Construct a hemisphere of radius r with center at ΔA. The radius must be taken large enough so that ΔA can be considered a point source. The intensity of the element in the direction of the shaded zone is

$$\Delta I_\theta = B_\theta \Delta A \cos\theta.$$

The area of the zone is

$$2\pi r^2 \sin\theta \, d\theta,$$

and the solid angle it subtends at ΔA is

$$d\omega = 2\pi r^2 \sin\theta \, d\theta / r^2 = 2\pi \sin\theta \, d\theta.$$

Hence from Eq. (13-9), the flux $d(\Delta F)_\theta$ crossing the shaded zone is

$$d(\Delta F)_\theta = (B_\theta \Delta A \cos\theta)\,(2\pi \sin\theta \, d\theta),$$

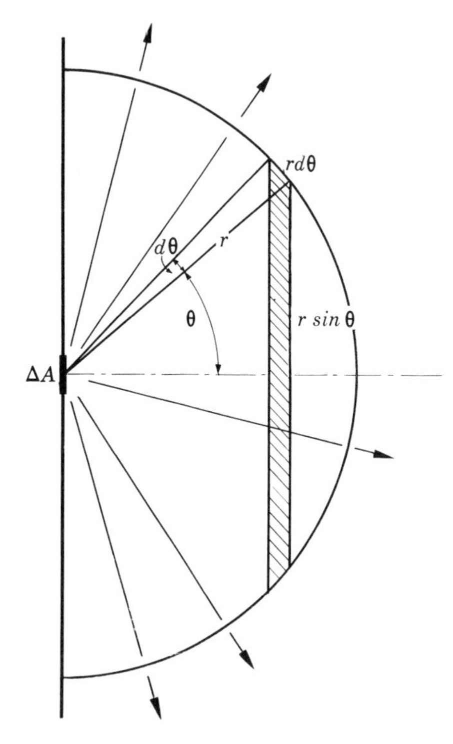

FIG. 13-14. Total flux emitted by an element of surface.

and the total flux crossing the hemisphere, which includes all of the flux emitted by the element is,

$$\Delta F = 2\pi \Delta A \int_0^{\pi/2} B_\theta \sin\theta \cos\theta \, d\theta.$$

In order to perform the indicated integration, the luminance B_θ must be known as a function of θ. In the special case of a perfectly diffuse surface, the luminance is a constant independent of θ. Let us represent the constant luminance by B. The preceding equation then integrates to

$$\Delta F = \pi B \Delta A. \quad (13\text{-}14)$$

The flux emitted by the source per unit area is

$$\frac{\Delta F}{\Delta A} = \pi B.$$

The ratio $\Delta F/\Delta A$, or the total luminous flux emitted in all directions, per unit area of the emitting surface, is called the *luminous emittance* of the surface and is represented by L. Notice that luminous emittance is exactly analogous to *radiant emittance* W, which is the radiant flux emitted per unit area of emitting surface.

$$L = \frac{\Delta F}{\Delta A} = \pi B. \quad (13\text{-}15)$$

That is, a perfectly diffuse surface whose luminance B is 1 candle/cm^2 has a luminous emittance L of π lumens/cm^2; a surface whose luminance is 1 candle/m^2 has a luminous emittance of π lumens/m^2; and a surface whose luminance is 1 candle/ft^2 has a luminous emittance of π lumens/ft^2.

It was stated in Sec. 13-2 that the maximum luminous efficiency of monochromatic radiant flux is 685 lumens/watt. We can now see how this figure is obtained. Let us for the moment consider it as unknown and represent it by K_m. The luminous efficiency at any wave length is then $K_m V$, where V is the relative luminosity at that wave length.

The fundamental photometric standard is the surface of a blackbody at the temperature of freezing platinum, 2042°K. From Planck's law, the radiant emittance of a blackbody (radiant flux emitted per unit area) at the wave length λ is

$$dW_{bb} = \frac{c_1\lambda^{-5}}{\epsilon^{c_2/\lambda T} - 1}\, d\lambda.$$

The luminous flux emitted per unit area, at this wave length, is the product of the radiant flux and its luminous efficiency, or

$$dL = K_m\, V\, dW_{bb}.$$

The total flux emitted per unit area, or the luminous emittance L, is

$$L = \int dL = \int K_m\, V\, dW_{bb} = K_m \int_0^\infty V\, \frac{c_1\lambda^{-5}}{\epsilon^{c_2/\lambda T} - 1}\, d\lambda.$$

The numerical value of the integral, for a temperature of 2042°K, is 2750 in mks units. (Numerical or graphical methods of integration must be used since the relative luminosity V cannot be expressed analytically as a function of wave length.) Then

$$L = 2750\, K_m \text{ lumens/m}^2.$$

By definition, the luminosity of the blackbody standard is

$$B = 60 \times 10^4 \text{ (new) candles/m}^2.$$

From Eq. (13-15), the luminous emittance of the standard is

$$L = \pi B = 60\pi \times 10^4 \text{ lumens/m}^2.$$

The two expressions for L may now be equated, giving

$$2750\, K_m = 60\pi \times 10^4.$$

Hence

$$K_m = \frac{60\pi \times 10^4}{2750}$$

$$= 685 \text{ (new) lumens/watt}.$$

In terms of the older international candle the luminosity of the blackbody standard is 58.9×10^4 candles/m² and the corresponding maximum luminous efficiency is

$$K_m = 673 \text{ (international) lumens/watt}.$$

Experimental uncertainties in the values of the constants c_1 and c_2 lead to an uncertainty in K_m of about 5 units in the last digit.

Luminance is often expressed in terms of another set of units each of which is $1/\pi$th as great as the units mentioned above. These units are the *lambert*, the *meterlambert*, and the *footlambert*. (To be perfectly consistent, the first should be called a *centimeterlambert*.)

$$1 \text{ lambert} = 1/\pi \text{ candle/cm}^2,$$

$$1 \text{ meterlambert} = 1/\pi \text{ candle/m}^2,$$

$$1 \text{ footlambert} = 1/\pi \text{ candle/ft}^2.$$

It follows that a perfectly diffuse surface whose luminance is 1 lambert has a luminous emittance of 1 lumen/cm^2, that is, the luminous emittance, in lumens/cm^2, is numerically equal to the luminance in lamberts. Similarly, the luminous emittance in lumens/m^2 is numerically equal to the luminance in meterlamberts and the luminous emittance in lumens/ft^2 is numerically equal to the luminance in footlamberts.

The "lambert" units are convenient when dealing with nonself-luminous surfaces, such as the diffusing shades around incandescent lamps or the walls or ceiling of an indirectly illuminated room. If such a surface does not absorb any of the light incident on it, the number of lumens *incident* per unit area equals the number of lumens *emitted* per unit area. (The term "emitted" is used here in a general sense to include the flux diffusely transmitted or reflected from a surface.) But the number of lumens incident per unit area is the illuminance on the surface and the number of lumens emitted per unit area is the luminous emittance of the surface. Since the latter is equal to the luminance of the surface in the appropriate "lambert" unit, it follows that for a surface which is perfectly diffuse, and which transmits or reflects 100% of the light incident on it, the *illuminance* (in lumens/cm^2), the *luminance* (in lamberts), and the *luminous emittance* (in lumens/cm^2), are all numerically equal. Similarly, the illuminance, luminance, and luminous emittance are numerically equal when expressed respectively in lumens/m^2 or lumens/ft^2, meterlamberts or footlamberts, and lumens/m^2 or lumens/ft^2.

If some light is absorbed by the surface, the luminance (in lamberts, meterlamberts, or footlamberts) is equal to the illuminance (in lumens/cm^2, lumens/m^2, or lumens/ft^2) multiplied by the fraction of the incident light transmitted or reflected. The luminous emittance (in lumens/cm^2, lumens/m^2, or lumens/ft^2) equals the luminance (in lamberts, meterlamberts, or footlamberts).

Examples. (1) A lamp whose intensity is known to be 20 candles is set up at one end of a photometer bar 3 meters long, with an unknown lamp at the other end. A photometer head indicates a balance when at a distance of 2 meters from the standard lamp. What is the intensity of the unknown lamp, in the direction of the photometer?

The illuminance of the photometer by the standard lamp is

$$\frac{20}{2^2} \text{ lumens/m}^2.$$

The illuminance by the unknown lamp is

$$\frac{I}{1^2} \text{ lumens/m}^2.$$

Since these are equal,

$$I = 5 \text{ candles.}$$

(2) An incandescent lamp is placed in a box, in one side of which is a circular hole 10 cm in diameter covered with diffusely transmitting opal glass. The box is set up at one end of the photometer bench with the opal glass in the position occupied by the unknown lamp in Example 1. The photometer now balances when distant 180 cm from the glass. What is the luminance of the glass?

Since the distance of the photometer from the glass disk is 36 times the radius of the disk the inverse square law may be applied. The illuminance of the photometer by the disk, from Eq. (13-8), is

$$E = BA/b^2 = BA/180^2 \text{ lumens/cm}^2.$$

The illuminance by the standard is

$$\frac{20}{120^2} \text{ lumens/cm}^2.$$

Since these are equal, $BA = 45$ lumens/steradian or 45 candles. Finally, since the area of the disk is 25π cm², its luminance is

$$B = \frac{45}{25\pi} = 0.57 \text{ candle/cm}^2 = 0.57\pi \text{ lamberts} = 1.79 \text{ lamberts.}$$

(3) What is the intensity of the disk, considering it a point source, in a direction at 45° with the normal to its surface?

From Eq. (13-11),

$$\begin{aligned} \Delta I_\theta &= B\,\Delta A \cos\theta \\ &= 0.57 \times 25\pi \times .707 \\ &= 32 \text{ candles.} \end{aligned}$$

(4) If the disk transmits 60% of the light incident on it, what is the illuminance on its inner surface?

The luminance of the outer surface, in lamberts, equals the illuminance on the inner surface, in lumens/cm², multiplied by the fraction of light transmitted. Hence

$$1.79 = E \times .6,$$

$$F = 3 \text{ lumens/cm}^2.$$

13-12 Light sources. The sun radiates very approximately like an ideal blackbody, the value of λ_m in the solar spectrum (about 500 mμ) corresponding to a surface temperature of about 5750°K. Every square centimeter of the earth's surface receives about 2 gram-calories of radiant energy per minute from the sun. This value is called the *solar constant.*

The solar spectrum is crossed by a large number of very fine dark lines which were first noted by Fraunhofer, who measured the wave lengths of many of them very carefully. They are called, after him, *Fraunhofer lines.* These dark Fraunhofer lines are found to occupy the same positions in the spectrum (i.e., to have the same wave lengths) as the *bright* lines observed in the laboratory in the emission spectra of various substances. It is found that the relatively cool vapors of a substance will absorb, from light passing through them, many of the same wave lengths they emit if excited by an electrical discharge. The conclusion is that the Fraunhofer lines are the *absorption lines* of relatively cool gases either in the sun's outer atmosphere or in the earth's atmosphere, and from their positions in the spectrum the elements responsible for their appearance may be determined. In this way, more than 60 of the elements known to us on the earth have been shown to be present in the atmosphere of the sun. In fact, one element, the now familiar helium, was "discovered" in the sun by the presence of its absorption lines in the solar spectrum, before it had been isolated in the laboratory. (The name helium is from "helios," the sun.)

The more prominent of the Fraunhofer lines serve as useful landmarks in the spectrum, and are designated by letters. The wave lengths of some, together with the elements responsible for them, are given below:

Fraunhofer Line	λ(mμ)	Element
A	759.38	O (Atmospheric)
B	686.72	O (Atmospheric)
C	656.28	H
D_1	589.59	Na
D_2	589.00	Na
F	486.13	H
G'	434.05	H
h	410.19	H
H	396.85	Ca
K	393.37	Ca

The first practical incandescent lamp was developed by Edison in 1879. It consisted of a filament of a carbonized strip of bamboo in an evacuated glass bulb. This was followed by lamps with metal filaments, first of tantalum and then of tungsten. The latter, when properly purified, can be drawn into wires as fine as 0.0005 inch in diameter. The rate of evaporation of the filament is reduced by introducing into the bulb some inert gas such as argon or nitrogen. The chief advantage of the gas-filled tungsten lamp over the earlier types is that it may be operated at a much higher temperature without unduly shortening its life, some gas-filled tungsten lamps being operated today at temperatures as high as 3400°K. As will be recalled from Fig. 13-4, the luminous efficiency of a thermal radiator is a maximum at about 6500°K. Since the melting point of tungsten is 3665°K and no suitable material having a higher melting point has been discovered, it seems unlikely that the luminous efficiency of incandescent sources will be greatly increased over the values attainable at present. The luminous efficiency of tungsten light at 3665°K is about 59 lumens/watt, but most incandescent lamps operate at considerably lower temperatures and their efficiencies are of the order of 15 lumens/watt.

The familiar "neon sign," the mercury vapor lamp, and the sodium vapor lamp, are electrical discharges in gases at low pressure. The light is emitted by atoms which have temporarily acquired energy from collisions with rapidly moving electrons. When the atoms return to their normal state the energy is released as radiant energy. The wave lengths emitted are characteristic of the gas in the tube and the energy distribution in the spectrum is not given by Planck's equation which is characteristic of thermal radiators only. Much of the radiant energy emitted by a gas discharge may lie in the ultraviolet portion of the spectrum where it contributes nothing to the luminous output. The luminous efficiency of gas discharges may be increased by coating the tube walls with a fluorescent material, that is, one that has the property of absorbing radiant energy in the ultraviolet and re-emitting the energy at wave lengths in the visible spectrum. The overall luminous efficiency of fluorescent lamps runs as high as 60 lumens/watt.

One occasionally sees reference to "cold light," implying a beam of visible light which would produce no heating of a body by which the light is absorbed. Such a beam is obviously an impossibility. Of course most light sources, and thermal radiators particularly, emit radiant energy which includes wave lengths both longer and shorter than the limits of the visible spectrum. These waves contribute nothing to the luminosity of the light beam, but the energy which they carry raises the temperature of any body by which the beam is absorbed. If a light source could be found which would convert all of the energy supplied to it into radiant energy in the visible spectrum, its luminous efficiency would be much greater than that of an incandescent source, and the energy associated with a given amount of visible light would be correspondingly smaller. From this point of view, the ideal source would be one emitting only monochromatic light of that frequency to which the eye is most sensitive, namely, about 555 mμ. It should be noted, however, that even with such a source, a power input of 0.00146 watt would be required for every lumen, corresponding to a luminous efficiency of 685 lumens per watt.

A monochromatic or nearly monochromatic light source is frequently found necessary in the laboratory. A Bunsen flame into which common salt (NaCl) may be introduced is a simple and fairly satisfactory source for some purposes. The D-lines of sodium vapor, which are responsible for the yellow color, are very pronounced in the emitted light. A sodium vapor lamp is a more intense source of these lines. The mercury arc and the hydrogen discharge tube contain a number of prominent and fairly well separated lines, so that all except the one desired may be eliminated by use of the proper filters. If a monochromatic source of any arbitrary wave length is desired, recourse is usually had to a *monochromator*, which is essentially a spectrometer (see Fig. 6-31) in which a narrow slit is placed in the plane of the image formed by the telescope objective. By moving the slit across the spectrum of some continuous source or by rotating the prism, a narrow spectral band in any portion of the spectrum may be isolated by the slit.

Summary

The photometric concepts defined in this chapter are summarized below:

Quantity	Symbol	Unit	Definition
Luminous flux	F	Lumen	1 lumen = 0.00146 watt of radiant flux of wave length 555 mμ, or its equivalent in evoking the sensation of brightness.
Luminous efficiency	F/P	Lumen/watt	Ratio of luminous flux to radiant flux.
Luminous intensity	I	Lumen/steradian or candle	Luminous flux emitted per steradian.
Luminous emittance	L	Lumen/cm^2 Lumen/m^2 Lumen/ft^2	Luminous flux emitted per unit area.
Illuminance	E	Lumen/cm^2 Lumen/m^2 Lumen/ft^2	Luminous flux incident per unit area.
Luminance	B	Candle/cm^2 Candle/m^2 Candle/ft^2	Luminous intensity per unit of projected area.

Problems—Chapter 13

(1) A sample of radiant flux consists of 20 watts of monochromatic light of wave length 500 mμ and 10 watts of monochromatic light of wave length 600 mμ. (a) What is the radiant flux in the sample? (b) What is the luminous flux in the sample? (c) What is the luminous efficiency of the sample?

(2) A point source of light is 2 m from a screen in which there is a circular hole 10 cm in diameter. The screen is at right angles to the line joining the center of the hole and the source. It is found that 0.05 lumen of luminous flux from the source passes through the hole. (a) What is the solid angle, in steradians, subtended by the hole at the source? (b) What is the intensity of the source in the direction of the hole? (c) If the source emits uniformly in all directions, find the total number of lumens it emits. (d) The overall luminous efficiency of the source is 20 lumens/watt. What is the power input to the source?

(3) A relay is to be controlled by a vacuum photocell, actuated by the light passing through an aperture measuring 15 mm by 40 mm. At least 0.2 lumen must strike the photocell to operate the relay. What is the maximum permissible distance from the aperture to a uniform point source of intensity 50 candles, if the light from the source is to operate the relay?

(4) A uniform point source is suspended 4 ft above a desk. The illuminance on the desk at a point directly below the source is 10 footcandles. (a) What is the illuminance on the desk at a point 3 ft from the first point? (b) At what distance from the first point is the illuminance 2 footcandles?

(5) A room is 16 ft high, 30 ft long, and 20 ft wide. Four sources of intensity 200 candles each are suspended from the ceiling on cords 4 ft long. Each cord is 5 ft from both side walls in its respective corner. Compute the direct illuminance on a table top in the center of the room, 30 inches above the floor.

(6) The standard photometer bar is 3 m long. A lamp whose intensity in the direction of the photometer is known to be 30 candles is set up at one end of the bar, and a lamp of unknown intensity at the other end. The field of view of a Lummer-Brodhun photometer appears of uniform brightness when it is 80 cm from the 30-candle lamp. What is the intensity of the other lamp, in the direction of the photometer?

(7) At what height above the center of a circular table of radius R should a point light source be suspended, to produce the maximum illumination at the edges of the table?

(8) A searchlight beam has an intensity of 90,000,000 lumens/steradian. (a) Neglecting atmospheric losses, what illumination would it produce on a ship 3000 m distant, assuming the ship to be broadside to the searchlight beam? (b) If the side of the ship reflects diffusely (according to Lambert's law) half of the light incident upon it, what is its luminance?

(9) A circular disk of white blotting paper, 10 cm in diameter, is 50 cm from a lamp whose intensity in the direction of the disk is 2000 candles. The plane of the disk is at right angles to the line joining the lamp and the center of the disk. (a) What is the illuminance on the disk? (b) If the disk reflected diffusely all the light falling on it, what would its luminance be, in candles/cm^2 and in lamberts? (c) What would be the intensity of the disk in a direction making an angle of 60° with the normal, at a distance sufficiently great for the disk to be treated as a small

element of area? (d) The intensity in this direction is measured and found to be 8 candles. What fraction of the light incident on the disk is reflected?

(10) A circular disk of diffusely transmitting opal glass 10 cm in diameter is set in the wall of an opaque box. There is a light source within the box of such intensity that the luminance of the outer side of the disk is 2 candles/cm². Compute the illuminance at points on the axis of the disk (on a screen at right angles to the axis, outside the box) at the following distances from the disk: 20 cm, 10 cm, 5 cm, 2 cm, 0 cm. Show the results in a graph.

(11) Compute the illuminance, on a screen parallel to the surface, due to a perfectly diffuse surface of infinite extent whose luminance is B candles/m².

(12) A uniform point source of intensity 100 candles is mounted 50 cm above a horizontal screen containing a hole 1 cm in radius located directly under the source. Light passing through this hole falls on a second horizontal screen 50 cm below the first. (a) Find the total number of lumens from the source that fall on the second screen. (b) The lower side of the first screen is illuminated by reflected light from the second. Compute the illuminance at a point 50 cm from the center of the hole, on the lower side of the first screen. Assume the second screen to be a perfectly diffuse reflector having a reflectance of 80%.

(13) The full moon is capable of producing an illuminance of 0.2 lumen per square meter on the surface of the earth. Assuming the full moon to be optically equivalent to a uniform circular disk 2200 miles in diameter and distant 250,000 miles from the surface of the earth, compute the luminance of the moon. Neglect absorption of light by the atmosphere.

(14) A carbon arc produces an illuminance of 1,000 lumens/m² on a normal surface 4 m distant. (a) What is the luminous intensity of the arc? (b) If the light is emitted exclusively by the positive carbon, which has an area of 1 cm², what is its luminance in candles/cm²? (c) What is its luminous emittance in lumens/cm²?

(15) An incandescent sphere of tungsten 2 mm in diameter has an intensity of 100 candles when regarded as a point source. The sphere is located at the focal point of a positive lens, of focal length such that the sphere subtends a solid angle of 1/100 steradian (plane angle = 6.40°) at the center of the lens. The size of the lens is such that it collects a solid angle of ⅕ steradian (plane angle = 32.4°) from the sphere, as shown in Fig. 13-15. (a) Find the total number of lumens falling on the lens. (b) Show by a careful drawing the paths of all the rays from points A and B which pass through the lens. Assume the lens to be free of spherical aberration. (c) What will be the "beam candlepower" of the system, considered as a searchlight? (d) Find the illumination produced by the searchlight on a surface normal to the searchlight beam at a distance of 100 ft from the lens.

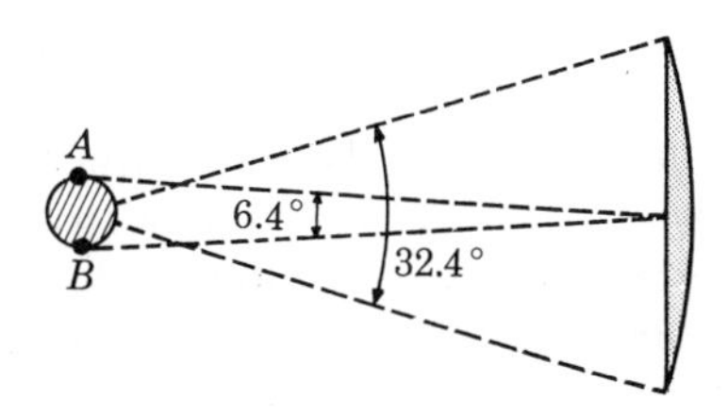

FIG. 13-15.

CHAPTER 14

COLOR

14-1 Colorimetry. A modern magazine with its wealth of illustrations in color, color photography for the amateur as well as the professional, colored plastics, goods in colored packages, all testify to the increasing importance of color in our daily lives. Color has become the concern not only of the artist, but of the physicist and chemist and of the engineer and industrialist as well.

The following statement of Lord Kelvin was quoted in Chap. 1 of the first book of this series and it is appropriate to repeat it here. "I often say that when you can measure what you are speaking about, and express it in numbers, you know something about it; but when you cannot express it in numbers, your knowledge is of a meagre and unsatisfactory kind; it may be the beginning of knowledge, but you have scarcely, in your thoughts, advanced to the stage of *science*, whatever the matter may be." We have today a science of color; color is something that can be measured and expressed in numbers. The science of color measurement is called *colorimetry*.

The word color is commonly used in several different senses. The psychologist uses the word with reference to the sensation in the consciousness of a human observer when the retina of his eye is stimulated by radiant energy. In an entirely different sense, the term is used to specify a property of an object, as, for example, when we say that the color of a book is red. The Committee on Colorimetry of the Optical Society of America has recommended the following definition: "*Color consists of the characteristics of light* other than spatial and temporal inhomogeneities; light being that aspect of radiant energy of which a human observer is aware through the visual sensations which arise from the stimulation of the retina of the eye."

The "characteristics of light" referred to in this definition are threefold. The first is *luminous flux*, which is a measure of the effectiveness of the light in evoking the sensation of brightness. The other two characteristics, which are referred to jointly as the *chromaticity* of the light, are *dominant wave length* and *purity*. Methods for determining dominant wave length and purity will be described in this chapter. The former corresponds to the attribute of *color sensation* called *hue*, the latter to the attribute called *saturation*.

14-2 Additive color mixture. Suppose that three projection lanterns, A, B, and C, are set up so as to project onto a white screen three overlapping circular patches of light, each of a different color, as in Fig. 14-1. The light from each lantern is referred to as a *component.* The regions lettered A, B, C, in Fig. 14-1, are illuminated by a single component. The regions lettered $A + B$, $B + C$, and $A + C$, are illuminated by two components, while all three components illuminate the central region lettered $A + B + C$. The light reflected from a region illuminated by more than one component is called a *color mixture.* The mixture is additive, because the reflected light is made up of the fraction of component A reflected by the screen, plus the fraction of component B reflected, plus the fraction of component C reflected.

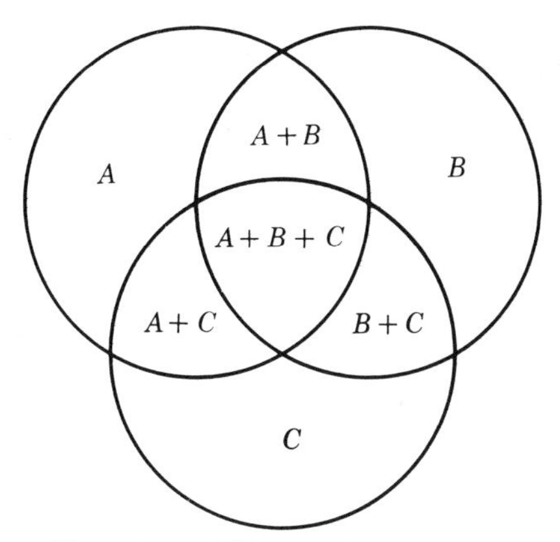

FIG. 14-1. Additive color mixture.

It is well known that each of the mixtures differs from the others in color, as well as differing from each of the three components. Furthermore, it is not possible to detect in any mixture the colors of the components of which it is composed. In this respect the eye differs from the ear. If two notes of different pitch are struck simultaneously on a piano, the resulting sensation is not that of a single pitch intermediate between the two, but both notes can be distinguished. Our sense of hearing is analytical, while our color sense is not.

Let us now adjust the projection lanterns A, B, and C, so that all three circles coincide as in Fig. 14-2, and with a fourth lantern X project onto the screen a second circle illuminated by light of any arbitrary color. If

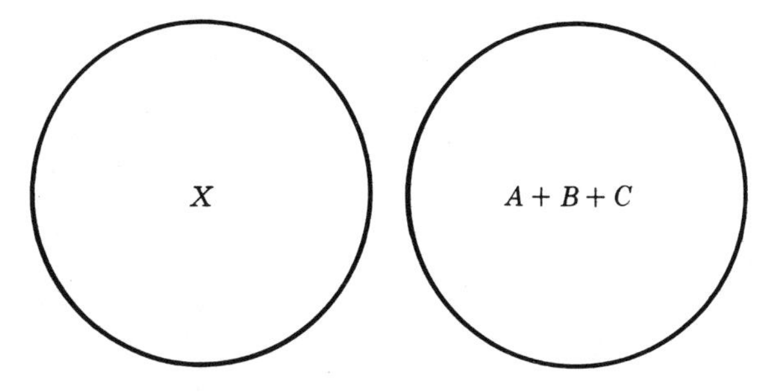

FIG. 14-2. An additive mixture of three components A, B, C, matches the color X.

lanterns A, B, and C are each provided with a device for controlling the quantity of luminous flux emitted (for example, a rheostat in series with the projection lamp) it is found that a wide gamut of colors at X can be matched by additive mixtures of the components A, B, and C in the proper proportions.

When used in this way, the apparatus constitutes one form of *colorimeter*. We can say that the color at X has been measured, in the sense that it can be specified by three numbers representing the quantities of the three components that are required for a color match.

Although a wide gamut of colors at X can be matched by the additive mixture of any three arbitrary components, it is not possible to match all colors. This is not because of a faulty choice of components. A second set of components, different from the first, could be combined to match some colors which the first set could not, but it would then be found that other colors, which could be matched by mixing the first set, could not be matched by the second. To extend the number of components to four or more would widen the gamut to some extent, but unless the number of components were infinite there would still be some colors for which a match could not be secured.

It is found that when a color at X can not be matched by an additive mixture of three components, a color match can be secured between a mixture of the unknown and one of the components, on one hand, and a mixture of the other two components on the other. For example, a mixture of X and A might be matched by a mixture of B and C. In some cases, it may be necessary to add two of the components to the color at X, and match this mixture with the third component. Thus even if it is not possible to *match* the original color with but three components, it is nevertheless possible to *specify* it by stating the quantity of the component (or components) which, when added to it, will result in a match with stated amounts of the remaining components. Thus *all colors can be measured in terms of any three components*, and the results of the measurement expressed by three numbers. Quantities of any component that must be added to a given color to secure a match are considered negative.

We see, therefore, that the common belief that all colors can be matched by a mixture of three properly chosen "primary colors" is incorrect, unless the concept of "matching" is extended in the sense described above. Furthermore, there are no three unique components that *must* be used for color matching. We shall show later that red, green, and blue components permit matching the widest gamut of colors without using negative quantities of a component, and in this sense red, green, and blue can be considered as the "primary colors."

Notice that the preceding statements are experimental facts, and are independent of any theories of color vision.

14-3 Three-color mixture data for matching spectrum colors. It is not necessary actually to set up a colorimeter and match a given color, in order to know the amounts of three arbitrary components that would be required for a match. A preliminary experiment is first performed in which one measures the amounts of the components needed to match all of the *spectrum colors.* (By a spectrum color is meant the color of a light beam comprising only a narrow range of wave lengths.) Then the amounts of the components that would be required to match any given color can be computed if the radiant flux at each wave length, in the given color, is known. The method of computation is explained in Sec. 14-4.

Fig. 14-3 shows, for one particular set of components, the amount of each required to match a spectrum color at any wave length. These components are themselves spectrum colors; a red of wave length 650 mμ, a green of wave length 530 mμ, and a blue of wave length 425 mμ. The ordinates of the curves give the number of lumens of each component such that an additive mixture matches 1 watt of radiant flux at the indicated wave length.

If other components had been used, a different set of curves would, of course, have been obtained. It is possible, however, by an algebraic transformation, to compute what the form of the curves would be for any other set of components, so experiments need be performed with one set of components only.

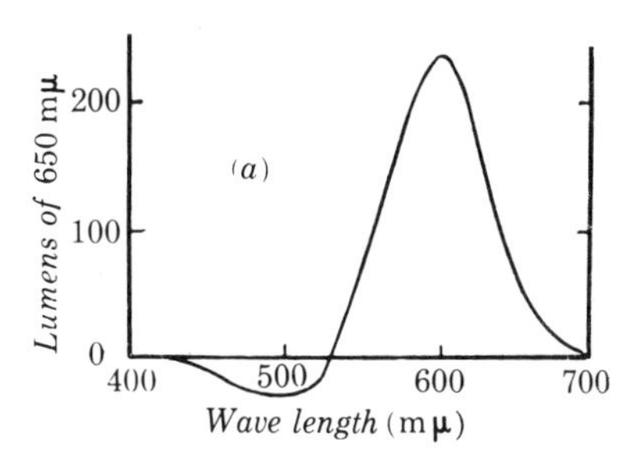

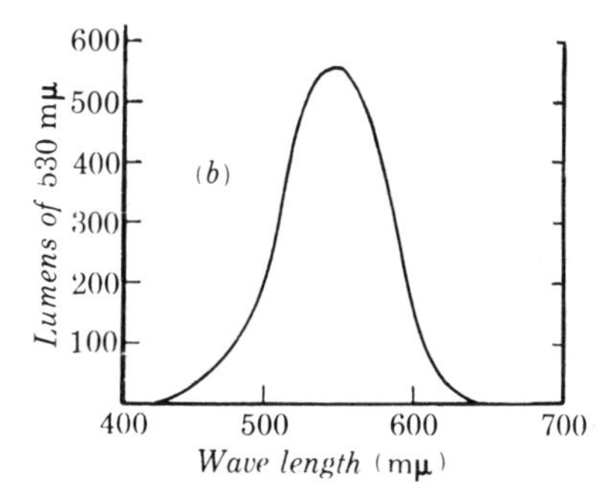

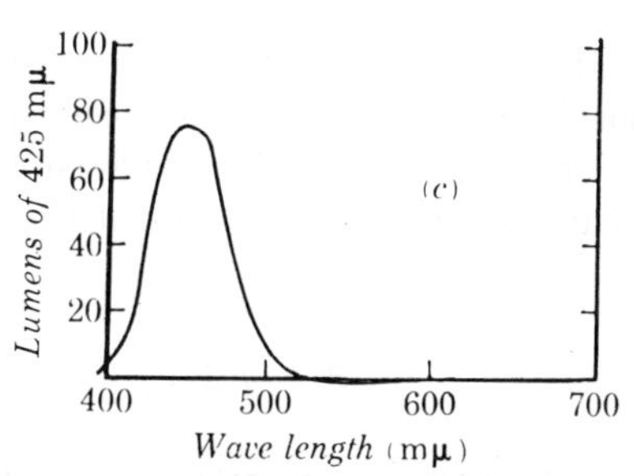

FIG. 14-3. Number of lumens of each of three monochromatic components required to match 1 watt of monochromatic radiant flux.

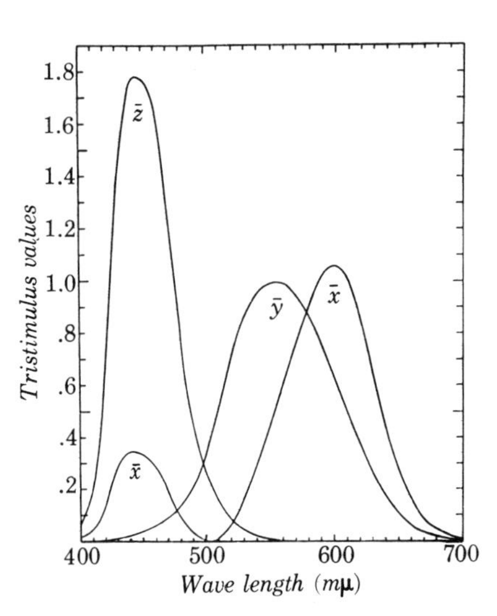

FIG. 14-4. Standard I.C.I. color mixture curves.

The International Commission on Illumination, in 1931, agreed to express all color mixture data in terms of three components so chosen that the curves corresponding to those in Fig. 14-3 were everywhere above the X-axis. This avoids the use of negative numbers in computations. The standard I. C. I. components lie outside the realm of real colors,[1] but this is not of importance since, as stated above, the amounts of the components required for a match can be obtained by mathematical methods without the necessity of setting up an actual colorimeter. As long as everyone agrees to specify colors in terms of the same set of components, one set is as good as another.

The standard I. C. I. color mixture curves are given in Fig. 14-4. For convenience in computation, the ordinates are expressed in arbitrary units so that the areas under all three curves are equal. The symbols $\bar{x}$, $\bar{y}$, and $\bar{z}$ are used for the ordinates of the respective curves, and the values of $\bar{x}$, $\bar{y}$, and $\bar{z}$, at any wave length, are called the *tristimulus values* of a spectrum color of that wave length. Thus the tristimulus values of green light of wave length 500 mμ are

$$\bar{x} = 0.0049, \quad \bar{y} = 0.3230, \quad \bar{z} = 0.2720.$$

Since three numbers (the three tristimulus values) are required to specify a color, a three-dimensional diagram would be needed to represent colors graphically. This difficulty is avoided by introducing three other quantities x, y, and z, defined (for spectrum colors) by the equations

$$x = \frac{\bar{x}}{\bar{x} + \bar{y} + \bar{z}}, \quad y = \frac{\bar{y}}{\bar{x} + \bar{y} + \bar{z}}, \quad z = \frac{\bar{z}}{\bar{x} + \bar{y} + \bar{z}}.$$

The quantities x, y, and z, are called *trichromatic coefficients*. Since from their definition $x + y + z = 1$, any two of these quantities are sufficient

[1] See Sec. 14-6.

to define a color. The quantities x and y are those usually chosen. Taking spectrum light of wave length 500 mμ as an example, we find:

$$x = \frac{.0049}{.5999} = 0.0082, \quad y = \frac{.3230}{.5999} = 0.5384, \quad z = \frac{.2720}{.5999} = 0.4534.$$

Hence this spectrum color may be represented in a two-dimensional rectangular coordinate system by a point whose coordinates are $x = 0.0082$, $y = 0.5384$. If the same computation is made for all spectrum colors, and the results plotted, the curve shown in Fig. 14-5 is obtained. This curve is called the *spectrum locus*, and the diagram a *chromaticity diagram*.

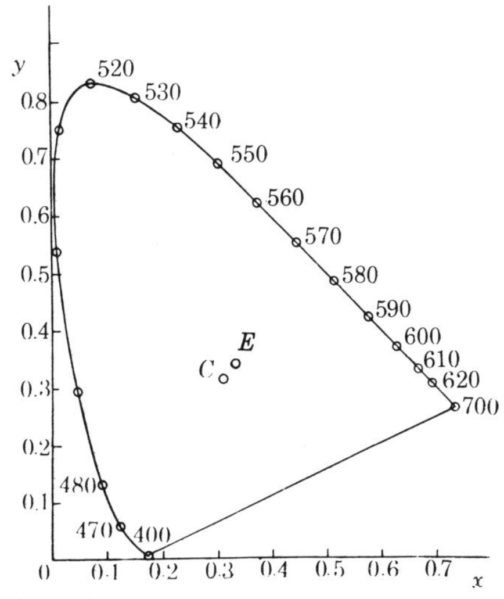

FIG. 14-5. Chromaticity diagram showing the spectrum locus and illuminants C and E.

The color names commonly associated with certain portions of the spectrum are listed below.

Designation	Wave length (mμ)
Violet	Shorter than 450
Blue	450–500
Green	500–570
Yellow	570–590
Orange	590–610
Red	Longer than 610

14-4 Trichromatic coefficients of light of any color. The three tristimulus values of any sample of light are defined as the amounts of the three

I. C. I. standard components which when added would match the sample. These values are denoted by X, Y, and Z, and may be computed by an integration process. The ordinates of the curves in Fig. 14-4 represent the amounts of the I. C. I. components required to match unit quantity of radiant flux at each wave length. Let $f(\lambda)d\lambda$ be the radiant flux at wave length λ in the given sample of light. The amount of the first primary required, for this wave length range, is then

$$\bar{x}f(\lambda)d\lambda,$$

and hence the total amount for a match is

$$X = \int_0^\infty \bar{x}f(\lambda)d\lambda. \tag{14-1}$$

In the same way,

$$Y = \int_0^\infty \bar{y}f(\lambda)d\lambda, \qquad Z = \int_0^\infty \bar{z}f(\lambda)d\lambda. \tag{14-2}$$

Since the curves in Fig. 14-4 cannot be expressed analytically, the integration must be performed by graphical, numerical, or mechanical methods. Instruments are available which perform the integration automatically and with sufficient precision.

Having found the X, Y, and Z tristimulus values, the trichromatic coefficients x, y, and z can be computed from the equations

$$x = \frac{X}{X+Y+Z}, \quad y = \frac{Y}{X+Y+Z}, \quad z = \frac{Z}{X+Y+Z}. \tag{14-3}$$

When the procedure above is carried out for a sample of light in which the radiant flux is the same in each wave length interval (a so-called *equal-energy* spectrum) one obtains $x = .333$, $y = .333$. The chromaticity of such a sample is represented by point E in Fig. 14-5. A standard illuminant known as *illuminant C* (see Sec. 14-6), which is a satisfactory substitute for average daylight, is plotted at point C.

Notice that the integration indicated by Eqs. (14-1) and (14-2) is of precisely the same form as that used to evaluate the luminous flux in a sample of light. (See Eq. (13-1).) A great deal of foresight was displayed by the I. C. I. in its selection of the standard I. C. I. components, when they chose the y-component so that the curve of $\bar{y}$ vs λ, in Fig. 14-4, has exactly the same form as the standard relative luminosity curve in Fig. 13-2. The tristimulus value Y is therefore directly proportional to the luminous flux in the sample.

14-5 Spectrophotometry. We see objects by means of the light they reflect. (For simplicity, the discussion will be limited for the present to opaque, nonself-luminous objects.) According to the Colorimetry Committee's definition of color, it is not strictly correct to attribute color to an object, but only to the light reflected from it. The color of the reflected light depends on the color of the incident light, and on the particular way the color is modified in the reflection process, since most objects do not reflect uniformly throughout the spectrum. Their reflectance (i.e., the fraction of incident light reflected) is a function of wave length and they are said to exhibit *selective reflection.*

The methods of measuring the reflectance of an object at each wave length constitute one branch of the science of *spectrophotometry.* The principle of one type of spectrophotometer is shown in Fig. 14-6. Light from a source A is dispersed by prism B, and a narrow range of wave lengths is isolated by slit C. The beam passing through the slit is divided at D into two beams of equal intensity by a half-silvered mirror or its equivalent. The transmitted beam strikes a standard white surface of

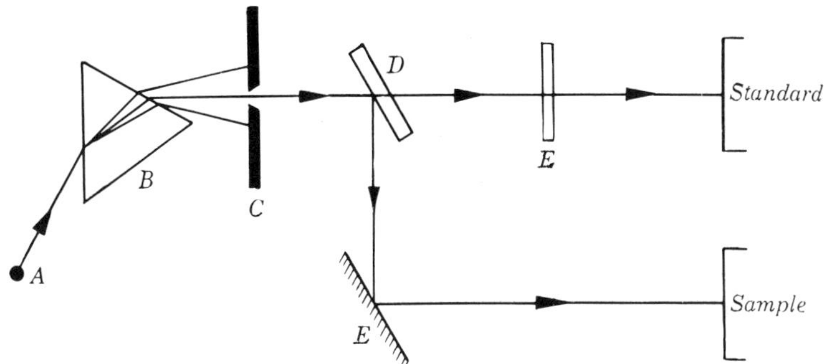

FIG. 14-6. Principle of the spectrophotometer used to measure spectral reflectance.

magnesium oxide, while the reflected beam, after reflection from mirror E, strikes the surface of the sample. The latter, in general, has a lower reflectance than the standard white, so it appears less bright than the standard. The quantity of light striking the standard may be reduced by a device shown schematically at F, until sample and standard appear equally bright. If, for example, the light incident on the standard must be reduced to 70% to secure a brightness match, the reflectance of the sample at this wave length is 70%. By repeating the measurement at other wave lengths, the complete reflectance curve of the sample may be obtained.

Fig. 14-7 is a photograph of an automatic recording photoelectric spectrophotometer developed by Professor A. C. Hardy of M. I. T. In

FIG. 14-7. Automatic recording photoelectric spectrophotometer. The sample and the standard white can be seen at the right, and a portion of the reflectance curve on the cylinder at the left.

(Courtesy of General Electric Co.)

this instrument the standard and sample are viewed by a photoelectric cell, the amplified current from which is used to adjust the light beams until equal quantities of light are reflected from standard and sample. The reflectance curve is drawn by the instrument on a sheet of coordinate paper.

The color version of Fig. 14-8, in the frontispiece, shows six colored rectangles. Beside each is its reflectance curve, replotted on a smaller scale from the curve drawn by the recording spectrophotometer. Notice carefully that each sample reflects to some extent throughout the spectrum. That is, it is not true that the yellow sample reflects only yellow, the green sample only green, and so on.

Let r_λ represent the reflectance of a sample at a wave length λ, and $F(\lambda)d\lambda$ the radiant flux incident on the sample in this wave length range. The reflected flux in the same wave length range, say $f(\lambda)d\lambda$, is evidently

$$f(\lambda)d\lambda = r_\lambda F(\lambda)d\lambda.$$

The tristimulus values of the reflected light are therefore

$$X = \int_0^\infty \bar{x} r_\lambda F(\lambda)d\lambda, \qquad Y = \int_0^\infty \bar{y} r_\lambda F(\lambda)d\lambda, \qquad Z = \int_0^\infty \bar{z} r_\lambda F(\lambda)d\lambda. \tag{14-4}$$

The trichromatic coefficients x, y, z, are computed by Eq. (14-3).

Evidently the trichromatic coefficients of the light reflected from a given object depend on the spectral composition $F(\lambda)d\lambda$ of the light by which it is illuminated. In other words, the color of the reflected light depends on the color of the illuminant and is not a unique property of the object. For purposes of standardization, the I. C. I. has recommended that the "colors of objects" be stated in terms of the color of the light reflected by them from one of three standard light sources produced by a specified arrangement of incandescent lamps and selectively absorbing solutions. The spectral distributions of these sources, which are designated as illuminants A, B, and C respectively, approximate closely to those of a blackbody at temperatures of 2848°K, 4800°K, and 6500°K. Illuminant C is a good approximation to average daylight, and illuminant A to the light from an incandescent tungsten lamp.

The points representing the chromaticities of the samples in Fig. 14-8, when illuminated by illuminant C, have been plotted in a chromaticity diagram beside each sample. If another illuminant had been used, the coefficients, and the position of the plotted point, would of course have been different.

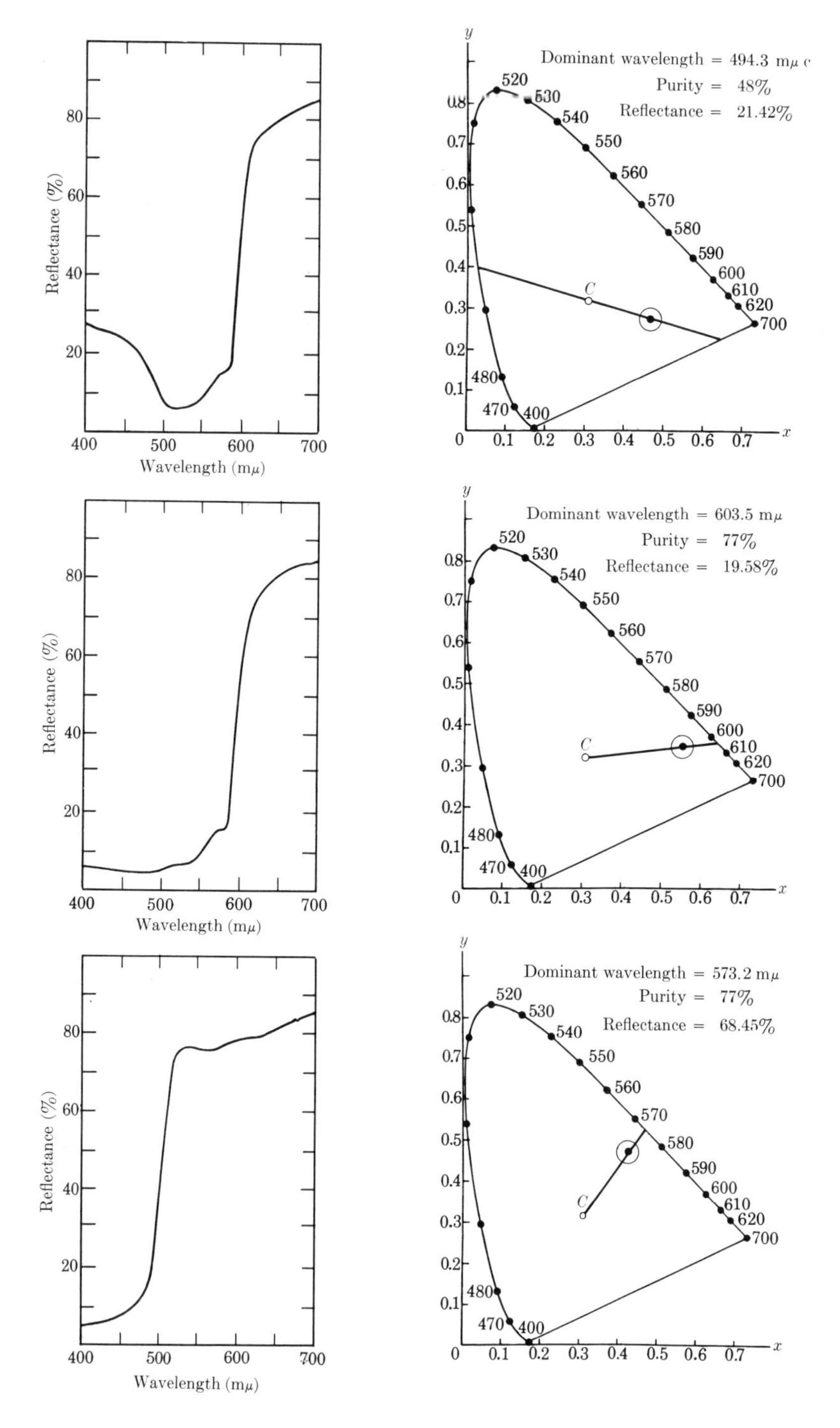

FIG. 14-8. These two pages show reflectance curv

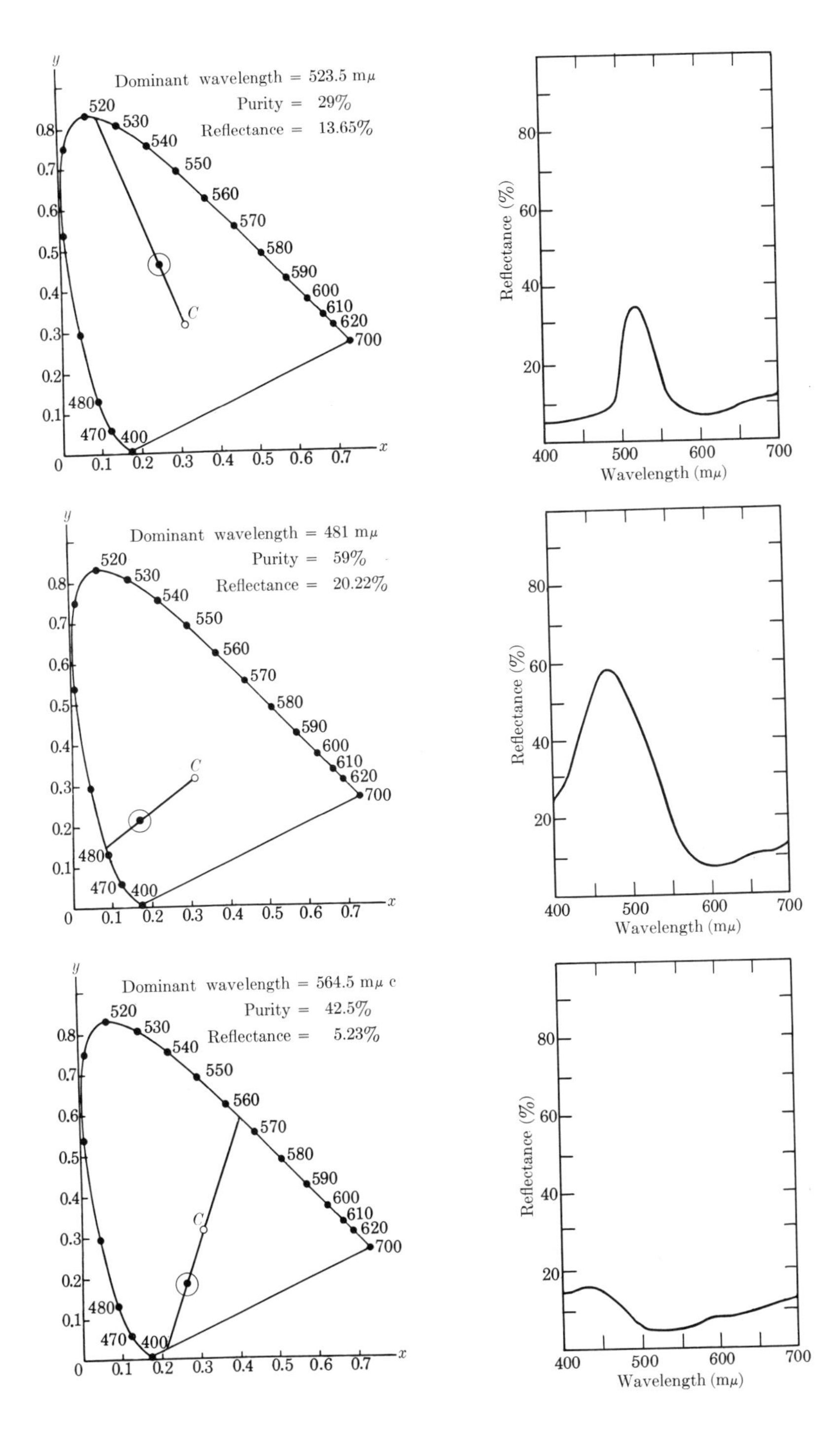

chromaticities of six color samples.

14-6 Dominant wave length and purity When two colors are mixed additively, the point representing the chromaticity of the mixture lies on a straight line connecting the points that represent the chromaticities of the components. Thus all additive mixtures of the colors represented by points D and E in Fig. 14-9 lie somewhere on the line DE. If each component is assigned a weight proportional to the sum of its tristimulus values, the point representing the chromaticity of the mixture lies at the center of gravity of these weights. In other words, the greater the proportion of, say, component D, the closer to point D lies the point representing the mixture. This property of the chromaticity diagram makes possible another means of specifying a color.

A word should first be said about "white" light. There is no unique definition of white light or, more technically, of *achromatic* light. Average daylight, sunlight, and skylight, although their spectral distributions differ widely, may all be considered "white." It has also been suggested that an equal-energy spectrum (point E, Fig. 14-5) should be considered "white." In the absence of an accepted definition, let us speak of illuminant C (average daylight) as white or achromatic light, and call point C in Fig. 14-9 the *white point*.

All colors that could be obtained or matched by a mixture of white light and the spectrum color G, in Fig. 14-9, are represented by points on the line CG. If the proportion of white light is large, the representative point of the mixture lies close to the white point. As the proportion of the spectrum color G in the mixture is increased, the representative point moves closer to the spectrum locus and the color approaches a pure spectrum color. Quantitatively, the *purity* of any color is defined as the distance of its representative point from the white point, expressed as a percentage of the distance from the white point to the spectrum locus along a straight line from the white point passing through the given point. For example, the distance CF in Fig. 14-9 is about 75% of the distance CG, and the purity of the color represented by point F is about 75%. The purity of any spectrum color is of course 100%, and the purity of white is zero.

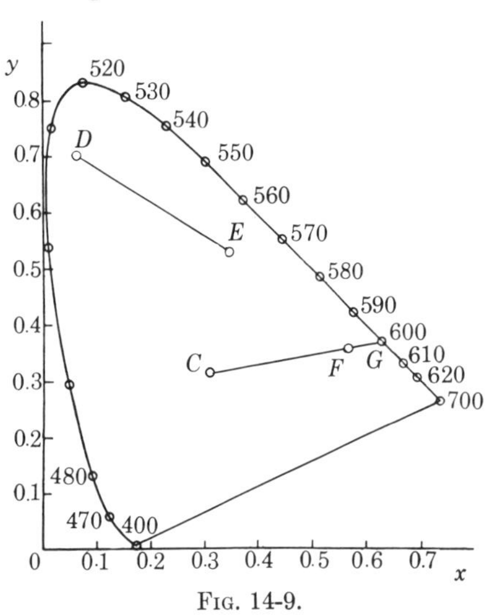

FIG. 14-9.

The *dominant wave length* of a given color is the wave length at which a line from the white point, passing through the point representing the color, intersects the spectrum locus. The dominant wave length of color F in Fig. 14-9 is 600 mμ. As stated in Sec. 14-1, the characteristics of dominant wave length and purity, taken together, constitute the *chromaticity* of a given color. The specification of a color in terms of dominant wave length and purity enables the appearance of the color to be visualized more readily than does its description in terms of tristimulus values or trichromatic coefficients.

The dominant wave lengths and purities of the samples in Fig. 14-8 are indicated on their respective chromaticity diagrams.

In the preceding discussion we assumed that the spectral distribution $F(\lambda)$ of the light incident on an object, was expressed in absolute units such as watts per millimicron. In practice, the values of $F(\lambda)$ for illuminants A, B, and C are tabulated in arbitrary units of such magnitude that if the reflectance of a sample, r_λ, is 100% at all wave lengths, (r_λ = constant = 1), the integral

$$Y = \int_0^\infty \bar{y} \times 1 \times F(\lambda) d\lambda = 1.$$

Then if the reflectance of any sample is less than unity, the value of Y is decreased in proportion. That is, the value of Y gives, not the luminous flux in the reflected light, but the *average reflectance of the sample* for the particular illuminant to which $F(\lambda)$ applies. The average reflectance is usually expressed as a percentage.

The "color of an object" is understood to include a statement of its average reflectance, together with the dominant wave length and purity of the light it reflects, all with reference to a stated illuminant. For example, the green sample in Fig. 14-8 has an average reflectance of 13.65%, a dominant wave length of 523.5 mμ, and a purity of 29%, all referred to illuminant C.

The dominant wave length of a color, although it is not the same thing as the sensation of hue evoked by the color, does correspond to hue in the same way that the wave length of a sound wave corresponds to the sensation of pitch evoked by the wave. Similarly, the purity of a color corresponds to the sensation of saturation and average reflectance corresponds to the sensation of brightness. Chromaticity, which includes the characteristics of dominant wave length and purity, corresponds to chromaticness which includes both hue and saturation.

Mixtures having chromaticities within the triangle HCJ in Fig. 14-10 are described as purples or magentas. Since lines from the white point through points in this triangle do not intersect the spectrum locus, purples cannot be matched by a mixture of white and a spectrum color. They are called *nonspectral* colors. The dominant wave length of a purple is obtained by extending a line from its representative point through the white point until it intersects the spectrum locus. The dominant wave length of the color at K, in Fig. 14-10, is that of the spectrum color at point M. Purple samples reflect more strongly in the red and blue, and less so in the green, and may be described as "minus greens." From the properties of the chromaticity diagram it can be seen that the purple at K in Fig. 14-10, and the spectrum color at M, could be combined in proper proportions to match illuminant C, or white light. When two colors can be added to obtain white they are called *complementary*. The spectrum color at M is a green, and is complementary to the purple or minus green at K. This is indicated by the suffix c following a statement of the wave length of the spectrum color at M. (See Fig. 14-8.)

The purity of a color in the region HCJ in Fig. 14-10 is defined as the distance of its representative point from the white point, expressed as a percentage of the distance from the white point to the line HJ joining the extremities of the spectrum locus. The purity of the color at K in Fig. 14-10 is about 45%.

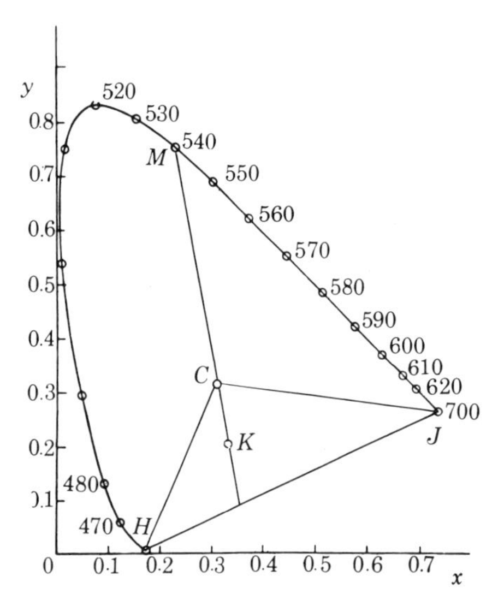

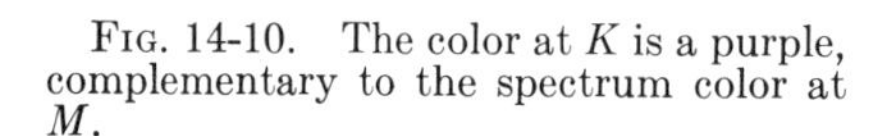
FIG. 14-10. The color at K is a purple, complementary to the spectrum color at M.

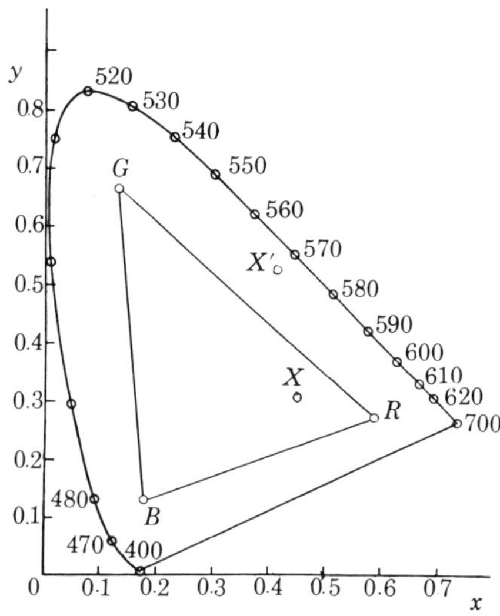

FIG. 14-11. All additive mixtures of components R, G, B, are represented by points within the triangle RGB.

Any real color can be considered an additive mixture of spectrum colors. It follows that the representative point of any real color must lie somewhere within the region bounded by the spectrum locus and the straight line joining its extremities. This region is called the *locus of real colors.*

It can now be understood why a wide gamut of colors, but not all colors, can be matched by an additive mixture of three properly chosen (real) components. Suppose the three components are represented by points R, G, and B in Fig. 14-11. All additive mixtures of B and G lie on the line BG. By adding component R to one of these mixtures, any color such as X within the triangle RGB can be matched. A color such as X' could not be matched by mixtures of R, G, and B, but if B were added to X' in such proportions that the mixture were represented by a point on the line RG, this color could then be matched by a mixture of R and G.

It will also be seen that there is no set of real primaries such that the triangle of which they form the corners will include all real colors, but that the widest gamut of colors can be matched if the components are a highly saturated (or a spectrum) red, green, and blue.

The x- and y-coefficients of the standard I. C. I. primaries are, $x = 0$, $y = 0$; $x = 1$, $y = 0$; $x = 0$, $y = 1$. As stated earlier, these points lie outside the realm of real colors, but since the triangle of which they form the corners includes all of the spectrum locus, all real colors can be matched by mixtures of them.

14-7 The subtractive method of color mixing. Curves A and B, Fig. 14-12, are the transmittance curves of a blue and yellow filter respectively. Suppose the two filters are placed in contact, and inserted in a beam of white light from a projection lantern. We wish to find the color of the transmitted light.

Let the light pass first through the blue and then through the yellow filter. At each wave length, curve A gives the fraction of the incident light transmitted by the blue

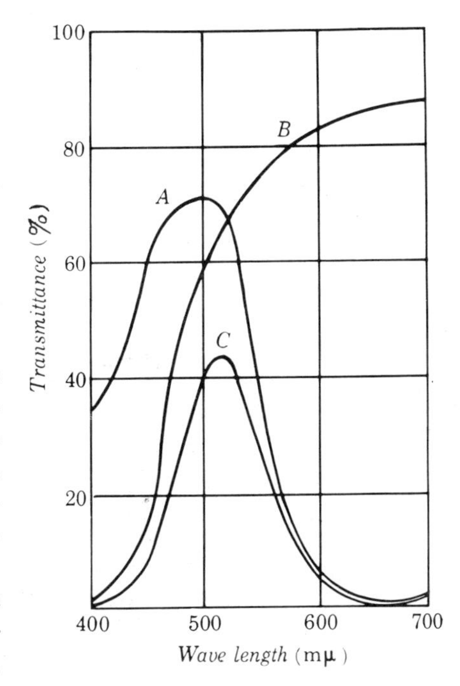

FIG. 14-12. Subtractive color mixing.

filter, and curve B gives the *fraction of this fraction* transmitted by the yellow filter. Thus at 500 mμ 69% of the incident light is transmitted by the blue filter, and 58% of 69%, or 40%, is transmitted by the yellow filter. Hence the transmittance of the combination is found by multiplying the transmittance curves of the filters together, wave length by wave length. The resulting curve is C, Fig. 14-12. Evidently the final result is the same if the light passes first through the yellow and then through the blue filter. The transmittance curve has a maximum in the central portion of the spectrum, and the color of the transmitted light will be green. Since each filter subtracts some energy from the light incident upon it, this method of mixing colors is called a *subtractive* process.

14-8 The color of paints and inks. The colors obtained by mixing paints and inks are due to a subtractive process. In the first place, let us consider an opaque white paint or ink. Its base is a liquid "vehicle," usually linseed oil. The vehicle is quite colorless and transparent. Suspended in it are tiny particles of equally colorless and transparent material, such as an oxide of lead, zinc, or titanium. The index of refraction of the suspended material must be as different as possible from that of the vehicle. We have seen that whenever light is incident on a surface bounding two media of different indices of refraction, some of the light is reflected at the surface. Consider a ray striking the surface of the white paint, Fig. 14-13. Some light will be reflected at the air-vehicle surface, since there is a change in index at this surface. The remainder penetrates into the paint and strikes a boundary between vehicle and suspended particle, where again a portion is reflected. The reflected part returns through the surface, and the part remaining penetrates further, a portion being reflected at each boundary surface which it crosses. Since reflection occurs whatever the wave length of the incident light, the paint reflects uniformly throughout the spectrum, or in other words, it is "white." Note that its white "color" is not produced by suspending white particles in the vehicle, but is due simply to a difference in index between particle and vehicle, both of which are transparent.

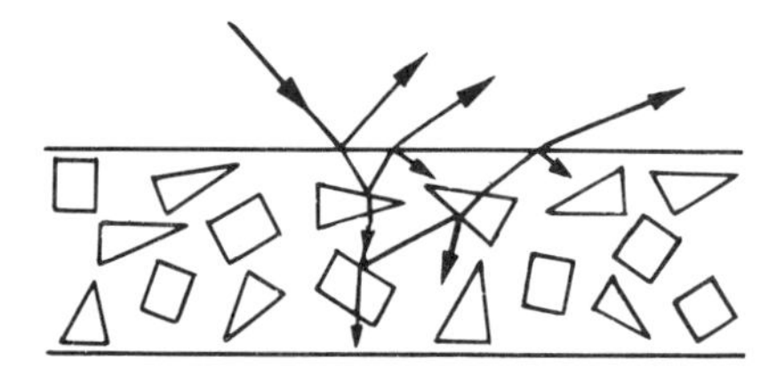

FIG. 14-13.

If a colored paint or ink is desired, the suspended particles are dyed the desired color, or other dyed particles are added to the white paint. The dyed particles then behave like tiny filters in the path of the light rays in the paint. The light reflected back out of the paint must pass through many of these filters on its way in and out, each filter absorbing some of the light incident upon it. The spectral distribution of the incident light is modified by this absorption, and we say that the paint is colored.

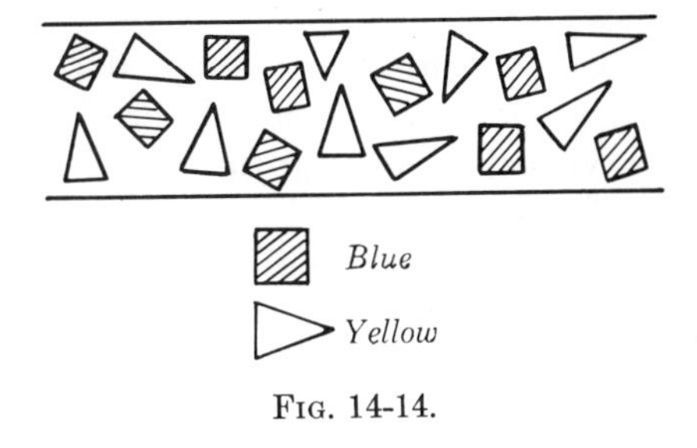

FIG. 14-14.

Suppose that a blue and a yellow paint are mixed. Then the light rays in the paint will pass through both blue and yellow filters before making their way out of the surface (Fig. 14-14). The effect is the same as that produced when a blue and yellow filter are placed "in series" in the path of a light beam. If the paint mixture is illuminated by white light, then green will predominate in the reflected light, and the color produced by this mixture of blue and yellow pigments is green.

In three-color printing processes where one color is printed over another, it is necessary that the inks be transparent. That is, only a small portion of the light should be reflected, and most of it transmitted. This result is secured by using suspended particles whose index is nearly the same as that of the vehicle. The inks are printed one over the other on white paper, and the incident light passes through them, is reflected from the white paper, and passes through the inks again on its way out.

14-9 Subtractive "primaries." As explained in Sec. 14-6, a wide gamut of colors can be matched by *additive* mixtures of a red, a green, and a blue component. The amounts of each component in the mixture are controlled, say, by rheostats in the circuit of each of three projection lan-

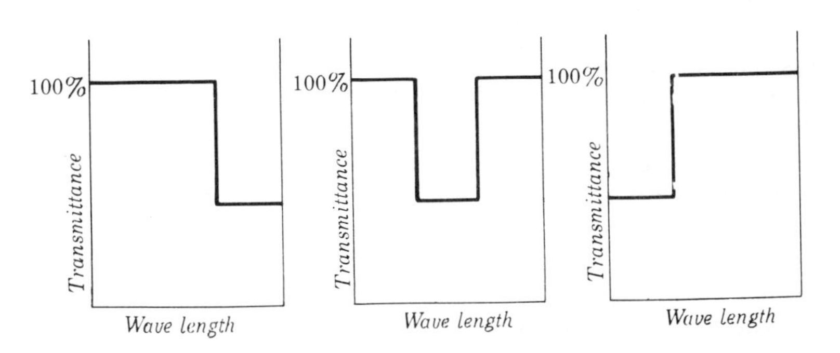

FIG. 14-15. Transmittance curves of ideal dyes for a three-color subtractive process.

terns. When colors are reproduced by printing inks, or by the dyes in a color photograph, the mixture is *subtractive* and control is obtained by varying the density with which each ink is printed, or the concentration of each dye.

The particular portion of the spectrum that each ink or dye controls is that portion in which it *absorbs*. The greater the density or concentration, the more light is absorbed; the smaller its concentration, the less light is absorbed.

The widest gamut of colors can be secured in a three-color subtractive process, i.e., one using three inks or dyes, if the same colors are *controlled* that give the widest gamut in an additive process, namely, a red, a green, and a blue. Note carefully that this does not imply that the colors of the

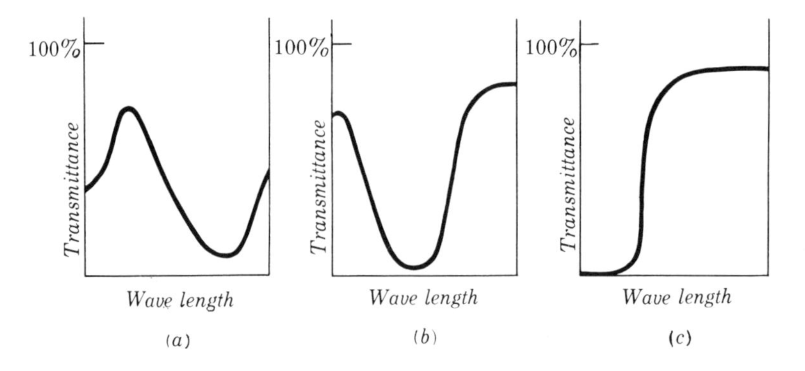

FIG. 14-16. Transmittance curves of the three dyes used in the Eastman "wash-off relief" process.

inks or dyes are red, green, and blue. The ideal dyes would have transmittances as shown in Fig. 14-15, which apply to one particular concentration. The dye in Fig. 14-15 (a) absorbs only in the red, that in (b) absorbs only in the green, that in (c) absorbs only in the blue. If the concentrations were greater, less light would be transmitted (or more absorbed) in the red, green, and blue respectively. If the concentrations were less, less light would be absorbed and more transmitted. Dyes or inks having the ideally sharp cut-offs of Fig. 14-15 are not available. Transmittance curves of the dyes used in the Eastman "wash-off relief" process for making color prints are shown in Fig. 14-16. It will be seen that they approximate the idealized curves. The three dyes in Fig. 14-16 can be described as a "minus-red," a "minus-green," and a "minus-blue." The color of the "minus-red" dye, when printed on a white surface and viewed by white light, is blue. The color of the "minus-green" is a reddish purple or magenta. That of the "minus-blue" is yellow. The "minus-green" or magenta is usually (but erroneously) described as "red."

There is a common belief that "the primary colors are red, yellow, and blue." This mistaken impression comes about in part as a result of the incorrect description of the magenta dye or ink used in color reproduction as "red," and in part because of the failure to appreciate the nature of the subtractive process, i.e., each dye or ink controls the particular region of the spectrum in which it absorbs. The three colors *controlled* by the three dyes in Fig. 14-16 are the same as those that would give a wide gamut of colors in an additive process, namely, red, green, and blue. The blue, or "minus-red" dye, controls the red; the magenta, or "minus-green" dye, controls the green; the yellow, or "minus-blue" dye, controls the blue.

Because the actual inks used in color printing, even when printed in maximum concentration, do not absorb to a sufficient extent to give good blacks, a fourth plate is usually used to accentuate the blacks. Hence the process is referred to as a "four-color process."

Three of the colored rectangles in Fig. 14-8 are printed with the "primary" inks of a four-color process. These are the magenta at the upper left (minus green), the yellow at the lower left (minus blue), and the blue or cyan at the center right (minus red). The orange at the center left is obtained by printing the magenta ink over the yellow ink, the green at the upper right by printing the yellow over the blue, and the purple at the lower right by printing the magenta over the blue. The reader will find it instructive to trace the reflectance curves for the yellow and blue inks on tracing paper and to place this composite curve over the reflectance curve of the green sample.

Problems—Chapter 14

(1) Suppose the three components A, B, and C, in Fig. 14-2 have the following trichromatic coefficients:

Component	x	y
A	0.20	0.20
B	0.10	0.70
C	0.50	0.30

Draw a careful diagram of the spectrum locus, show the positions of these components, and indicate the loci of colors that can be matched by mixtures of A and B, B and C, and of A and C. Where are colors which, if mixed additively with component A, can be matched by a mixture of B and C? Where are colors which, if mixed with both A and B, can be matched by component C?

(2) The tristimulus values of two painted surfaces when illuminated by illuminant C are, respectively

$$X_1 = 12.0, \quad Y_1 = 24.0, \quad Z_1 = 18.0.$$

$$X_2 = 18.0, \quad Y_2 = 36.0, \quad Z_2 = 27.0.$$

How do the surfaces differ with respect to (a) dominant wave length? (b) purity? (c) average reflectance?

(3) Light falls nearly normally on a table top from two sources whose tristimulus values are respectively $X = 20$, $Y = 60$, $Z = 20$, and $X' = 20$, $Y' = 10$, $Z' = 20$. (a) What are the trichromatic coefficients of the two sources? (b) If the distances of the two sources above the table are adjusted until the light incident on the top is white (assume illuminant E), what must be the ratio of their distances?

(4) A spectrum color whose wave length is 400 mμ is mixed additively with a spectrum color whose wave length is 580 mμ. (a) What is the hue of each before they are mixed? (b) What changes in hue will take place as the proportions of the mixture are changed progressively from 100% of the first to 100% of the second?

(5) Two pieces of colored glass, A and B, have spectral transmission curves as shown in Fig. 14-17. Sketch the transmission curve for this pair when bound together so that the light must traverse both.

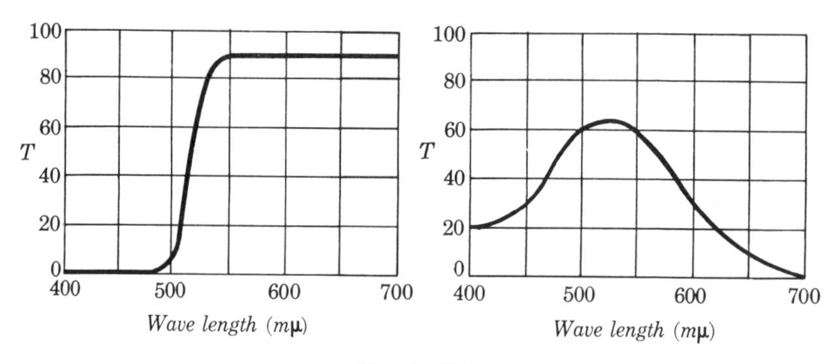

FIG. 14-17.

(6) Let the reflectance curves of the blue and yellow samples in Fig. 14-8 represent the transmittance curves of a blue and a yellow filter. (a) Find the transmittance curve of the composite filter obtained by placing the two in contact. (b) If white light is incident on this filter, what will be the hue of the transmitted light?

(7) What is the wave length of monochromatic light complementary to (a) the red cadmium line at 644 mμ? (b) the sodium D-line at 589 mμ? Consider illuminant C as white light.

ANSWERS TO PROBLEMS

CHAPTER 1

1. 232,000 miles; approximately equal

3. 187,000 miles/sec

5. (a) 2×10^8 m/sec
(b) 333 mμ

7. 2 cm

9. 5×10^{-9} sec

11. 1.57×10^{-5} to 2.76×10^{-5} inch

CHAPTER 2

1. 3.48 rev/sec

5. (a) The one containing water
(b) 1.22

7. 4.67 inches below top face of plate

9. 6 ft from vertical line through cork

11. (a) 13.5°, 27°, 42°, 59°
(b) $y' = 1.4$ inches
$x'' = 0.16$ inch
$y'' = 1.36$ inches

13. 1.03 cm

15. 32°

17. 1.88

19. A plane angle of 97°

21. 3.39 inches

23. 120°

25. (a) 54°
(b) 48°
(c) 9 minutes

27. 1°

29. (b) 2°
(c) 8°

31. 5.5°

CHAPTER 3

1. (a) $s' = -1$ in; $m = +2$
(b) $s' = -0.2$ in; $m = +1.2$

3. (a) $s' = -0.5$ ft
(b) $m = +1.33$

5. (a) $s' = +30$ cm
(b) $m = -1$

7. (a) -20 cm
(b) Virtual
(c) 8 cm to right of second vertex
(d) Real, inverted
(e) -0.6 mm

9. (a) $f = +4$ cm; $f' = +6$ cm
(c) $s' = +12$ cm; $m = -1$

11. (a) 4.67 inches below top of plate
(b) $m = +1$

13. $s' = -6$ inches; $m = -2$

15. $y' = -0.108$ ft

17. (a) $s = 18.75$ cm, $s = 31.25$ cm
(b) $s' = 75$ cm, $s' = -125$ cm
(c) Virtual, real

19. (a) 50 cm beneath surface
(b) 250 m

21. (a) 0.247 inch beneath surface of bearing
(b) 0.0247 inch

23. At vertex of silvered surface

25. 21 inches to the right of his eye

CHAPTER 4

1.

R_1	R_2	f
10 cm	20 cm	40 cm
10	−20	13.3
−10	20	−13.3
−10	−20	−40

3. (b) $s' = 2$ inches, ∞, 0, 0.5 inch (relative to second lens)

$m = +1, \infty, -1, -\frac{1}{2}$

(relative to original object)

7. (a) $\frac{1}{f} = \left(\frac{n_1}{n_2} - 1\right)\left(\frac{1}{R_1} - \frac{1}{R_2}\right)$
(b) 35.2 inches

9. (a) 90 cm to right of lens
(b) Virtual
(c) Inverted

11. 0.01 inch away from lens

13. 120 mm farther from film

15.

Lens	Focal length (mm)	Distance from first vertex (mm)			
		F	F′	H	H′
(a)	+101.7	− 98.3	+108.3	+ 3.4	+ 6.6
(b)	+387.1	−393.5	+384.2	− 6.4	− 2.9
(c)	−413.8	+427.6	−396.9	+13.8	+16.9
(d)	− 98.4	+101.6	− 91.6	+ 3.2	+ 6.8

17. $f = \frac{3}{2} R$, principal points coincide at center of sphere.

19. F' = 0.67 in to right of second lens
F = 0.67 in to left of first lens
H' = 0.67 in to left of second lens
H = 0.67 in to right of first lens

21. (a) ∞
(b) At infinity

23. (a) $x' = +180$ mm
(b) $x' = -180$ mm
(c) $x' = -40$ mm

25. $s' = 10.67$ cm

27. 2.4×; 0.41×

29. (a) 0.167 inch to right of previous focus
(b) 23.3 ft

CHAPTER 6

1. (a) $R = 6.3$ mm
(b) $y' = -0.94$ mm

3. (a) 50 cm
(b) 200 cm

5. (a) -1 diopter
(b) 20 cm

7. 2.5 cm

9. (a) 1.74 cm
(b) -11.3
(c) $-113\times$

11. (a) $-5\times$
(b) $y' = 1.49$ inches

13. (b) $1000\times$
(c) $500\times$

15. (a) $m = \frac{1}{3}$
(b) $\gamma = 3\times$

17. (a) $f = 26.7$ mm
(b) 40 mm to left of first lens

19. Negative lens, $f = -20$ cm; crude telescope length 120 cm compared to Galilean telescope length of 80 cm

21. (a) $f = 6$ inches
(b) 50 inches $\times$ 50 inches
(c) Complete image would be formed, one-half as bright

23. (a) $d\phi = \dfrac{L\,dx}{x^2 + L^2}$
(b) 43.6 ft; 4360 ft
(c) 4.36 ft; 436 ft

CHAPTER 7

1. (a) 37°
(b) E-vector horizontal

3. (a) 53°
(b) 11.4°

5. (a) 17.8%
(b) -0.356

7. (a) 45°
(b) 60°
(c) 69°

9. (a) Linearly polarized
(b) Elliptically polarized

11. (a) Quarter-wave plate
(b) Long wave length end of spectrum transmitted to greater extent

13. $t_2 = 1.981$ mm

CHAPTER 8

1. 113 mμ; 451 mμ

3. 480 mμ

5. 2.95×10^{-4} radian

7. 4.5×10^{-5} radian

9. (a) 500 mμ
(b) 0.87 mm

11. 2.5 mm

13. 1.47×10^{-4} radian

15. 2 mm; 3.5 mm

CHAPTER 9

1. (a) 16.6°
(b) 30°

3. 25 cm

5. (a) 0.02 mm
(b) 6 mm

7. 12.5°

9. $d = 0.01$ mm

11. 0.022 cm

13. (a) 1.55 mm
(b) 3.10 mm

15. 0.069 cm

CHAPTER 10

1. (a) 4.7×10^{-4} cm
(b) 6.3×10^{-4} cm
(c) 4500 m

3. (a) 5×
(b) 0.76λ

5. 5×10^{-5} cm

7. (a)

Magnification of ocular	5×	10×	20×
Focal length of ocular	5 cm	2.5 cm	1.25 cm
Angular magnification of telescope	5×	10×	20×
Radius of exit pupil	2 mm	1 mm	0.5 mm
Radius of diffraction disks	5.72×10^{-4} cm	5.72×10^{-4} cm	11.4×10^{-4} cm
Distance between centers of diffraction disks on retina	2.81×10^{-4} cm	5.63×10^{-4} cm	11.3×10^{-4} cm
Relative brightness of retinal image	1	1	¼

7. (b) 10×, (c) Objects can be resolved with 10× and 20× oculars

9. About 900 cm for green light

11. 70%

13. (a) 4 times
(b) ¼ times
(c) 16 times

15. (a) 20,000 lines
(b) 35,000 lines

17. (a) About 0.06 mμ
(b) About 2 cm

CHAPTER 11

1. Lyman: 121.5 mμ, 91.2 mμ
Paschen: 1870 mμ, 820 mμ
Brackett: 4050 mμ, 1460 mμ

3. (a) 0.0075 mμ
(b) 27,000 lines

5. (a) 620 mμ
(b) 8.67×10^{7} cm/sec

7. (a) ¼ that of the first
(b) The same

9. (a) 0.0413 mμ
(b) 0.000248 mμ

CHAPTER 12

1. (a) $r = 0.6$
   (b) $a = 0.4$
   (c) $P = 0.5$ watt
   (d) $W = 20$ watts/m^2
   (e) 173°K
   (f) $W_{bb} = 50$ watts/m^2

3. (a) 5794 mμ
   (b) 579.4 mμ
   (c) 5220°K

5. 3.1

7. 2020 watts

9. 10.2 times

11. 45%

13. (a) 368 watts/m^2
    (b) 264°K
    (c) 18.5 watts/m^2

CHAPTER 13

1. (a) 30 watts
   (b) 9100 lumens
   (c) 303 lumens/watt

3. 38.7 cm

5. 2.41 lumens/ft^2

7. $R/\sqrt{2}$

9. (a) 0.8 lumen/cm^2
   (b) 0.8 lambert
   (c) 10 candles
   (d) 0.8

11. πB lumens/m^2

13. 3.3×10^3 candles/m^2

15. (a) 20 lumens
    (c) 2000 candles
    (d) 0.2 lumen/ft^2

CHAPTER 14

1. AB, BC, AC; colors in area B-C-600-520; colors in area C-600-700-512c

3. (a) 0.2, 0.6, 0.2
   (b) 0.4, 0.2, 0.4
   (c) 2:1

7. (a) 492 mμ
   (b) 485 mμ

N	0	1	2	3	4	5	6	7	8	9
0	....	0000	3010	4771	6021	6990	7782	8451	9031	9542
1	0000	0414	0792	1139	1461	1761	2041	2304	2553	2788
2	3010	3222	3424	3617	3802	3979	4150	4314	4472	4624
3	4771	4914	5051	5185	5315	5441	5563	5682	5798	5911
4	6021	6128	6232	6335	6435	6532	6628	6721	6812	6902
5	6990	7076	7160	7243	7324	7404	7482	7559	7634	7709
6	7782	7853	7924	7993	8062	8129	8195	8261	8325	8388
7	8451	8513	8573	8633	8692	8751	8808	8865	8921	8976
8	9031	9085	9138	9191	9243	9294	9345	9395	9445	9494
9	9542	9590	9638	9685	9731	9777	9823	9868	9912	9956
10	0000	0043	0086	0128	0170	0212	0253	0294	0334	0374
11	0414	0453	0492	0531	0569	0607	0645	0682	0719	0755
12	0792	0828	0864	0899	0934	0969	1004	1038	1072	1106
13	1139	1173	1206	1239	1271	1303	1335	1367	1399	1430
14	1461	1492	1523	1553	1584	1614	1644	1673	1703	1732
15	1761	1790	1818	1847	1875	1903	1931	1959	1987	2014
16	2041	2068	2095	2122	2148	2175	2201	2227	2253	2279
17	2304	2330	2355	2380	2405	2430	2455	2480	2504	2529
18	2553	2577	2601	2625	2648	2672	2695	2718	2742	2765
19	2788	2810	2833	2856	2878	2900	2923	2945	2967	2989
20	3010	3032	3054	3075	3096	3118	3139	3160	3181	3201
21	3222	3243	3263	3284	3304	3324	3345	3365	3385	3404
22	3424	3444	3464	3483	3502	3522	3541	3560	3579	3598
23	3617	3636	3655	3674	3692	3711	3729	3747	3766	3784
24	3802	3820	3838	3856	3874	3892	3909	3927	3945	3962
25	3979	3997	4014	4031	4048	4065	4082	4099	4116	4133
26	4150	4166	4183	4200	4216	4232	4249	4265	4281	4298
27	4314	4330	4346	4362	4378	4393	4409	4425	4440	4456
28	4472	4487	4502	4518	4533	4548	4564	4579	4594	4609
29	4624	4639	4654	4669	4683	4698	4713	4728	4742	4757
30	4771	4786	4800	4814	4829	4843	4857	4871	4886	4900
31	4914	4928	4942	4955	4969	4983	4997	5011	5024	5038
32	5051	5065	5079	5092	5105	5119	5132	5145	5159	5172
33	5185	5198	5211	5224	5237	5250	5263	5276	5289	5302
34	5315	5328	5340	5353	5366	5378	5391	5403	5416	5428
35	5441	5453	5465	5478	5490	5502	5514	5527	5539	5551
36	5563	5575	5587	5599	5611	5623	5635	5647	5658	5670
37	5682	5694	5705	5717	5729	5740	5752	5763	5775	5786
38	5798	5809	5821	5832	5843	5855	5866	5877	5888	5899
39	5911	5922	5933	5944	5955	5966	5977	5988	5999	6010
40	6021	6031	6042	6053	6064	6075	6085	6096	6107	6117
41	6128	6138	6149	6160	6170	6180	6191	6201	6212	6222
42	6232	6243	6253	6263	6274	6284	6294	6304	6314	6325
43	6335	6345	6355	6365	6375	6385	6395	6405	6415	6425
44	6435	6444	6454	6464	6474	6484	6493	6503	6513	6522
45	6532	6542	6551	6561	6571	6580	6590	6599	6609	6618
46	6628	6637	6646	6656	6665	6675	6684	6693	6702	6712
47	6721	6730	6739	6749	6758	6767	6776	6785	6794	6803
48	6812	6821	6830	6839	6848	6857	6866	6875	6884	6893
49	6902	6911	6920	6928	6937	6946	6955	6964	6972	6981
50	6990	6998	7007	7016	7024	7033	7042	7050	7059	7067
N	0	1	2	3	4	5	6	7	8	9

N	0	1	2	3	4	5	6	7	8	9
50	6990	6998	7007	7016	7024	7033	7042	7050	7059	7067
51	7076	7084	7093	7101	7110	7118	7126	7135	7143	7152
52	7160	7168	7177	7185	7193	7202	7210	7218	7226	7235
53	7243	7251	7259	7267	7275	7284	7292	7300	7308	7316
54	7324	7332	7340	7348	7356	7364	7372	7380	7388	7396
55	7404	7412	7419	7427	7435	7443	7451	7459	7466	7474
56	7482	7490	7497	7505	7513	7520	7528	7536	7543	7551
57	7559	7566	7574	7582	7589	7597	7604	7612	7619	7627
58	7634	7642	7649	7657	7664	7672	7679	7686	7694	7701
59	7709	7716	7723	7731	7738	7745	7752	7760	7767	7774
60	7782	7789	7796	7803	7810	7818	7825	7832	7839	7846
61	7853	7860	7868	7875	7882	7889	7896	7903	7910	7917
62	7924	7931	7938	7945	7952	7959	7966	7973	7980	7987
63	7993	8000	8007	8014	8021	8028	8035	8041	8048	8055
64	8062	8069	8075	8082	8089	8096	8102	8109	8116	8122
65	8129	8136	8142	8149	8156	8162	8169	8176	8182	8189
66	8195	8202	8209	8215	8222	8228	8235	8241	8248	8254
67	8261	8267	8274	8280	8287	8293	8299	8306	8312	8319
68	8325	8331	8338	8344	8351	8357	8363	8370	8376	8382
69	8388	8395	8401	8407	8414	8420	8426	8432	8439	8445
70	8451	8457	8463	8470	8476	8482	8488	8494	8500	8506
71	8513	8519	8525	8531	8537	8543	8549	8555	8561	8567
72	8573	8579	8585	8591	8597	8603	8609	8615	8621	8627
73	8633	8639	8645	8651	8657	8663	8669	8675	8681	8686
74	8692	8698	8704	8710	8716	8722	8727	8733	8739	8745
75	8751	8756	8762	8768	8774	8779	8785	8791	8797	8802
76	8808	8814	8820	8825	8831	8837	8842	8848	8854	8859
77	8865	8871	8876	8882	8887	8893	8899	8904	8910	8915
78	8921	8927	8932	8938	8943	8949	8954	8960	8965	8971
79	8976	8982	8987	8993	8998	9004	9009	9015	9020	9025
80	9031	9036	9042	9047	9053	9058	9063	9069	9074	9079
81	9085	9090	9096	9101	9106	9112	9117	9122	9128	9133
82	9138	9143	9149	9154	9159	9165	9170	9175	9180	9186
83	9191	9196	9201	9206	9212	9217	9222	9227	9232	9238
84	9243	9248	9253	9258	9263	9269	9274	9279	9284	9289
85	9294	9299	9304	9309	9315	9320	9325	9330	9335	9340
86	9345	9350	9355	9360	9365	9370	9375	9380	9385	9390
87	9395	9400	9405	9410	9415	9420	9425	9430	9435	9440
88	9445	9450	9455	9460	9465	9469	9474	9479	9484	9489
89	9494	9499	9504	9509	9513	9518	9523	9528	9533	9538
90	9542	9547	9552	9557	9562	9566	9571	9576	9581	9586
91	9590	9595	9600	9605	9609	9614	9619	9624	9628	9633
92	9638	9643	9647	9652	9657	9661	9666	9671	9675	9680
93	9685	9689	9694	9699	9703	9708	9713	9717	9722	9727
94	9731	9736	9741	9745	9750	9754	9759	9763	9768	9773
95	9777	9782	9786	9791	9795	9800	9805	9809	9814	9818
96	9823	9827	9832	9836	9841	9845	9850	9854	9859	9863
97	9868	9872	9877	9881	9886	9890	9894	9899	9903	9908
98	9912	9917	9921	9926	9930	9934	9939	9943	9948	9952
99	9956	9961	9965	9969	9974	9978	9983	9987	9991	9996
100	0000	0004	0009	0013	0017	0022	0026	0030	0035	0039
N	0	1	2	3	4	5	6	7	8	9

Angle	Sine	Cosine	Tangent	Angle	Sine	Cosine	Tangent
0°	0.000	1.000	0.000				
1°	.018	1.000	.018	46°	.719	.695	1.036
2°	.035	0.999	.035	47°	.731	.682	1.072
3°	.052	.999	.052	48°	.743	.669	1.111
4°	.070	.998	.070	49°	.755	.656	1.150
5°	.087	.996	.088	50°	.766	.643	1.192
6°	.105	.995	.105	51°	.777	.629	1.235
7°	.122	.993	.123	52°	.788	.616	1.280
8°	.139	.990	.141	53°	.799	.602	1.327
9°	.156	.988	.158	54°	.809	.588	1.376
10°	.174	.985	.176	55°	.819	.574	1.428
11°	.191	.982	.194	56°	.829	.559	1.483
12°	.208	.978	.213	57°	.839	.545	1.540
13°	.225	.974	.231	58°	.848	.530	1.600
14°	.242	.970	.249	59°	.857	.515	1.664
15°	.259	.966	.268	60°	.866	.500	1.732
16°	.276	.961	.287	61°	.875	.485	1.804
17°	.292	.956	.306	62°	.883	.470	1.881
18°	.309	.951	.325	63°	.891	.454	1.963
19°	.326	.946	.344	64°	.899	.438	2.050
20°	.342	.940	.364	65°	.906	.423	2.145
21°	.358	.934	.384	66°	.914	.407	2.246
22°	.375	.927	.404	67°	.921	.391	2.356
23°	.391	.921	.425	68°	.927	.375	2.475
24°	.407	.914	.445	69°	.934	.358	2.605
25°	.423	.906	.466	70°	.940	.342	2.747
26°	.438	.899	.488	71°	.946	.326	2.904
27°	.454	.891	.510	72°	.951	.309	3.078
28°	.470	.883	.532	73°	.956	.292	3.271
29°	.485	.875	.554	74°	.961	.276	3.487
30°	.500	.866	.577	75°	.966	.259	3.732
31°	.515	.857	.601	76°	.970	.242	4.011
32°	.530	.848	.625	77°	.974	.225	4.331
33°	.545	.839	.649	78°	.978	.208	4.705
34°	.559	.829	.675	79°	.982	.191	5.145
35°	.574	.819	.700	80°	.985	.174	5.671
36°	.588	.809	.727	81°	.988	.156	6.314
37°	.602	.799	.754	82°	.990	.139	7.115
38°	.616	.788	.781	83°	.993	.122	8.144
39°	.629	.777	.810	84°	.995	.105	9.514
40°	.643	.766	.839	85°	.996	.087	11.43
41°	.656	.755	.869	86°	.998	.070	14.30
42°	.669	.743	.900	87°	.999	.052	19.08
43°	.682	.731	.933	88°	.999	.035	28.64
44°	.695	.719	.966	89°	1.000	.018	57.29
45°	.707	.707	1.000	90°	1.000	.000	∞

CONSTANTS AND CONVERSION FACTORS

$\pi = 3.1416$

$\epsilon = 2.7183$

$\log_e 10 = 2.3026$

1 Angström unit = A = 10^{-8} cm
1 micron = 0.001 mm
1 centimeter = 0.39370 in
1 inch = 2.5400 cm
1 foot = 30.480 cm
1 radian = 57.2958 degrees

1 gram = 15.432 grains
1 ounce = 28.350 gm
1 newton = 0.224 lb = 10^5 dynes
1 pound (wt.) = 445,000 dynes

1 atmosphere = 14.697 lb per sq in
1 joule = 10,000,000 ergs
1 calorie = 4.186 joules
1 sq inch = 6.4516 sq cm
1 sq foot = 929.03 sq cm
1 cu inch = 16.387 cu cm
1 liter = 1000 cu cm
1 gallon = 3.785 liters
1 gallon = 231 cu in

1 pound = 453.59 gm
1 kilogram = 2.2046 lb
1 slug = 14.6 kgm

1 foot-pound = 1.3549 joules
1 Btu = 252.00 cal
1 Btu = 778 ft-lb
1 horsepower = 746 watts

Greek Alphabet

Α	α	Alpha
Β	β	Beta
Γ	γ	Gamma
Δ	δ	Delta
Ε	ϵ	Epsilon
Ζ	ζ	Zeta
Η	η	Eta
Θ	θ	Theta
Ι	ι	Iota
Κ	κ	Kappa
Λ	λ	Lambda
Μ	μ	Mu
Ν	ν	Nu
Ξ	ξ	Xi
Ο	ο	Omicron
Π	π	Pi
Ρ	ρ	Rho
Σ	σ	Sigma
Τ	τ	Tau
Υ	υ	Upsilon
Φ	φ	Phi
Χ	χ	Chi
Ψ	ψ	Psi
Ω	ω	Omega

Periodic Table of the Elements.

Atomic weights are based on the most recent values adopted by the International Union of Chemistry. (For artificially produced elements, the approximate atomic weight of the most stable isotope is given in brackets.)

Group→		I	II	III	IV	V	VI	VII	VIII			0
Period	Series											
1	1	1 H 1.0080										2 He 4.003
2	2	3 Li 6.940	4 Be 9.013	5 B 10.82	6 C 12.011	7 N 14.008	8 O 16.0000	9 F 19.00				10 Ne 20.183
3	3	11 Na 22.991	12 Mg 24.32	13 Al 26.98	14 Si 28.09	15 P 30.975	16 S 32.066	17 Cl 35.457				18 A 39.944
4	4	19 K 39.100	20 Ca 40.08	21 Sc 44.96	22 Ti 47.90	23 V 50.95	24 Cr 52.01	25 Mn 54.94	26 Fe 55.85	27 Co 58.94	28 Ni 58.71	
	5	29 Cu 63.54	30 Zn 65.38	31 Ga 69.72	32 Ge 72.60	33 As 74.91	34 Se 78.96	35 Br 79.916				36 Kr 83.80
5	6	37 Rb 85.48	38 Sr 87.63	39 Y 88.92	40 Zr 91.22	41 Nb 92.91	42 Mo 95.95	43 Tc [99]	44 Ru 101.1	45 Rh 102.91	46 Pd 106.4	
	7	47 Ag 107.880	48 Cd 112.41	49 In 114.82	50 Sn 118.70	51 Sb 121.76	52 Te 127.61	53 I 126.91				54 Xe 131.30
6	8	55 Cs 132.91	56 Ba 137.36	57–71 Lanthanide series*	72 Hf 178.50	73 Ta 180.95	74 W 183.86	75 Re 186.22	76 Os 190.2	77 Ir 192.2	78 Pt 195.09	
	9	79 Au 197.0	80 Hg 200.61	81 Tl 204.39	82 Pb 207.21	83 Bi 209.00	84 Po 210	85 At [210]				86 Rn 222
7	10	87 Fr [223]	88 Ra 226.05	89– Actinide series**								

*Lanthanide series:	57 La 138.92	58 Ce 140.13	59 Pr 140.92	60 Nd 144.27	61 Pm [147]	62 Sm 150.35	63 Eu 152.0	64 Gd 157.26	65 Tb 158.93	66 Dy 162.51	67 Ho 164.94	68 Er 167.27	69 Tm 168.94	70 Yb 173.04	71 Lu 174.99
**Actinide series:	89 Ac 227	90 Th 232.05	91 Pa 231	92 U 238.07	93 Np [237]	94 Pu [242]	95 Am [243]	96 Cm [245]	97 Bk [249]	98 Cf [249]	99 E [253]	100 Fm [255]	101 Md [256]	102 No	103

INDEX